PLC and HMI Development with Siemens TIA Portal

Develop PLC and HMI programs using standard methods and structured approaches with TIA Portal V17

Liam Bee

Packt>

BIRMINGHAM—MUMBAI

PLC and HMI Development with Siemens TIA Portal

Copyright © 2022 Packt Publishing

Group Product Manager: Rahul Nair
Publishing Product Manager: Meeta Rajani
Senior Editor: Athikho Sapuni Rishana
Content Development Editor: Sayali Pingale
Technical Editor: Arjun Varma
Copy Editor: Safis Editing
Associate Project Manager: Neil Dmello
Proofreader: Safis Editing
Indexer: Rekha Nair
Production Designer: Sinhayna Bais
Marketing Coordinator: Nimisha Dua and Sanjana Gupta

First published: April 2022

Production reference: 2060522

Published by Packt Publishing Ltd.
Livery Place
35 Livery Street
Birmingham
B3 2PB, UK.

978-1-80181-722-6

www.packt.com

To my boys, Lewis and Ryan – keep learning! I love you both very much.

– Liam Bee

Contributors

About the author

Liam Bee has worked in automation for over 16 years, after starting his career at 16 years old as an instrument technician in the water industry. He began his automation journey by maintaining PLCs and the instruments connected to them. He found very early on that he had an interest in PLCs and automation, taking the time to learn PLC programming in his own time, as well as exposing himself as much as possible to automation while working.

After 8 years of working in maintenance, Liam started his own side business, providing bespoke controls using lower-range PLC solutions. This experience proved invaluable to his progression; he learned quickly, often through failure, and his knowledge of control design improved significantly.

12 years into his career, he moved roles again; this time, he was looking for something to fill knowledge gaps and target Siemens as he was yet to use Siemens extensively. Liam started at Aquabio Ltd and quickly found himself immersed in Siemens SIMATIC Manager and TIA Portal. Over the next 3 years, he worked hard to understand Siemens' tools and development environments, calling on knowledge from other PLC environments that he had previously worked with.

Over his years working with automation, he has learned many different languages and development environments. He has worked with Siemens, Allen Bradley, Schneider, Mitsubishi, and a host of other PLC platforms, all of which have helped shape the design concepts that he uses today. Liam has also taught himself computer programming languages such as VBA, VBS, VB.NET, C#, Java, and more. Closing the space between IT and industrial automation is important as time moves forwards and he has always tried his hardest to be at the forefront of innovation.

I would like to thank my lovely wife, Ellie, for putting up with my constant ramblings about PLCs and this book and for giving me the space and support to achieve this.

I would also like to thank everyone at Aquabio Ltd, for helping me to learn as much as I have and for allowing me the access to TIA Portal that has enabled me to write this book.

Thank you to Packt and the great team involved in this process; it's been very rewarding.

About the reviewer

Anna Goncharova lives near Boston, USA. She attended college at Vanderbilt University, where she received her degree in electrical engineering. She has over 10 years of experience working with PLCs and HMIs.

Table of Contents

Preface

Section 1 – The TIA Portal – Project Environment

1

Starting a New Project with TIA Portal

Windows and panes – layout of the development environment	**4**	Details view	18
		Overview view mode	19
Portal view – windows and panes	6	**Instructions and libraries**	**19**
Project view – windows and panes	8	Instructions	20
Getting started with a new project in the Project view	**9**	Libraries	22
		Project library	23
Starting a new project	9	Global library	27
Changes to the project tree	11	**Online testing environment**	**31**
Adding devices	12	CPU operator panel	33
Configuration of devices	14	Call environment	33
The Reference and Details views – tools to aid development	**15**	Breakpoints	34
		Call hierarchy	34
Reference projects	16	**Summary**	**35**

2

Creating Objects and How They Fit Together

Creating a project's structure	**38**	**Hierarchy in TIA Portal**	**45**
Data management	38	Call structure	45

Dependency structure   46
Parent/child relationships   47

Instance and global data   48

Using instance and global data   49
Accessing data   51
Configuration options   52

Interfaces and the effects on running PLCs   54

Optimized data/non-optimized data   54
Mixing optimized and non-optimized data   56
Passing data through interfaces   56

Summary   62

3
Structures and User-Defined Types

What are structs and UDTs?   64

Structs   64
UDTs   66

Creating struct/UDTs – best practices   69

Understanding what is required   69
Defining structure variables   70
Finding commonalities between assets   72
Naming conventions   73

Simplifying interfaces with structs/UDTs   74

Passing inputs as a single struct   74
Passing outputs as a single struct   78

Passing InOut data as a single struct   79
Structures in static and temporary memory   79
Creating static declarations of UDTs or structs 80
Creating temporary instances of UDTs or structs 80

Drawbacks of structs and UDTs   81

Libraries   81
Lack of open protocol support   86
Cross-referencing   86
Overusing UDTs/structs   89

Summary   89

Section 2 – TIA Portal – Languages, Structures, and Configurations

4
PLC Programming and Languages

Getting started with languages   94

Available languages   94
Languages in program blocks   99
Different language types   102

Selecting the best language   105

Understanding the use case   105
Memory management   112

Differences between Siemens and other PLCs 112

Timers 112

Valid networks in ladder logic 113

GRAPH is not SFC 114

Bit access in the byte, Word, and Dword variables 115

Summary 115

5

Working with Languages in TIA Portal

The control scenario 118

Control overview 120

Using the HMI 123

Languages used in TIA Portal 123

Ladder logic 123

Function Block Diagram 135

Structured Control Language 145

GRAPH 154

Cause and effect matrix 168

Summary 181

6

Creating Standard Control Objects

Planning standard interfaces 184

Defining variables in an interface 184

Large variables in the interface 187

Planning standard control interfaces 189

Creating control data 193

Improving control data accessibility with UDTs 195

Example 196

Creating HMI data 201

Setpoints/parameters 201

Structuring logic 202

General layout 203

Supportive methods 203

Considerations that have an impact on usability 205

How flexible does the control object need to be? 205

How likely is it that the control object will need to be modified? 206

What does the control object interact with? 208

Summary 211

7

Simulating Signals in the PLC

Running PLC/HMI in simulate mode 213

Starting a PLC simulation 214

Managing simulated inputs 220

Using watch tables to change inputs 220

Using an input mapping
layer to change inputs 223

Creating a simulation interface 228

Safeguarding outputs when
in simulation mode 235
Summary 236

8
Options to Consider When Creating PLC Blocks

Extending standard functions 238
Extending standard data 241
Managing data through
instance parameters 243
Principle to this approach 244

TIA Portal example 245
Asynchronous data
access considerations 249
The correct method 251
Summary 252

Section 3 – TIA Portal – HMI Development

9
TIA Portal HMI Development Environment

TIA Portal Comfort Panel 256

Adding an HMI to a project 256
HMI development
environment overview 259
Runtime settings 259
Screens 261

Screen objects 264
Special objects 266
Elements 267
Controls 268
Graphics and Dynamic widgets 268
Summary 270

10
Placing Objects, Settings Properties, and Events

Setting static properties 272
Types of static values 274
Key properties 275

Setting dynamic properties 276
Assigning tags to dynamization
properties 277

Using scripts 283
Raising events 287
Event scripts 290
Summary 290

11

Structures and HMI Faceplates

What are faceplates?	**292**	Property interface	299
TIA Portal V17 faceplates	293	**Creating and handling events in faceplates**	**301**
Creating a faceplate	**293**		
Available objects and controls	295	Accessing tags	302
Creating interfaces	**296**	**Summary**	**304**
Tag interface	296		

12

Managing Navigation and Alarms

HMI navigation	**306**	**Alarm tags**	**321**
Managing page changes	307	**PLC-driven alarming**	**323**
HMI alarm controls	**310**	Supervision categories	325
		Types of supervision	326
Configuration of HMI alarms	311	Alarm texts	327
The configuration of classes	313	Setting global alarm class colors	330
Configuration of alarm controls	317		
Setting filters on alarm controls	318	**Summary**	**330**

Section 4 – TIA Portal – Deployment and Best Practices

13

Downloading to the PLC

Downloading to a PLC	**336**	**Retaining data in optimized and non-optimized blocks**	**343**
Initiating a download	337		
Setting load actions	338	Retaining data in instance data	344
Downloads requiring the PLC to be stopped	342	Downloads without reinitialization	348
		Snapshots	350

Uploading from a PLC   352
Considerations   357
Data segregation   357

Using functions   358
Summary   358

14
Downloading to the HMI

Connection parameters   360
Creating connections   360
Devices and networks   362

Downloading to an HMI   364

Simulating a unified HMI   365
Accessing a unified HMI simulation 366

Security considerations   369
Summary   370

15
Programming Tips and Additional Support

Simplifying logic tips   372
Delay timers   372
AT constructor   374
IF statements   375
Serializing   376
Refactoring   380
Consolidating blocks   381

Sequences – best practices   382
Using constants instead of
numerical values   382
Managed transitions   383

Managing output requests   386

Naming conventions
and commenting   388
Comments in SCL   390

Additional Siemens support   394
Using TIA Portal's help system   394
Siemens forum   398
Siemens documentation archive   399

Further support – Liam Bee   400
Summary   400

Index

Other Books You May Enjoy

Preface

This book is designed to teach the fundamentals behind TIA Portal as well as the latest features that V17 offers.

TIA Portal V17 is the latest installment of Siemens' flagship development environment. It is designed to encompass all parts of project development, from PLC program writing to HMI development. TIA Portal V17 offers a complete solution to manage a project through design, development, commissioning, and maintenance.

Who this book is for

This book is aimed at anyone who wants to learn about the Siemens TIA Portal environment. No prior knowledge of Siemens is required, however, a basic understanding of PLCs/HMIs would be beneficial.

What this book covers

Chapter 1, Starting a New Project with TIA Portal, shows how to get started with a new project.

Chapter 2, Creating Objects and How They Fit Together, looks at how to create new PLC objects and how to use them together in a project.

Chapter 3, Structures and User-Defined Types, gives an introduction to UDTs and structures and how they benefit projects.

Chapter 4, PLC Programming and Languages, explores PLC programming in TIA Portal.

Chapter 5, Working with Languages in TIA Portal, discusses each programming language that TIA Portal offers, including the latest Cause and Effect language.

Chapter 6, Creating Standard Control Objects, looks at the benefits of standardizing control objects.

Chapter 7, Simulating Signals in the PLC, covers a writing pattern that creates an easy method to simulate all signals in the PLC.

Chapter 8, Options to Consider When Creating PLC Blocks, takes a look at different options that require some thought before and during the writing of code.

Chapter 9, TIA Portal HMI Development Environment, gives an introduction to TIA Portal's latest HMI environment, using the new Unified hardware and design platform built into TIA Portal.

Chapter 10, Placing Objects, Settings Properties, and Events, shows how to create objects in an HMI and set their properties and events.

Chapter 11, Structures and HMI Faceplates, looks at the benefits of using structures and faceplates in the HMI.

Chapter 12, Managing Navigation and Alarms, shows how to manage navigation between pages and how to manage alarms.

Chapter 13, Downloading to the PLC, details how to download a PLC program to PLC hardware.

Chapter 14, Downloading to the HMI, shows how to download an HMI design to HMI hardware.

Chapter 15, Programming Tips and Additional Support, provides information on programming tips and where additional support from Siemens can be found.

To get the most out of this book

In order to get the best out of this book, the basic concepts of what a PLC and HMI are and what they are used for should be understood. A keen interest in advancing learning beyond what this book offers will help to solidify the learning gained from this book.

Software/Hardware covered in the book	OS Requirements
TIA Portal V17	Windows 8,10,11

Trial licenses can be obtained from the official Siemens website. Search for `TIA Portal V17 Trial License` and you should find them.

Download the color images

We also provide a PDF file that has color images of the screenshots/diagrams used in this book. You can download it here: `https://static.packt-cdn.com/` `downloads/9781801817226_ColorImages.pdf`.

Conventions used

There are a number of text conventions used throughout this book.

`Code in text`: Indicates code words in text, database table names, folder names, filenames, file extensions, pathnames, dummy URLs, user input, and Twitter handles. Here is an example: "If `On_Rising_Edge` is True and `Data.Status_Data.Light_Flashes` is over `100000`, then `Data.Status_Data.Maintenance_Required` will be `True`."

A block of code is set as follows:

```
Outlet_Valve_Position_Request =
(((Max Tank Level - Min Tank Level)
/
(Max Valve Position - Min Valve Position))
*
(Current Tank Level - Min Tank Level))
+
Min Valve Position
```

Bold: Indicates a new term, an important word, or words that you see onscreen. For example, words in menus or dialog boxes appear in the text like this. Here is an example: "Double-click **Add new device** in the **Project tree** pane. This will open the **Add new device** window, as illustrated in the following screenshot."

> **Tips or Important Notes**
> Appear like this.

Get in touch

Feedback from our readers is always welcome.

General feedback: If you have questions about any aspect of this book, mention the book title in the subject of your message and email us at customercare@packtpub.com.

Errata: Although we have taken every care to ensure the accuracy of our content, mistakes do happen. If you have found a mistake in this book, we would be grateful if you would report this to us. Please visit www.packtpub.com/support/errata, selecting your book, clicking on the Errata Submission Form link, and entering the details.

Piracy: If you come across any illegal copies of our works in any form on the Internet, we would be grateful if you would provide us with the location address or website name. Please contact us at copyright@packt.com with a link to the material.

If you are interested in becoming an author: If there is a topic that you have expertise in and you are interested in either writing or contributing to a book, please visit authors.packtpub.com.

Share Your Thoughts

Once you've read *PLC and HMI development with Siemens TIA Portal*, we'd love to hear your thoughts! Scan the QR code below to go straight to the Amazon review page for this book and share your feedback.

https://packt.link/r/1801817227

Your review is important to us and the tech community and will help us make sure we're delivering excellent quality content.

Section 1 – The TIA Portal – Project Environment

Learn about the TIA Portal development environment and how to get started with the basics of programming with Siemens PLCs.

This part of the book comprises the following chapters:

- *Chapter 1, Starting a New Project with TIA Portal*
- *Chapter 2, Creating Objects and How They Fit Together*
- *Chapter 3, Structures and User-Defined Types*

1
Starting a New Project with TIA Portal

This chapter covers the core requirements to get started with TIA Portal. This includes the physical layout of the environment and different viewpoints, available tools, adding/configuring devices, and library management. The differences between online and offline views are also covered.

After reading this chapter, programmers should feel comfortable with the basic navigation of TIA Portal and have enough knowledge to be confident about the following:

- Adding devices to a new project
- Knowing where instructions and other tools are located
- Using the project library
- Using a global library
- Viewing online and offline views of connected devices

The following topics will be covered in this chapter:

- Windows and panes – layout of the development environment
- Getting started with a new project in theProject view
- The Reference and Details views – tools to aid development
- Instructions and libraries
- Online testing environment

Windows and panes – layout of the development environment

Before jumping in and creating a project, let's get familiar with the development environment. In TIA Portal **Version 17 (V17)**, there are two different ways to view and use the application, as outlined here:

- **Portal view**
  - Allows for the quick setup of hardware
  - Easy navigation of all TIA areas of development
  - Easy access to diagnostic and other online tools
- **Project view**
  - More familiar **Block & Code** view
  - Required for **programmable logic controller** (**PLC**) programming
  - Required for access to more advanced setting dialogs

Depending on the view that TIA is running in, this changes the look and feel of the development environment and also where/how objects are interacted with.

When opening TIA Portal for the first time, the **Portal** view will be presented, as illustrated in the following screenshot:

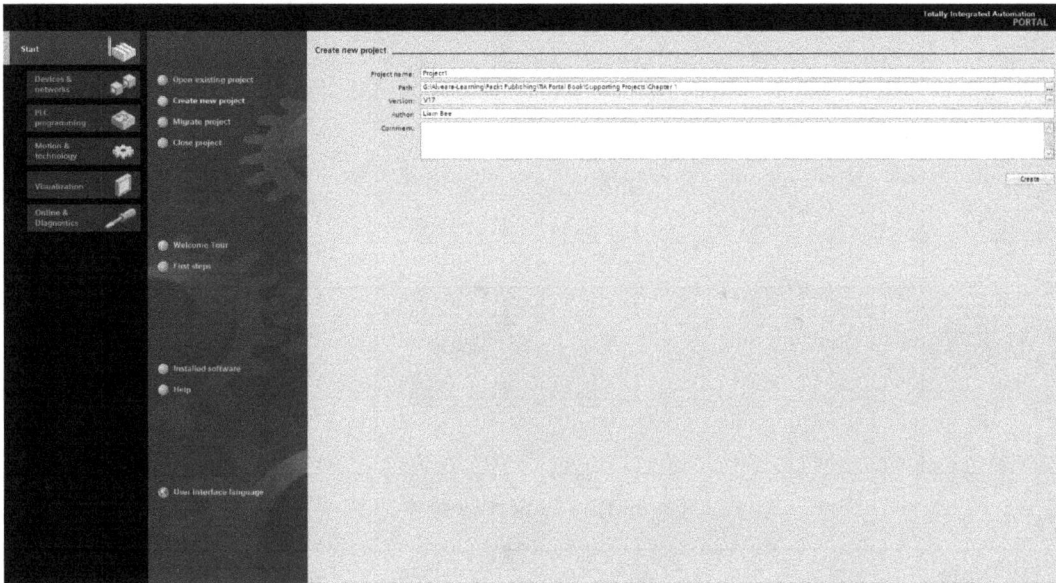

Figure 1.1 – Portal view

The Portal view is TIA's default setting and will be the mode that is in use every time TIA Portal V17 is opened (unless changed in the settings).

This view is best for quickly gaining access to different areas of a project.

When required, TIA will automatically switch from the Portal view to the Project view. The following screenshot shows the Project view:

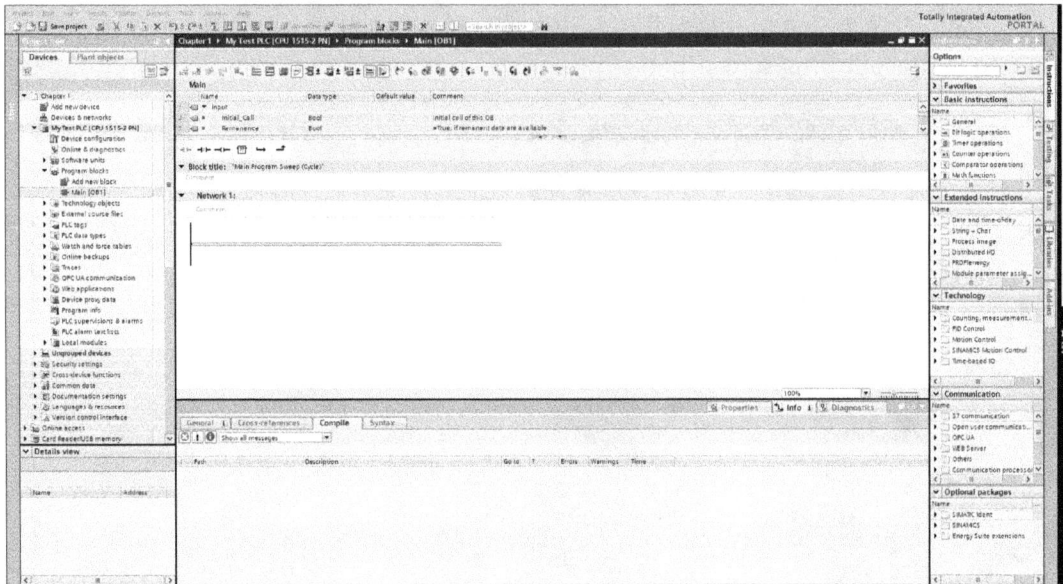

Figure 1.2 – Project view

> **Note**
> The Project view is required in order to actually program information in program blocks.

Portal view – windows and panes

The Portal view is much more simplistic than the Project view. This view is for high-level tasks such as adding equipment, and the layout of the view does not change much between different areas.

The windows in the Portal view can be split up into three different key areas, as illustrated in the following screenshot:

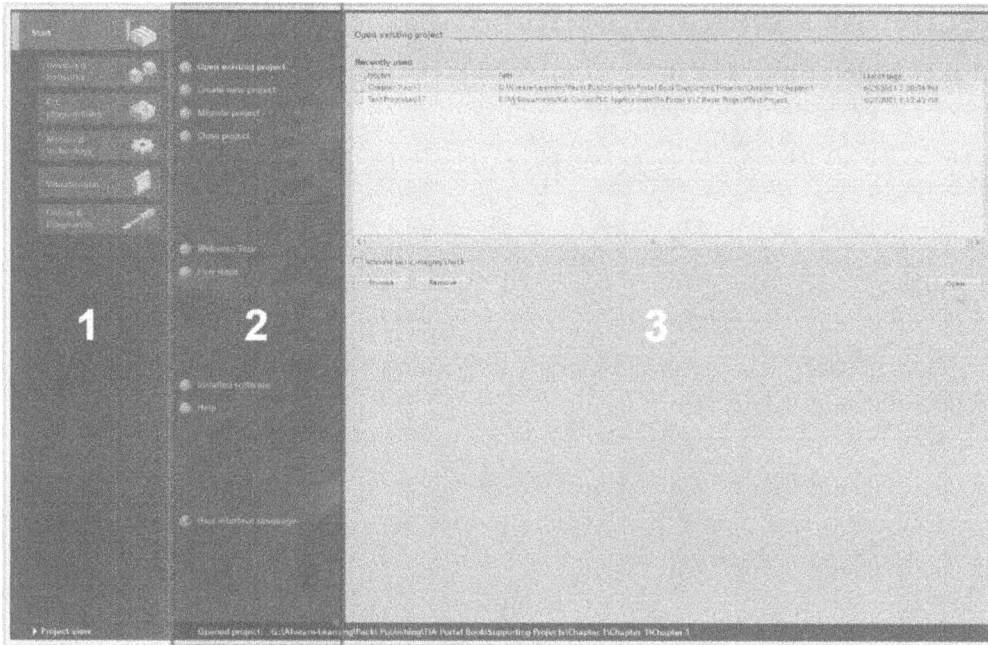

Figure 1.3 – Portal view key areas

The three areas are described in more detail here:

1. **Main menu**: These are the top-level areas of TIA Portal. Nearly all aspects of development fall into one of these categories.

2. **Submenu**: These are the navigational items that relate to the main menu area. For example, switching between **Start** and **PLC programming** would change the options available in this menu.

3. **Activity area**: This is the area in which interaction with selected processes occurs, such as creating a new project or navigating PLC programming blocks.

> **Note**
>
> Be careful in the submenu area of the Portal view! In some locations (such as **PLC programming**), a device must also be selected. It's very easy to accidentally leave the last selected item when returning. Make sure to check which device is active when in these types of menus!

Project view – windows and panes

The Project view is a little more involved than the Portal view. It is not as simplistic, and the menus and navigation can change depending on which area of the project is in view.

> **Note**
>
> The Project views shown in *Figure 1.4* are shown with the default layout. In TIA Portal, windows/panes can be moved around in the Project view to create a custom environment.

You can select the default layout by navigating to **Window | Default Window Layout**. The screen should then look like this:

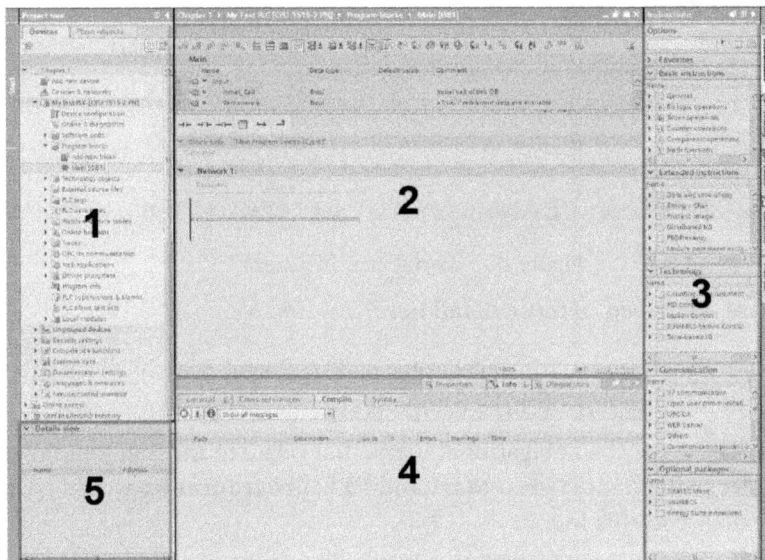

Figure 1.4 – Project view key areas

The four key areas are outlined here:

1. **Project tree**: This is the pane in which navigation of devices and associated objects (such as function blocks) is done. The addition of new devices and objects is also started here.

2. **Main activity area**: This area is where activities such as PLC programming and network configuration are completed.

3. **Supportive tools**: This pane contains a collection of tabs that offer tools relative to the activity that is being undertaken. The tabs that are offered also depend on the current activity.

4. **Properties**, **Info**, and **Diagnostics** pane: This pane changes often and has a multi-tab approach to display information in different categories. During development, this pane will most often be displaying information such as **Compilation Status** and **Properties**, which allow the setup of different objects. **Details view** pane—mini-project tree that only shows details of the object currently selected in the main project tree.

5. **Details view**: This area lists child objects when an object is selected in the project tree. It is also a useful method of exploring variables within a data block or tag list without having to open the object itself.

Getting started with a new project in the Project view

The Project view is the view that most programmers will spend their time in. While the Portal view has its advantages in simplicity, the Project view is a necessity and offers methods to do almost everything that the Portal view can do.

With this in mind, creating a brand-new project in the Project view may be easier for people new to TIA Portal as it follows similar steps to other development platforms.

> **Note**
>
> Remember, on a clean install with default options, TIA Portal will open in the Portal view. In order to get into the Project view, click the text (**Project view**) in the bottom left of the screen to change modes.
>
> This shows the view that will be switched to when clicked, not the current view.

Starting a new project

When in the Project view, TIA expects a project to be available in order to show anything project-related. At this point, though, no project is available, as none has been created.

A new project can be started by clicking the **New project** icon in the toolbar at the top of the window, as illustrated in the following screenshot:

Figure 1.5 – New project in the Project view

Once the **New project** icon has been clicked, the following dialog box is presented:

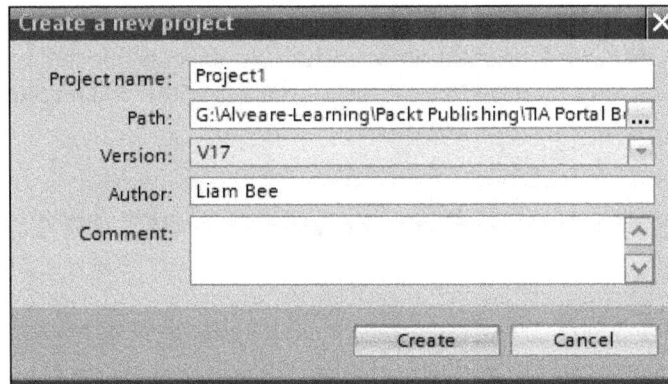

Figure 1.6 – Create a new project dialog box

This allows the setting of the project name, save path, and author, as well as some comments about the project.

Once the **Create** button has been clicked, TIA Portal will create (and save) the new project.

Changes to the project tree

Now that a project has been loaded into TIA (the one just created), the project tree is updated with additional objects that relate to the project, as shown in the following screenshot:

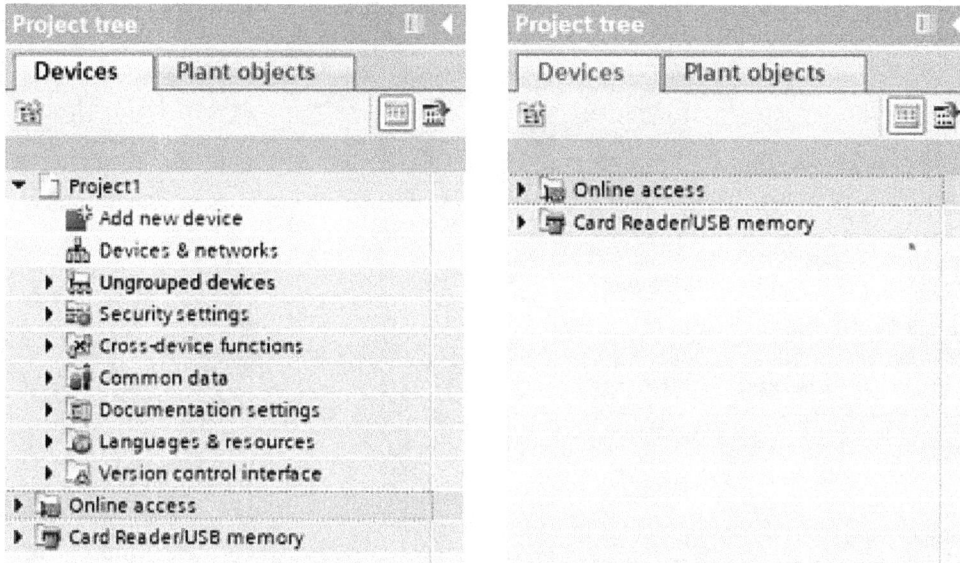

Figure 1.7 – Project tree differences when a device is added (left) compared to no device added (right)

These objects relate only to this project and may affect any devices or further objects that are added to the project (**Security Settings**, for example).

Adding devices

In the Project view (once a project has been opened), a device such as a PLC can be added by double-clicking the **Add new device** option in the project tree.

By double-clicking the **Add new device** option, a new dialog box will open, allowing the selection of supported devices. This is illustrated in the following screenshot:

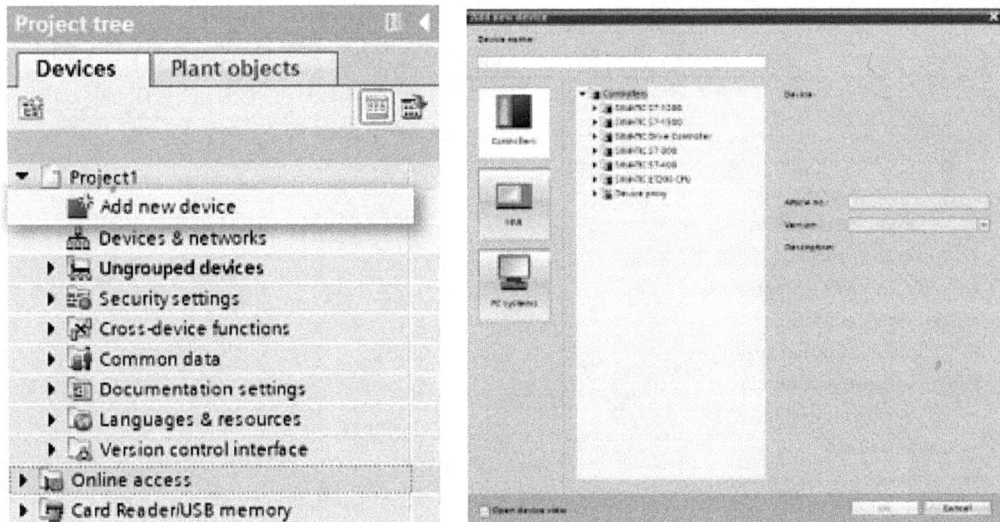

Figure 1.8 – Add new device

Once a suitable device has been selected and a name assigned, the **OK** button can be clicked, and the new PLC (or other hardware) will appear in the project tree, as illustrated in the following screenshot:

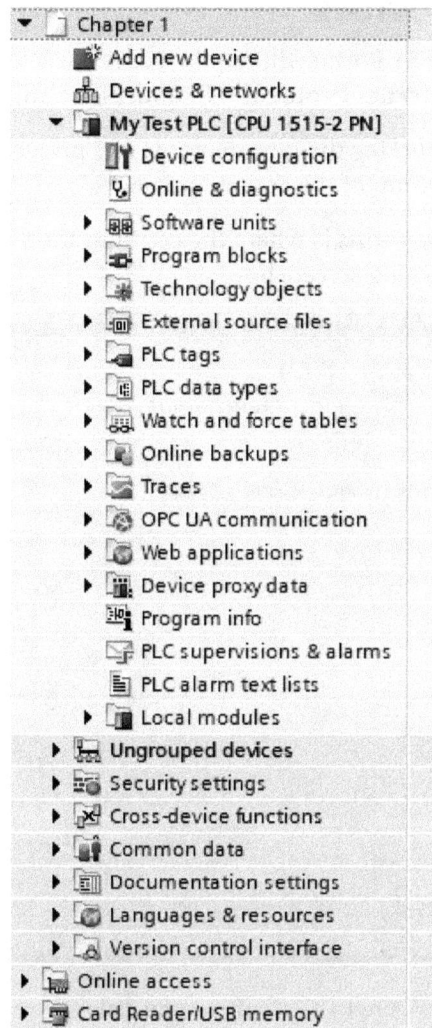

Figure 1.9 – Project tree with new device expanded

If the new device is expanded, all of the available child objects that are relative to the device are displayed.

> **Note**
>
> This is a necessary step to complete before software programming takes place.
>
> The **Programming Blocks** object (where software such as **ladder logic** is written) is not available until a device has been added.

Configuration of devices

Every device that is added in TIA Portal will require some basic level of configuration, such as setting the network **Internet Protocol** (**IP**) address or adding slave modules.

This can be accessed by *right-clicking* the new device in the project tree. The **Properties** item will be available at the bottom of the context menu.

The **Properties** dialog will open—this is where the configuration of the device can take place, as illustrated in the following screenshot:

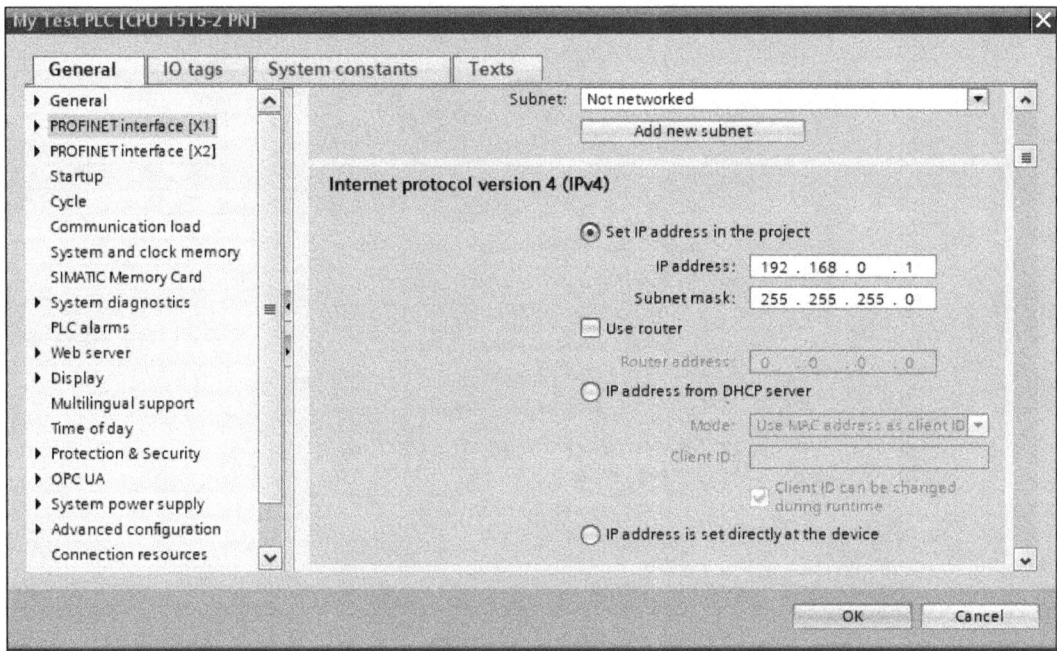

Figure 1.10 – Configuration dialog

The same configuration settings can also be accessed by opening the device configuration from the project tree. Once the Device view opens, the properties are displayed in the **Properties**, **Info**, and **Diagnostics** pane, as illustrated in the following screenshot:

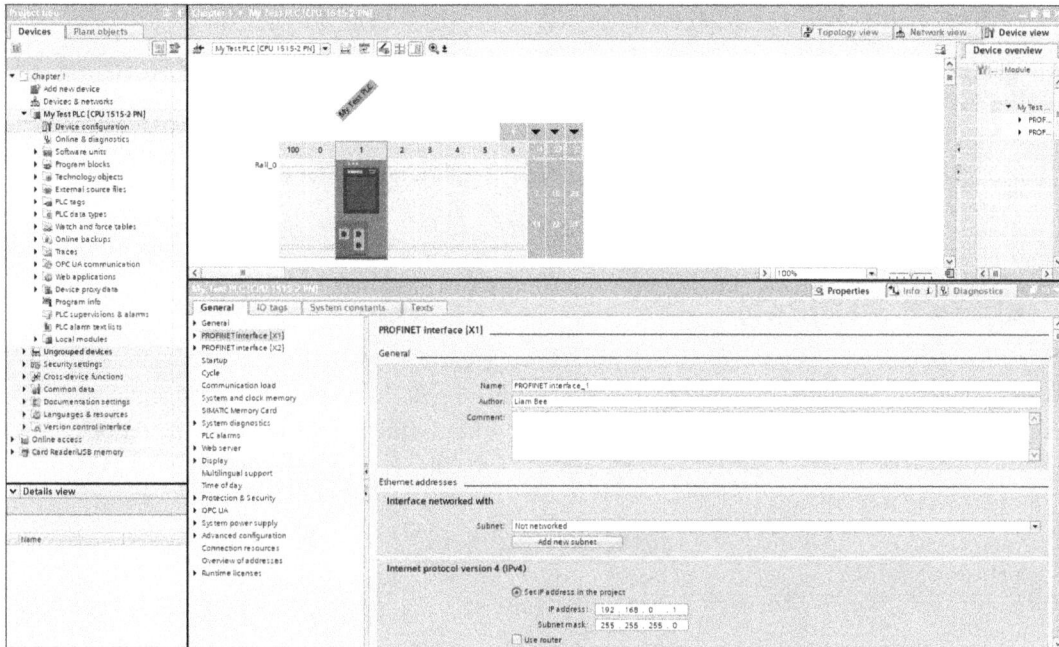

Figure 1.11 – Configuration dialog via the Device view

This pane then allows for the configuration of the device's properties and options.

The Reference and Details views – tools to aid development

TIA portal has some handy views that aren't immediately obvious as to their usefulness. Two views are particularly useful, as outlined here:

- **Reference project view**

  - Allows TIA to open a second project in **Read-Only** mode, to be used as a reference or guide to the current project
  - Blocks and other objects can be copied from the reference project

- **Details view**

  - Offers a *quick-peek*-style view of the selected object in the project tree
  - Allows the selection of internal object data for dragging and dropping into the programming view

Both of these views appear as **panels** on the left-hand side of the TIA Portal view.

> **Note**
>
> By default, TIA Portal has the **Reference projects** panel disabled. Both panels can be enabled (or disabled) by clicking **View** from the top menu and then checking the panels to be made visible.

Reference projects

Making use of a previous project as a template is a great way to maintain consistency between projects. Loading two instances of TIA Portal should be avoided if possible as TIA Portal can be resource-hungry; lighter development machines may lack the resources to maintain two or more instances of TIA Portal. You can see the **Reference projects** panel in the following screenshot:

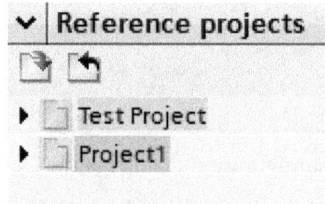

Figure 1.12 – Reference projects panel

With reference projects, multiple projects can be open at the same time in **Read-Only** mode.

Reference projects can be opened and closed via the buttons immediately below the title banner.

Once the project is expanded in the reference project view, it resembles exactly the same view as the project tree, with all objects available as they would be if the project were to be opened normally.

Once a reference project is open and expanded, items within it (such as program blocks) can be opened and read as normal; however, a gray hue will appear on the object icon to indicate it is a reference object, as illustrated in the following screenshot:

Figure 1.13 – Reference object indication example

The difference is very subtle—ensure that the correct blocks are being worked with (although it is not possible to write to a reference block).

Uses of reference projects

There are many different reasons why a reference project may be used to aid development. In most cases, a reference project is used to serve as a guide; however, the following functionalities are also available:

- **Copying information between the reference project and main project**: Objects from the reference view can be copied and pasted into the main project tree or dragged and dropped. This is a useful way to move blocks quickly between two projects.

- **Comparison of objects**: Objects can be compared between the reference project and the main `project.mp` file. This is a useful feature that allows programmers to quickly check the differences between two offline projects.

> **Note**
>
> To compare the reference project with the main project, *right-click* on an object and select **Quick compare | Select as left object**. Then, select the object to compare with and right-click, and then select **Quick compare | Compare with <previously selected left object>**.
>
> TIA Portal will then compare the objects and display any differences.

Details view

The **Details view** pane is best described as a mini-project tree that shows only the child objects of the currently selected object in the project tree.

This is particularly useful when programmers want to drag and drop objects into the main activity area, as in the following screenshot example:

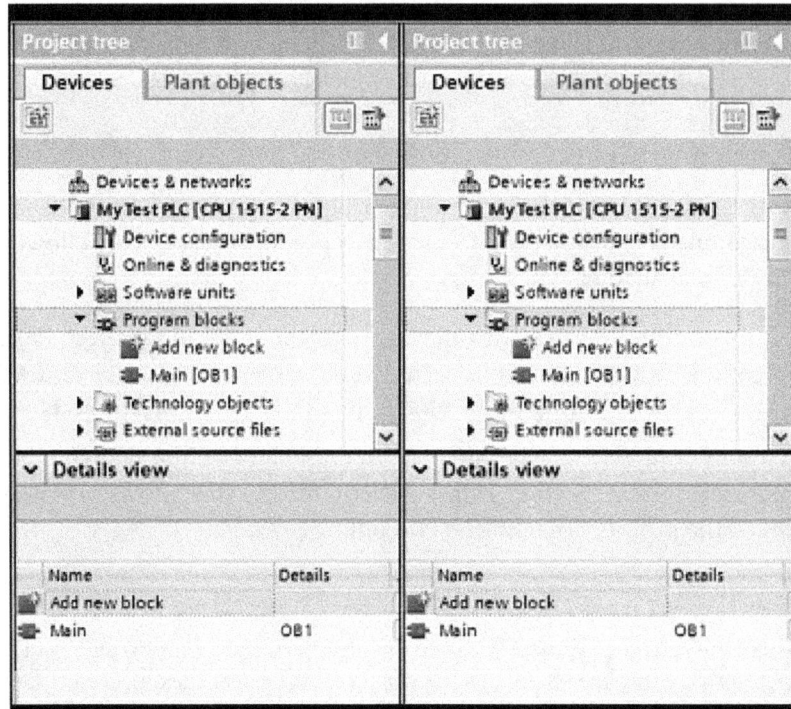

Figure 1.14 – Example of Project tree and Details view displaying the same child objects

> **Note**
>
> Not all options are available for an object when selected in the Details view. For example, an object can be opened, but cannot be renamed from the **Details view** pane.
>
> Right-click options are also unavailable.

Uses of the Details view

The **Details view** pane is only capable of showing child objects and will not show any information when there are no child objects available.

The **Details view** pane would typically be used to do the following:

- Preview child objects in a selected parent object.
- Directly open child objects in a selected parent object.
- Check basic details such as the block number or comments assigned to objects.

Overview view mode

The **Details view** pane also works in the **Overview** view mode, behaving in the same manner as if the object had been selected from the project tree directly.

The **Overview** view mode is similar to the Details view but uses the main activity area to display the results and shows all objects within the previously opened object.

The **Overview** view mode is entered by clicking the persistent **Overview** button at the bottom left of TIA Portal.

> **Note**
>
> Users of SIMATIC Manager (Siemens older development software) may find the **Overview** mode familiar when in the Details view (using the tabs at the top of the view). This displays blocks in a familiar way to how SIMATIC Manager displays blocks in the Details view.

Instructions and libraries

On the right-hand side of TIA Portal are the **Tool** panes used to develop application logic. Here, we find two important tabs called **Instructions** and **Libraries**, which are outlined in more detail here:

- **Instructions**
  - Contains instructions for use when developing code
  - Contains selection methods for selecting different versions of groups of instructions
  - Contains any optional package instructions that have been loaded into TIA Portal

- **Libraries**
  - Contains access to the project library
  - Contains access to global libraries

Instructions

The **Instructions** tab is the area in TIA Portal where programming instructions are stored. You can see an overview of the **Instructions** pane here:

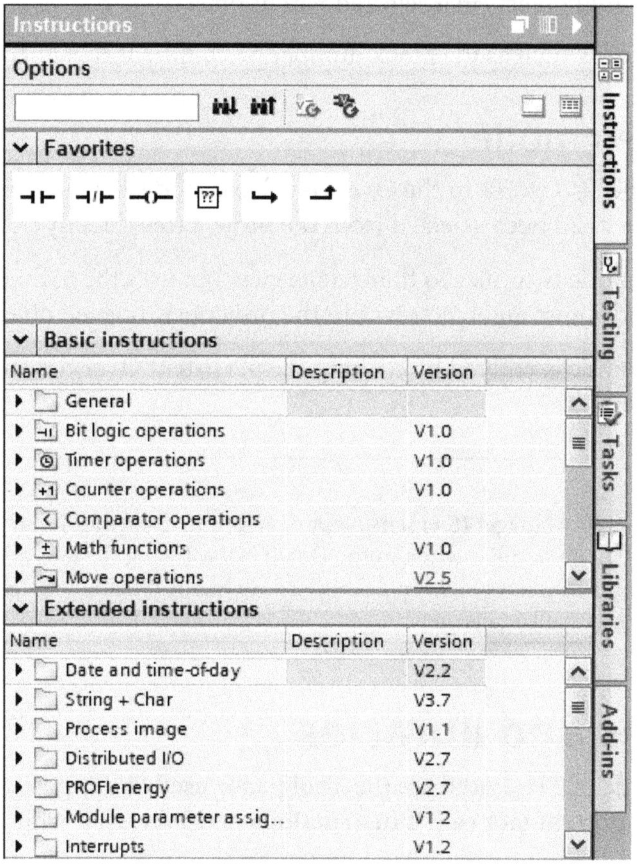

Figure 1.15 – Instructions pane

There are a lot of items here that are organized into categories (folders).

Some categories contain a selectable version. This allows programmers to use previous versions of operations/instructions in a newer environment if required by the project. The default version will always be the most recent, and it is not advised to change it unless there is a specific reason to do so.

Adding instructions from the Instructions tab

Instructions are the objects that are used in the programming window—the building blocks of the logic. These are used together to build the logical part of the program. An example is shown in the following screenshot:

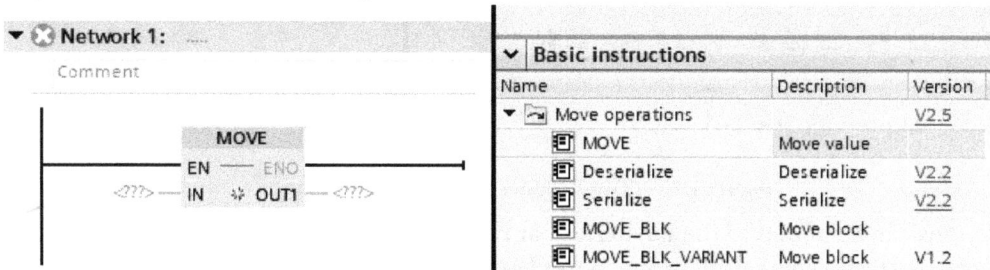

Figure 1.16 – Example of instruction added from the Instructions pane

Instructions are easily added to the program block in use through one of the two following methods:

- Double-clicking on the instruction in the **Instructions** pane. This will add the instruction to the current location in the **Program block** window.

- Dragging the instruction from the **Instructions** pane and placing it in the desired location in the **Program block** window.

Box instructions

A third option is also available for **box instructions**. Adding an **empty box instruction** (from the general category) will result in the following being added to the **Program block** window:

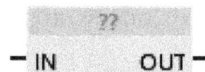

Figure 1.17 – Empty box instruction

?? can be replaced with the name of an instruction that uses the box notation, such as the previously shown MOVE instruction, as illustrated in the following screenshot:

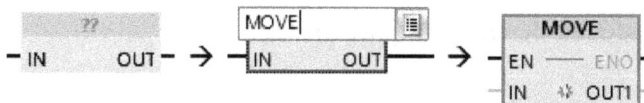

Figure 1.18 – Example of an empty instruction box being defined as a MOVE instruction

Adding favorite instructions

Favorites can be added to a **Favorites** bar in both the **Instructions** pane and in the **Program block** window below the **Block interface** pane, as illustrated in the following screenshot:

Figure 1.19 – Two instances of the Favorites bar

Instructions can be added to the **Favorites** bar by simply dragging and dropping them into place. There are some limitations to the **Favorites** bar, as outlined here:

- Only instructions from the basic instructions set can be added.

- Favorites cannot be re-organized once placed; however, they can be slotted between existing items. This means that if an instruction needs to be moved, it must be deleted and re-added.

Libraries

The library feature in TIA Portal is an extremely well-developed and feature-rich part of the development environment but is often underused and not appreciated by programmers.

Library management allows programmers to keep versions of program blocks so that they can be distributed throughout a project or shared between other projects. This offers the ability to use consistent and pre-tested code, wrapped up in a very well-controlled environment.

> **Note**
> It's best to start using libraries right at the beginning of a project (and throughout its life cycle) when defining a project structure is easier.

The **Libraries** tab consists of the following two areas:

- **Project library**

  - This contains **Types** and **Masters** that are relative to the current project

  - Can only be used by the currently opened project

- **Global library**

  - This also contains **Types** and **Masters** that are used to update or populate the project library
  - Can be updated via the project library

Types and Masters

Throughout the library system in TIA Portal, two common terms are used: **Types** and **Masters** (master copies). These refer to *how* objects placed inside the library behave.

Types: Program blocks placed in the Types folder will be explicitly connected to the project library. This means that they cannot be edited without first confirming the requirement to be edited. If a type is edited, it *must* be saved as a new version; TIA Portal will manage this appropriately.

Masters: Any object placed in the master copies will be a copy of the object placed. Unlike **Types**, these serve as templates and can be edited at will in the project without updating the library. If more than one object is placed in the Master copies folder at once, TIA Portal will group them together as a single object to ensure they are copied back together.

Project library

The **Project library** is a locally held collection of program blocks that directly relate to the current project, as illustrated in the following screenshot:

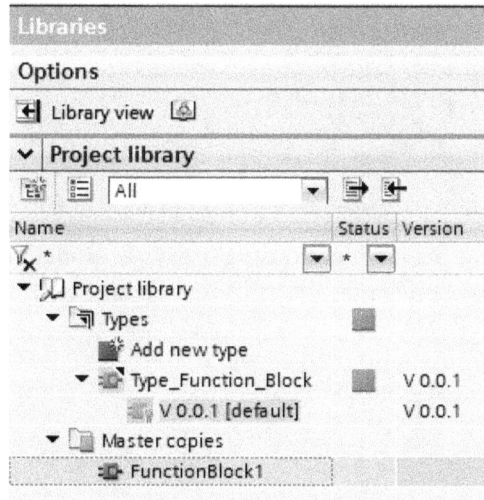

Figure 1.20 – Example of typed objects and master copies

This library is used to help programmers ensure that the program blocks used throughout the project are strictly version-controlled, or to provide templates in order to get started with a program block quickly.

The **Libraries** tab is found on the right-hand side of the TIA Portal application, below the **Instructions** tab.

Blocks that are used in the project library can only be opened if they are instantiated in the project itself. This means that in order to edit or open a project library object, it must at least exist in the project tree.

> **Note – Master Copies**
>
> Master copies cannot be opened at all from the **Libraries** pane and must be copied (dragged) into the project tree before an object can be opened.

Types

Types are version-controlled program blocks. Only **function blocks**, **functions**, and **user-defined types** (**UDTs**) can exist as a type in a library.

Types can be in one of the following two modes:

- **Released**
- **In test**

The following screenshot shows an example of a program block in test:

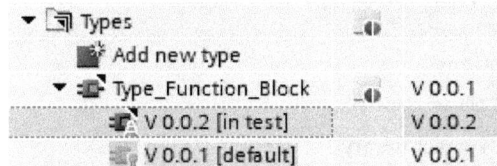

Figure 1.21 – Example of a program block in test

When a type is released, it has been added to the library, versioned, and is now available to be used in the project tree. At this point, it is protected and cannot be edited without creating a new version.

When a type is in test, it is available to be edited, versioned, and then committed back to the library. A type must be put into test to create a new version. The test version can be downloaded into the PLC via the project tree so that tests on the new version can be performed.

Editing Library Blocks (In Test)

The block to be edited *must* exist in the project tree before it can be edited. A block can be edited by right-clicking and choosing **Edit type** in the **Libraries** pane.

The project library should be used in order to release multiple in-test blocks at once. The status indication in the library will show the following symbols for different types of inconsistencies between the project library and project tree:

- **Multiple inconsistencies**: More than one object is **in test** (including dependents)

- **Non-Default version instantiated**: The project tree has an **in-test** version instantiated

- **Default dependent not used**: The default version of this block does not use the same default version of a dependent block (inconsistent versioning)

- **Consistent**: No inconsistencies detected

In order for library objects to be used correctly, all items should be consistent unless testing is actively occurring.

There is more than one method for releasing versions, as outlined here:

- Via the yellow information banner at the top of a typed block that is in test

- Via the right-click menu when clicking an object in test in the project library

When either is clicked, the following dialog box is presented:

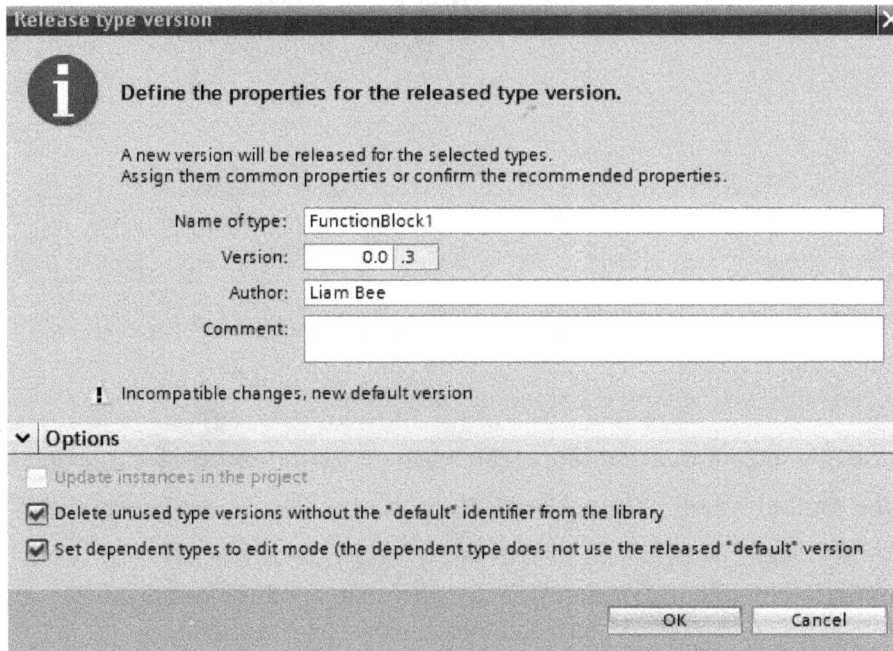

Figure 1.22 – Release type version dialog box

In this dialog box, information about the type can be entered, including an author and a comment.

The version number will automatically increase the minor revision number; however, this can be overridden to make larger version increments.

Checkboxes at the bottom of the dialog box allow programmers to change the behavior of the release.

Deleting unused types is a good way of keeping the project library clean and easy to use. Setting dependent types to edit mode will be pre-selected if the block being released has dependent types that are not in test yet (these dependent blocks will need to be updated to maintain library consistency).

Discarding Changes

Be careful! Clicking **Discard changes** will immediately delete the test version, without confirmation!

Master copies

Master copies work very differently from types. While master copies still reside in the library, they are not version-controlled, and modifying a master copy does not cause other blocks to need to adapt.

Master copies are also not restricted to just function blocks, functions, and UDTs. Other types—such as data blocks, tag lists, and even groups of objects and folders—can have master copies.

> **Editing Master Copies**
> Unlike types, master copies cannot be edited at all. In order to update a master copy, a new instance of the object must replace it. This can be done by simply deleting the existing version and dragging the new version (with the same name) into the `Master copies` folder.

Usage

A master copy serves as a template and can simply be dragged from the **Libraries** pane into the project tree in the position required. If the object dragged into the project tree contains more than one object (is a group of objects), all of the objects will be unpacked into their single instances, retaining any structural layout.

Icons

Different icons for master copies symbolize different meanings, as outlined here:

- **Standard singular objects**: Objects retain their normal icons
- **Group of data blocks**: A folder containing only data blocks
- **Group of function blocks**: A folder containing only function blocks
- **Group of mixed objects**: A folder containing more than one type of object
- **Empty folder**: A folder/group with no content (useful for templating project-tree layouts)

Global library

The **global library** is a library that can be used by more than one project. The global library behaves slightly differently from the project library, but still maintains a familiar approach to updating and maintaining objects within it.

> **Libraries Pane**
>
> Just as with the project library, the global library is found on the left-hand side of TIA Portal in the **Libraries** pane.

Creating a new global library

Global libraries exist outside of the project and can be saved in different directories from the active project. To create a new global library, click the ⌷ icon.

A familiar dialog box will open that is similar to the one used to create a new project, as shown here:

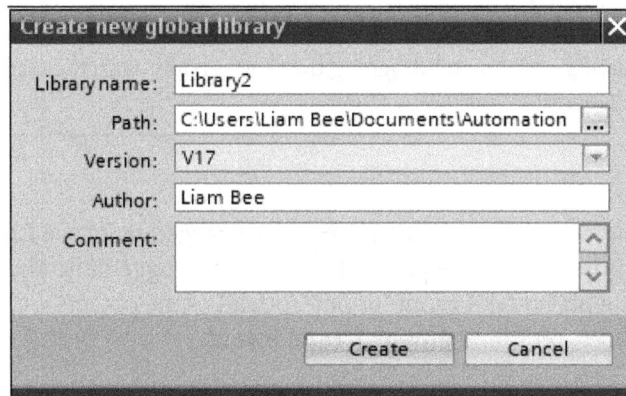

Figure 1.23 – Create new global library dialog box

By default, global libraries are saved in the `Automation` folder; however, this can be changed to any location.

It's important to note the version number of the global library—the version of TIA Portal must match in order to use the library. If an older version library is opened in a newer version of TIA Portal, a conversion to the new version takes place. You cannot downgrade TIA Portal libraries to lower versions.

Opening a global library

Just as with a project, a global library must be browsed for and opened. This can be done in the same place as when creating a new global library, by clicking the 🔾 icon.

A new dialog will open that allows a global library to be browsed for and opened, as illustrated in the following screenshot:

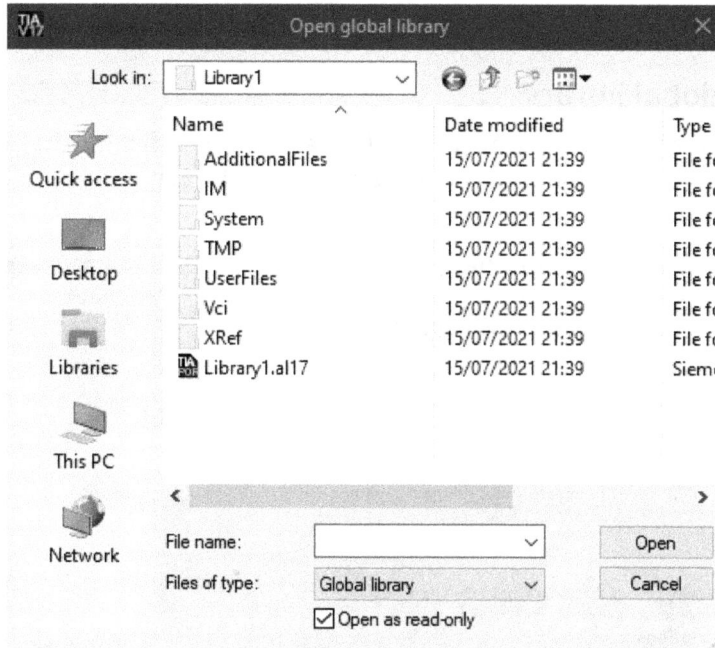

Figure 1.24 – Open dialog for a global library

If a global library is to be edited, remember to uncheck the **Open as read-only** checkbox. If this is left checked, objects can be read from the global library but cannot be written to the global library.

> **Note**
> A project doesn't have to be opened to open a global library; however, without a project, the use of a global library is limited!

Using a global library

Before any objects in the global library can actually be used, it is important to understand the differences between a global library and a project library. Almost all of the differences relate only to types, with master copies still behaving as templates.

If a type is required from a global library, it will automatically be added to the project library if dragged into the project tree, as will all of its required dependents.

Upgrading a global library

A global library is easily updated from a project library by right-clicking the object to be upgraded in the project tree and selecting **Update types | Library**. A new dialog will open, as illustrated in the following screenshot:

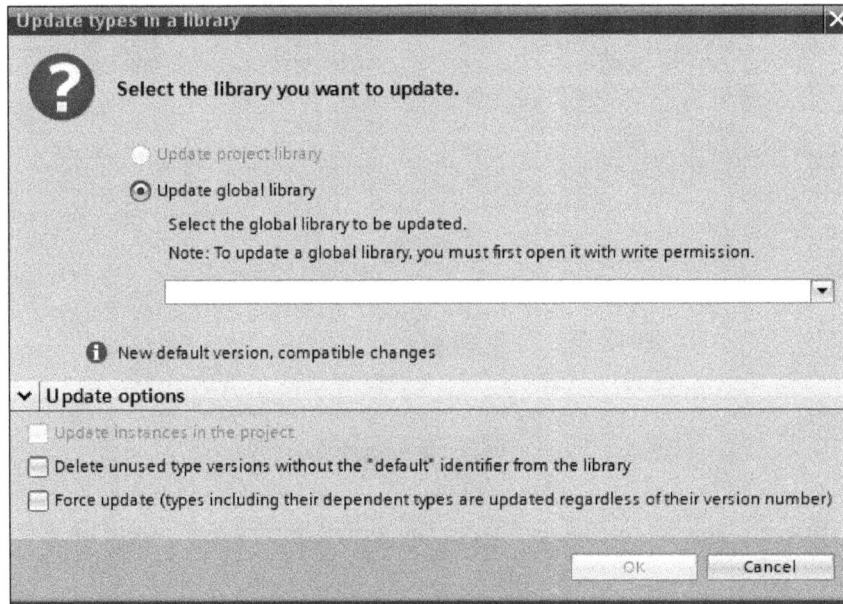

Figure 1.25 – Updating types in a library

From the dropdown, the open (and writable) global library can be selected. When **OK** is clicked, the global library will be updated with all new versions.

This also works the other way around, for updating project library types from the global library.

The global library must be saved after updating by clicking the **Save** icon (🔲) in the **Global library** pane.

> **Note**
>
> Checkbox options should only be selected if there is an explicit requirement to do so. It's good practice to keep previous versions available in the global library to support older projects as time and versions progress.
>
> Periodically, very old versions may be removed once no projects are making use of them (the project library will still contain the relative version, though, if one in use is removed from the global library).

Online testing environment

The **Testing** tab on the right-hand side of the TIA Portal application contains tools that are used only when a device is connected via the **Go online** function.

If no device is connected to, the **Testing** tab will simply display **No online connection** in the **central processing unit** (**CPU**) operator panel, and other functions are not available.

The **Go online** function can be found on the top menu bar—it looks like this:

Figure 1.26 – Go online function

Once clicked, a connection dialog will open that allows CPUs and other devices to be searched for and connected to, as illustrated in the following screenshot:

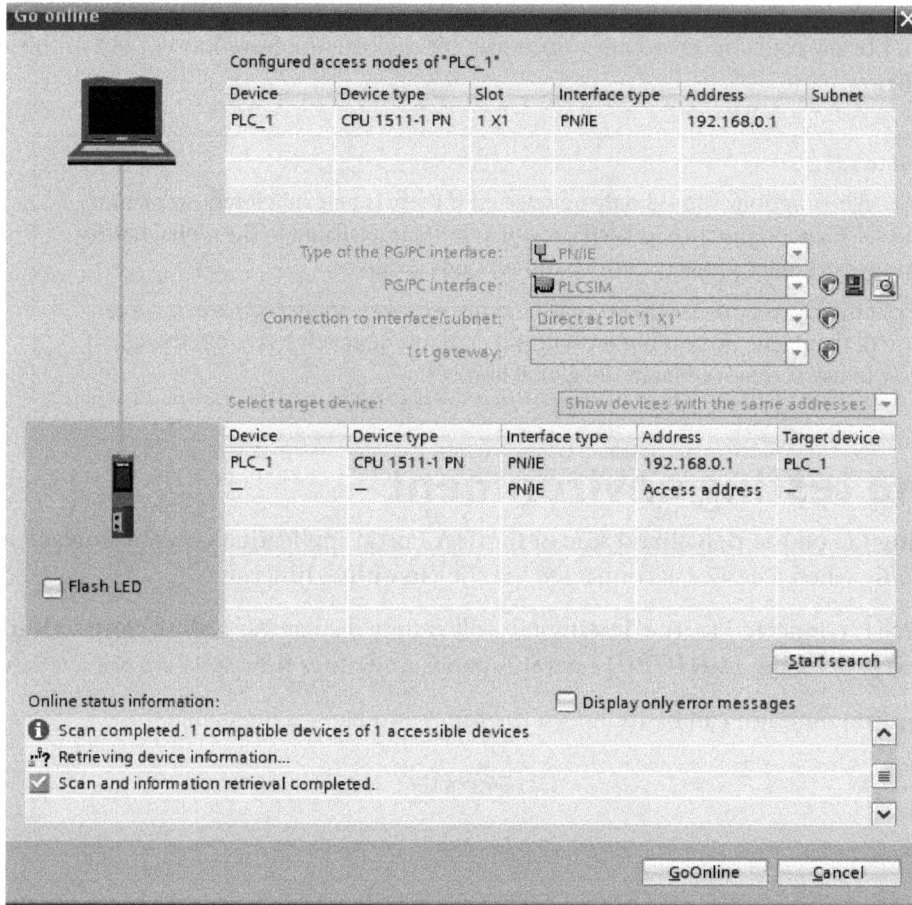

Figure 1.27 – Go online dialog that allows connection to devices

In the **Go online** dialog (*Figure 1.27*), devices can be searched for and connected to. Different interfaces can be selected from drop-down lists, and different subnets can be specified if required. Once the desired device is found, highlight it and click **GoOnline**. This will then connect the development environment to the device.

CPU operator panel

Once online, the **Testing** tab—specifically, the CPU operator panel—now contains data (if connected to a CPU in the project), as illustrated in the following screenshot:

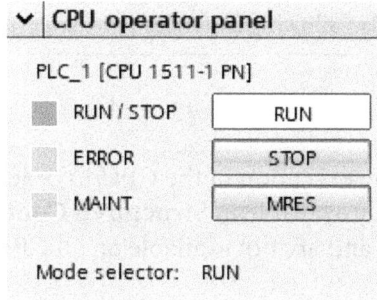

Figure 1.28 – CPU operator panel once online and connected to a CPU

These controls allow the CPU (PLC) to be stopped via the **STOP** button, memory reset via the **MRES** button, and placed back in to run via the **RUN** button.

The status of the CPU can also be read in this panel via the **RUN/STOP**, **ERROR**, and **MAINT** status LEDs. This is useful if the CPU is not local to the programming device running TIA Portal (connected via a **virtual private network** (**VPN**) connection, for example).

Call environment

The call environment is an important part of viewing and monitoring online code. In TIA Portal, a program block may be called more than once and via a different call environment (called by a different parent or instance). The **Call environment** panel in the **Testing** tab allows the call path to be defined, as illustrated in the following screenshot:

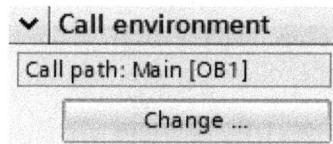

Figure 1.29 – Call environment panel

Figure 1.29 demonstrates a call path to a program block. When the **Change ...** button is pressed, a list of available call paths is displayed. A call path can be selected, and TIA Portal will then monitor the selected call path.

> **Note – SIMATIC Manager**
>
> Programmers who have worked with Siemens SIMATIC Manager may remember that multiple calls to the same function block on the same call path mean that only the first instance is available to be monitored. TIA Portal rectifies this behavior by also allowing instances to be selected.

Breakpoints

Breakpoints are used to pause the execution of the CPU in order to help debug programming logic. They are only available in **Structured Control Language** (**SCL**) and **Statement List** (**STL**) languages and are not available on all CPUs.

> **Warning – Outputs**
>
> If a breakpoint halts the execution of the CPU, outputs may remain in their current state until the breakpoint is released! This could cause damage to assets or personnel. It is important to understand the risks associated with using breakpoints in a live equipment environment.

Breakpoints and their usage are advanced topics that relate only to textual languages. Using breakpoints effectively is covered in more detail in *Chapter 5, Working with Languages in TIA Portal*.

Call hierarchy

When in offline mode (not online with a CPU), the **Call hierarchy** pane always displays **No call path available**.

When in online mode with a CPU and actively monitoring a program block via the **Monitor** function ⌨, the **Call hierarchy** pane displays the relative call path to the block being monitored, as illustrated in the following screenshot:

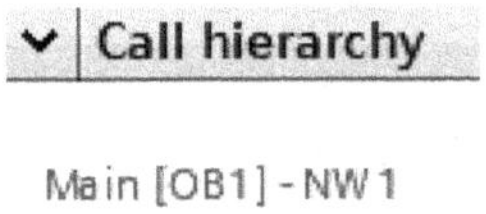

Figure 1.30 – Example of an object called by OB1

Figure 1.30 shows an example of a function block that is called by `Main [OB1]`. Should the block being monitored have multiple parents, the entire call hierarchy will be displayed.

Summary

This chapter has covered the required areas to start a project with a good understanding of TIA Portal's tools and layout, environment setup, and library management systems.

Understanding where and how to use the tools that TIA Portal offers is important. As a further learning exercise, it is recommended that users familiarize themselves with these tools. It is important to remember that TIA works best when all of the tools are utilized together.

The next chapter expands on the inner workings of TIA Portal , on how objects are created, and how these fit together to create a software structure of a program in a hierarchical manner.

2
Creating Objects and How They Fit Together

This chapter explains how, when creating objects, programmers should be mindful of how the hierarchy in TIA Portal operates. It helps programmers understand how to create a project's structure, how programming objects work with each other, how different instances of data in different spaces affect the project, and how interfaces that contain objects work.

We will cover the following topics in this chapter:

- Creating a project's structure
- Hierarchy in TIA Portal
- Instance and global data
- Interfaces and the effects of running PLCs

Creating a project's structure

When a new project is started, there's usually a good idea of the direction the project will take, such as whether control elements for equipment such as pumps will be added, what parameters will be created, and how data will be sent to and from a SCADA system. This information makes up most of the project, with logic and control only possible once the required assets and parameters have been established.

It is generally a good idea to enforce a workflow or project structure when developing with TIA Portal (or any PLC vendor). Being able to map out exactly where actions such as data exchange take place and where control logic takes place helps visualize a project, especially if the project that's being developed is large.

Data management

Data management is important in TIA Portal and Siemens has taken a very strong approach to ensure that the data and program cannot be compromised in an online environment (due to modifications) while it's running. This means that data blocks and instance data regularly need to be reinitialized to their default values if the structure of the program block changes. This ensures that while the data blocks are being modified, no incorrect values can be used in the program. With this in mind, it is important to protect data from being unnecessarily grouped with data that it has no relationship with. The following diagram shows project structure and data management:

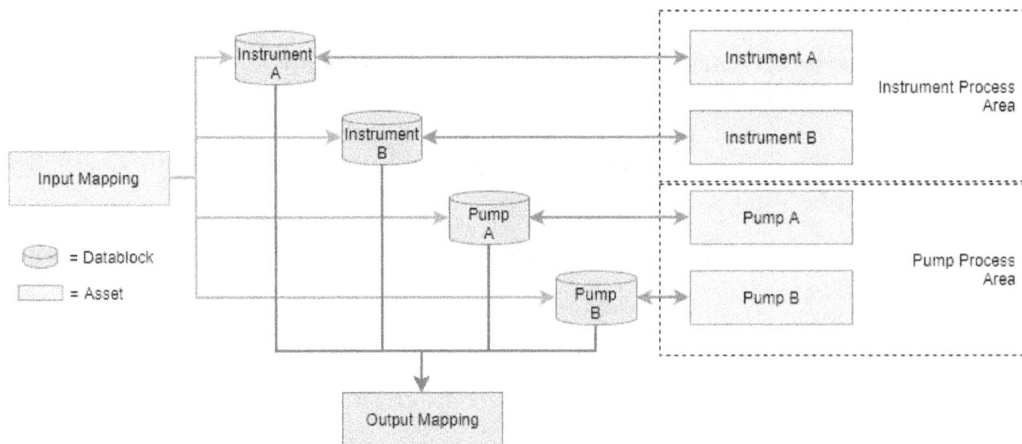

Figure 2.1 – Example of project structure and data management

The preceding diagram shows an example of managing data across four different assets, as well as how the data block that holds the data for that asset is interacted with:

- *Input mapping*: A logic area that maps input memory data into its relative data blocks for use in the project.

- *Data blocks*: The data blocks hold *all* the data that's required for the asset. This includes items such as parameters, SCADA interface variables, and control data.

- *Asset manager*: The asset manager is simply a function block that manages an asset completely, including items such as alarms, control logic, SCADA indication, and output commands.

- *Output mapping*: A logic area that maps data from the relative data blocks to their final output memory data.

This approach means that the project has naturally segregated development into the preceding key areas, with the fifth development area being to develop any interlocks and controls that are required to have assets interact/interface with each other. This serves as the workflow for the project.

> **Note**
>
> This approach purposely discourages the use of the %I, %O, and %M memory registers in code logic (outside of an asset manager). This is because the only places %I and %O should be used are in input mapping and output mapping, respectively.
>
> Memory registers (%M) should never be used in this design as programmers cannot guarantee that they are not already in use in a different project that the asset manager blocks may get used in. You're encouraged to use data blocks and instance data to store data and work with logic.

Example of use

In the *Structured project example 1* project, an example of this type of management is provided.

This example is for a simple use case where there's a red light on top of a windmill – the sort that warns low aircraft of its presence at night. This is a theoretical use case.

Project tree objects

Using the method laid out in *Figure 2.1*, the project tree in this example contains everything that's needed to create the `Pole_Light_Manager` (the asset manager), its respective data, and the mapping from its inputs and to its outputs:

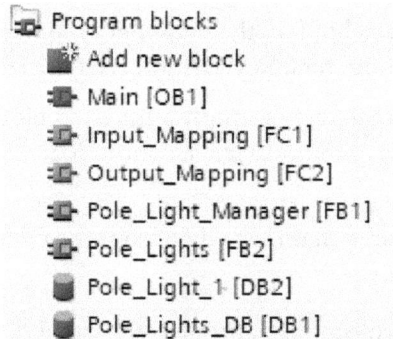

Figure 2.2 – Program blocks

The `Main [OB1]` block is used to call all of the required objects, in the desired sequence order. The `Pole_Lights` block is a function block that is used to hold different instances of the `Pole_Light_Manager` function block.

Main [OB1]

`Main [OB1]` calls the required program blocks in order from top to bottom, left to right:

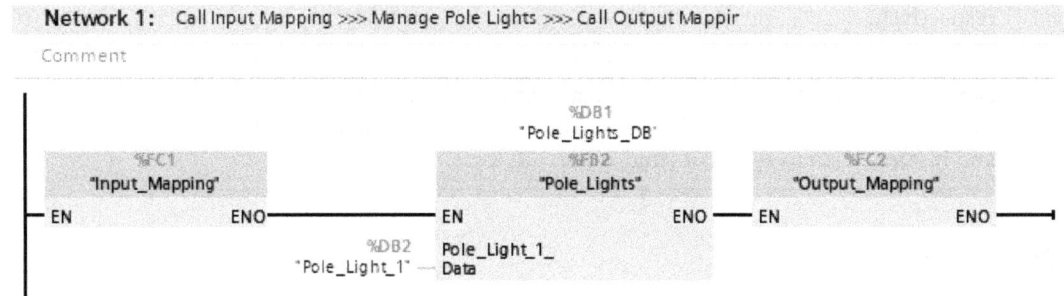

Figure 2.3 – Project structure – Main [OB1]

Because all the blocks are on the same network line in the preceding screenshot, they are executed from left to right. This calls `Input Mapping` to process raw inputs from the I/O, then the `Pole_Lights` function block, which manages all of the `Pole_Light_Manager` instances, and finally `Output Mapping`, which handles outputs at the I/O level.

Input mapping

The `Input_Mapping` block maps data from physical I/O to areas that are designated in the program (data blocks, for example):

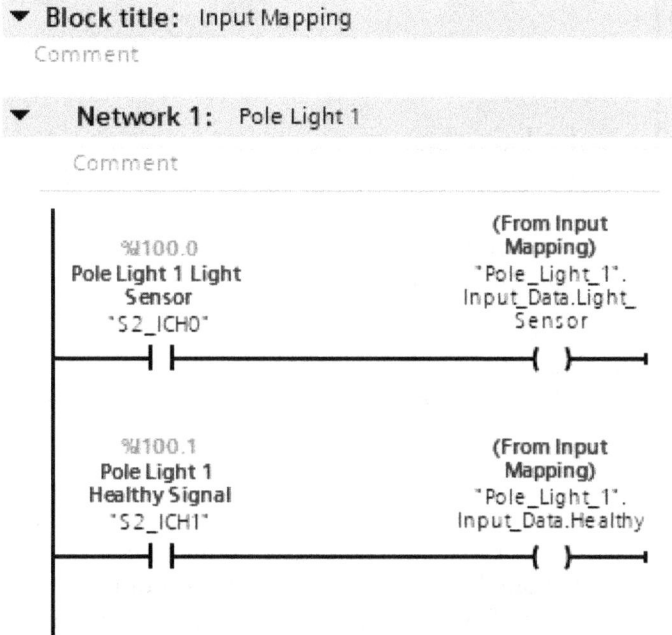

▼ **Block title:** Input Mapping

Comment

▼   **Network 1:**   Pole Light 1

Comment

```
        %I100.0                        (From Input
   Pole Light 1 Light                   Mapping)
       Sensor                        "Pole_Light_1".
      "S2_ICH0"                     Input_Data.Light_
                                        Sensor
   ─────┤ ├─────                   ─────( )─────


        %I100.1                        (From Input
    Pole Light 1                        Mapping)
   Healthy Signal                    "Pole_Light_1".
     "S2_ICH1"                      Input_Data.Healthy
   ─────┤ ├─────                   ─────( )─────
```

Figure 2.4 – Input mapping

This is the first step toward data management. By declaring an explicit place where all of the **inputs** for the project are mapped to the appropriate datasets (data blocks with a defined structure of data), the project naturally sets itself up to be *asset oriented* and all the signals can be deemed safe and correct for future use from the dataset.

By adopting this approach, raw signals can be modified *before* they are committed to the dataset, which is designed to be utilized by a management function block (`Pole_Light_Manager`, in this example).

For example, if the pole light hardware was a different model that had a low signal (*false*) when active and a high signal (*true*) when inactive, `S2_ICH0` could be inverted in the mapping layer without it affecting the normal use of the pole light later in the project.

Similarly, if the pole light was a newer version that produced an **analog input** for a light intensity measurement, the S2_ICH0 input referenced in the preceding screenshot could be replaced with the following:

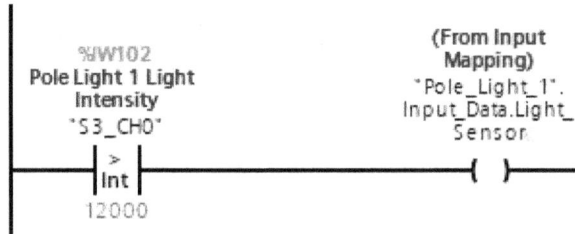

Figure 2.5 – Analog alternative – when above 12mA, consider the light sensor as active and write it to the appropriate variable in the dataset

Note that S2_ICH0 has been replaced with S3_CH0, the right-hand side **coil** has not changed, and the variable that's in use has not changed. This means that the function block that manages Pole_Light does not need to change, despite the environment changing around it.

Process area

The process area is everything that sits between the input mapping and the output mapping – essentially, it's the main bulk of the project. At this point, all the datasets should contain the input I/O data from the input mapping layer and can be deemed safe to work with.

In the Pole_Lights function block, there is a single instance of Pole_Light_Manager:

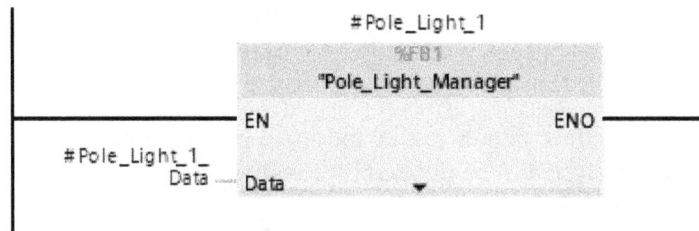

Figure 2.6 – The Pole_Light_Manager parent function block

It is this function block that manages all of the calls to the different instances of Pole_Light. This would make Pole_Light_Manager the child of Pole_Lights and Pole_Lights the parent of Pole_Light_Manager. This is called a **call structure**.

When `Pole_Light_Manager` is called, it is safe to assume the following:

- `Input_Mapping` has been called:

  - All the required input data is present and correct.

- All the required variables that exist in the dataset have been passed into (and out of) the function block via the data interface pin.

The preceding conditions are all satisfied via the project structure that has been created through the workflow and data management that's been designed for the project.

Dataset

A dataset is simply a data block that contains a known and explicit data structure.

If the data structure is known and the variables within it are placed in structures with valid symbolic meaning, it becomes very easy to use the data inside and outside the function blocks that manage the data.

`Pole_Light_1` is an example of a dataset:

	Name	Data type
	Pole_Light_1	
1	▼ Static	
2	▼ Input_Data	Struct
3	Light_Sensor	Bool
4	Healthy	Bool
5	▼ Control_Data	Struct
6	Light_Sensor_Active	Bool
7	▼ Status_Data	Struct
8	Light_Flashes	LInt
9	Healthy	Bool
10	Maintenance_Req...	Bool
11	▼ SCADA_Data	Struct
12	▶ From_SCADA	Struct
13	▶ To_SCADA	Struct
14	▼ Output_Data	Struct
15	Output	Bool

Figure 2.7 – Dataset structure using the Struct data type

Notice that a design that's similar to the project workflow in *Figure 2.1* exists in the dataset:

Input Data –> Process Data –> Output Data

The process data is made up of different types of signals, such as SCADA- and status-related ones. This means that this `Pole_Light_1` data block could be referenced directly from a SCADA system by referencing the `SCADA_Data` internal structure.

> **Note**
>
> Sometimes, it's better to have duplicated data in different structure areas within the dataset, which enables the dataset's structured areas to be used completely compartmentalized from any other area in the dataset.
>
> Take note of this when you're designing a dataset to try and reduce its overall size.

Output mapping

Once the management function blocks have processed the assets by reading inputs, executing logic, and setting outputs, the associated dataset will contain the current output values, ready to be sent to the I/O layer.

This is performed by the `Output_Mapping` function, which acts as a mapping layer:

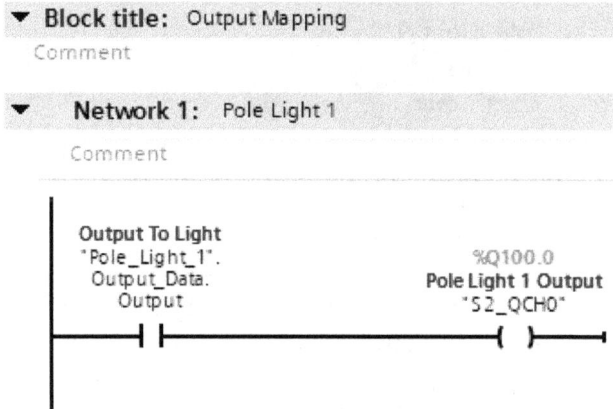

Figure 2.8 – Output mapping

The `Output_Mapping` function does the same thing as the `Input_Mapping` layer, only vice versa. It takes data from the dataset and sends it to the I/O so that assets can be controlled correctly.

As with the `Input mapping` layer, this provides us with the opportunity to modify signals to suit the output requirements, without changing the function block that provides the control (the asset manager).

Hierarchy in TIA Portal

TIA Portal works by using a **parent/child hierarchy**, where parent objects call child objects that they depend on to function correctly. The following is an example of a parent/child hierarchy:

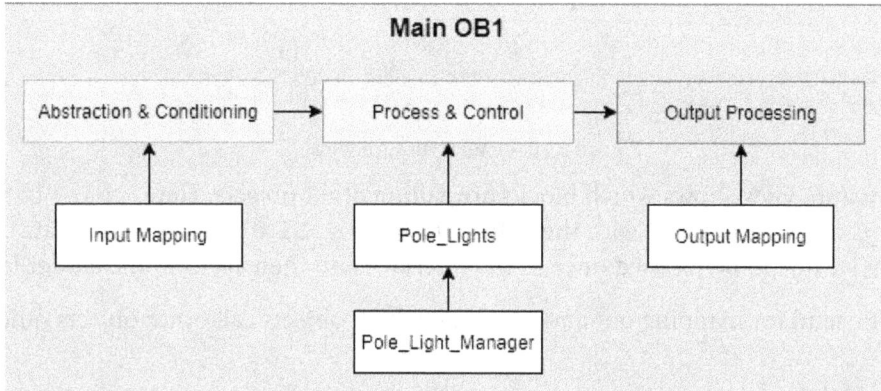

Figure 2.9 – Example of a parent/child hierarchy within the example project

The preceding diagram shows how the `Pole_Lights` function block is dependent on `Pole_Light_Manager`. If `Pole_Light_Manager` is modified, then the `Pole_Lights` function block would also need to be modified to utilize the new version. This is because the **interface** of the `Pole_Lights` function block would need to change its size to accommodate any new data that's added to `Pole_Light_Manager`.

This example also shows that input mapping and output mapping do not have any dependents, so they do not rely on any other function blocks or functions to process their logic.

Call structure

In TIA Portal, a view window can be opened that displays the call structure. This is a special view that helps map out the logical hierarchy of the project and the order in which blocks will be processed when the parent is called.

> **Note**
> To open the call structure view, from the project tree, right-click on an object in the `Program blocks` folder and choose **Call structure**. It doesn't matter which object you choose; the same call structure windows will open.

The following is the call structure view:

Call structure of PLC_1				
Call structure	I	Address	Call freq..	Details
▼ 🔲 Main		OB1		
▶ 🔲 Input_Mapping		FC1	1	@Main ► NW1 (Call Input Mapping >>> Manage P...
▶ 🔲 Output_Mapping		FC2	1	@Main ► NW1 (Call Input Mapping >>> Manage P...
🔲 Pole_Light_1 (Data block derived from UDT_Pole_Light)		DB2	1	@Main ► NW1 (Call Input Mapping >>> Manage P...
▼ 🔲 Pole_Lights, Pole_Lights_DB	🔲	FB2, DB1	1	@Main ► NW1 (Call Input Mapping >>> Manage P...
🔲 Pole_Light_Manager, #Pole_Light_1	🔲	FB1	1	@Pole_Lights ► NW1
🔲 Pole_Light_Manager, #Pole_Light_1	🔲	FB1	1	Block interface

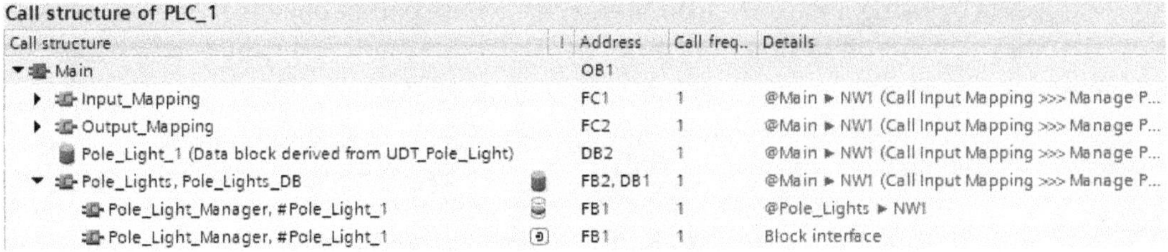

Figure 2.10 – Call structure view

The call structure view shows which blocks are calling other objects. Here, `Pole_Lights` calls `Pole_Light_Manager` with the data from `#Pole_Light_1` (instance data). It is displayed twice due to being used once in the interface and then again in the ladder logic.

This view is useful for mapping out and verifying which objects call other objects quickly.

Dependency structure

The dependency structure view is similar to the call structure view but shows objects that have other objects that depend on them.

The dependency structure view can be accessed by visiting the call structure view and then selecting **Dependency structure** from the tabs at the top of the screen:

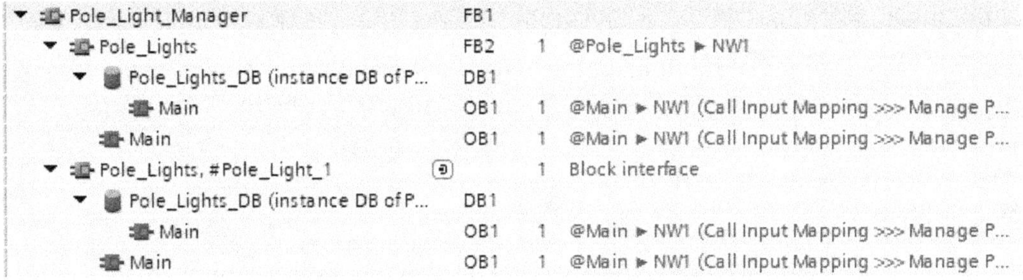

▼ 🔲 Pole_Light_Manager	FB1		
▼ 🔲 Pole_Lights	FB2	1	@Pole_Lights ► NW1
▼ 🔲 Pole_Lights_DB (instance DB of P...	DB1		
🔲 Main	OB1	1	@Main ► NW1 (Call Input Mapping >>> Manage P...
🔲 Main	OB1	1	@Main ► NW1 (Call Input Mapping >>> Manage P...
▼ 🔲 Pole_Lights, #Pole_Light_1 🔲		1	Block interface
▼ 🔲 Pole_Lights_DB (instance DB of P...	DB1		
🔲 Main	OB1	1	@Main ► NW1 (Call Input Mapping >>> Manage P...
🔲 Main	OB1	1	@Main ► NW1 (Call Input Mapping >>> Manage P...

Figure 2.11 – Dependency structure view

The preceding screenshot shows `Pole_Light_Manager` as the object with no dependencies. All of the objects listed below it depend on the `Pole_Light_Manager` object. As the hierarchy is expanded, more blocks that depend on child objects are displayed.

For example, `Main` depends on `Pole_Lights_DB`, which depends on `Pole_Lights`, which depends on `Pole_Light_Manager`. If `Pole_Light_Manager` were to be removed, the entire hierarchy would fail back to `Main`.

> **Note**
>
> As the structure of the project develops, it may become more and more important to understand what objects will be affected if changes are introduced. Remember that if a child object's data structure changes, the parent could be affected too and also need to be modified or recompiled.
>
> This becomes even more important when libraries are being used across multiple projects.

Parent/child relationships

It is important to understand the call and dependency structure in TIA Portal, especially when it comes to modifying items that have other items that depend on them:

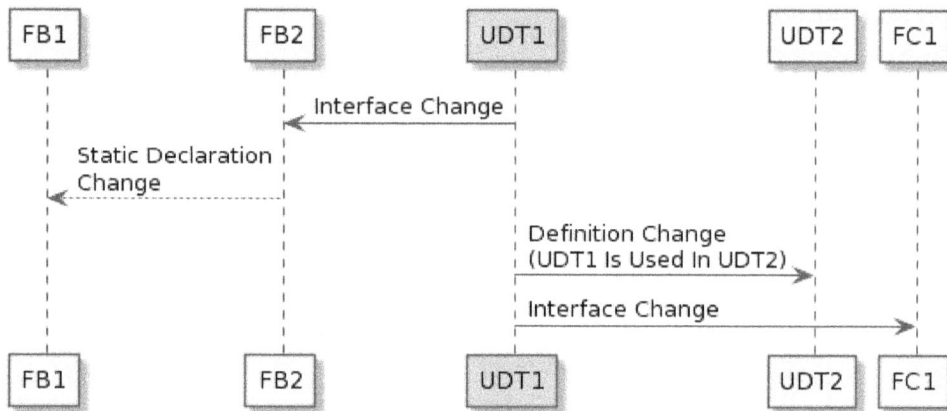

Figure 2.12 – Example of dependency changes

Program objects in TIA work at a hierarchical level, where parent objects are affected when child objects and their dependencies are changed.

In the preceding example, all the objects are typed in the project library. **UDT1** is being modified (**in test**). Once **UDT1** has been selected for modification, the objects that depend on it also switch to in test mode and are issued a minor revision increment automatically. This is because of the hierarchy and call structure. The preceding diagram explains that because **UDT1** is a **child** or indirect **dependent** of other objects, it must also be updated to preserve the library's integrity.

FB1, **FB2**, **UDT2**, and **FC1** all need to be modified because **UDT1** has been modified:

- **FC1**: Changes to the interface need to be made since **UDT1** is used in the interface of **FC1**.

- **UDT2**: Changes to the internal data of **UDT2** need to be made since **UDT1** is a dependency that is used in **UDT2**.

- **FB2**: Changes to the interface need to be made since **UDT1** is used in the interface of **FB2**.

- **FB1**: Because the interface of **FB2** has changed and **FB2** is a child of **FB1**, the interface of **FB1** needs to be updated to accommodate the new data.

Instance and global data

As with most PLCs, TIA Portal has two types of data – **global** and **instance**. What defines the data as a particular type is where and how it is used. What is classed as global or instance data depends on what information is being held and how it interacts or interfaces with logic:

- **Global data**:

  - Data that can be accessed anywhere in the project, at any hierarchy level.

  - Data can be freely defined by the programmer, including the creation of sub-structures that contain data or UDT definitions.

  - It cannot hold instance data for a function block, even in a sub-structure.

- **Instance data**:

  - Data that exists explicitly to be used with a function block or UDT.

  - Data is automatically defined by the requirements of the function block/UDT it is associated with.

  - It will contain any sub-instances of function blocks that have been called within the parent function block.

> **Note**
>
> There is an additional global data block type called **Array DB**. This allows a data block to be explicitly configured as an array and allows no elements other than those that fit the defined array element type.
>
> This is useful for systems that use queue patterns or recipe management.

Using instance and global data

Instance and global data are fundamental to any PLC application. Without them, it's not possible to store, move, read, or write data in different areas of the program. Being able to use instance and global data depends on where it's used initially.

Typically, a project will exist in an **organization block (OB)**, which calls a series of function blocks that require instance data. Usually, a few global data blocks are required to interface with other project environments such as SCADA, as shown in the following diagram:

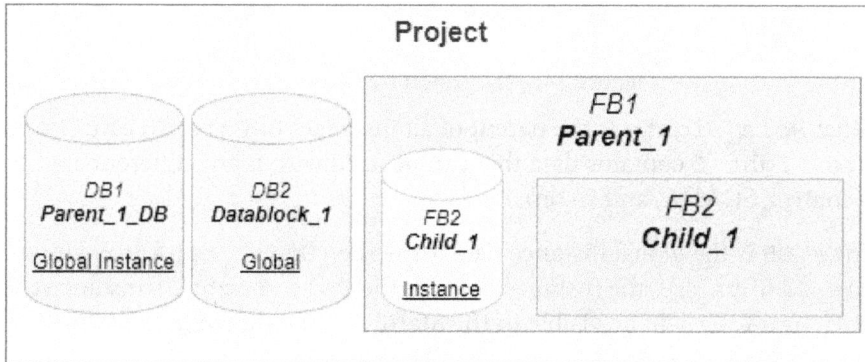

Figure 2.13 – Example of global and instance data

The preceding diagram shows how a project may lay data out. In this example, there are three data objects:

- `Parent_1_DB`: A **global instance data block**. This is a data block that holds instance data for **FB1**, accessible globally in the project.

- `Data block_1`: A **global data block**. This is a standard data block that simply holds any data for use globally in the project.

- `Child_1 instance`: An instance of FB2's data, which is stored in **Parent_1_DB**.

The *pole light* example shows what this layout looks like in the project tree:

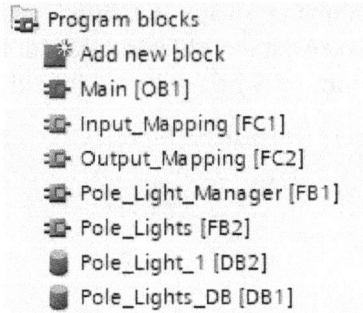

Figure 2.14 – The project tree's layout

Remember that `Pole_Lights` is the parent of all instances of `Pole_Light_Manager` and that `Pole_Light_1` contains data that can be used with many different aspects of the project (control, SCADA, and so on).

`Pole_Lights_DB` is the global instance data for `Pole_Light_Manager`. Upon opening `Pole_Lights_DB`, the instance data for the `Pole_Light_1` instance of `Pole_Light_Manager` will be visible in the **Static** area, as shown in the following screenshot:

		Name	Data type
		Pole_Lights_DB	
1		Input	
2		Output	
3	▼	InOut	
4	■	Pole_Light_1_Data	"UDT_Pole_Light"
5	▼	Static	
6	■ ▼	Pole_Light_1	"Pole_Light_Manager"
7	■	Input	
8	■ ▼	Output	
9	■	Ref_Light_Active	Bool
10	■	Ref_Light_Healthy	Bool
11	■ ▼	InOut	
12	■	Data	"UDT_Pole_Light"
13	■ ▼	Static	
14	■ ▶	Light_Sensor_Delay	TON_TIME
15	■ ▶	On_Duration	TOF_TIME
16	■ ▶	Off_Duration	TON_TIME
17	■	On_Rising_Edge	Bool

Figure 2.15 – Pole_Lights_DB

While this may be confusing at first, the principle behind this becomes simple when it's laid out in a graphical format, as follows:

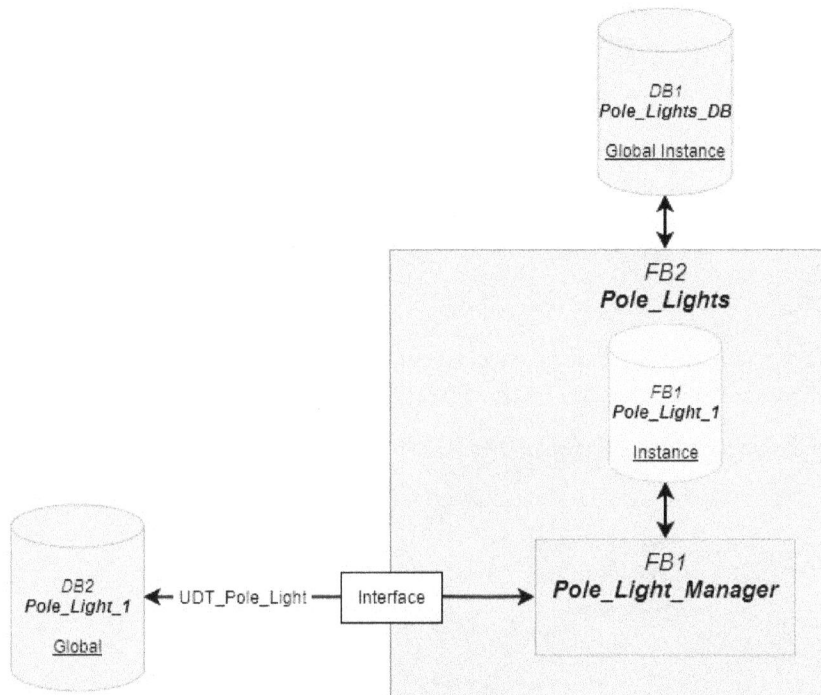

Figure 2.16 – Graphical example of global, global instance, and instance data

Remember that every function block needs an accompanying dataset. Whether it's global, global instance, or instance data, it must exist somewhere.

Accessing data

It is perfectly acceptable to use the same names in different locations. In the preceding diagram, the name `Pole_Light_1` is given both to the global data that houses the data for use with a `Pole_Light_Manager` asset function block, and the actual instance of `Pole_Light_Manager` (`Pole_Light_1`) inside `Pole_Lights`.

Because `Pole_Light_1` is a function block instance, inside the `Pole_Lights` function block, its full path is `Pole_Lights_DB.Pole_Light_1`, which is different from the data block for `Pole_Light_1`, which is simply `Pole_Light_1` and is not a function block instance.

Remember that these two data objects that are named the same hold very different information but are both common to `Pole_Light_Manager`.

If the global data of `Pole_Light_1` (DB2) is accessed, the following will be available:

			Name	Data type
			Pole_Light_1	
			Name	Data type
1	◀	▼	Static	
2	◀	▶	Input_Data	Struct
3	◀	▶	Control_Data	Struct
4	◀	▶	Status_Data	Struct
5	◀	▶	SCADA_Data	Struct
6	◀	▶	Output_Data	Struct

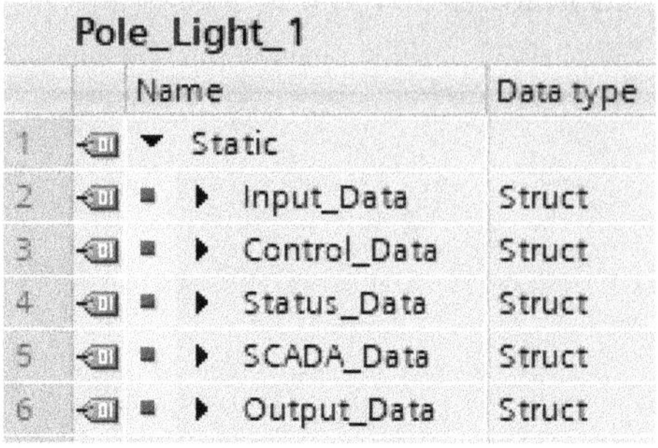

Figure 2.17 – The Pole_Light_1 global data block

This is simply information that can be used with or concerning `Pole_Light_1` as an asset. It is not instance data and can be used freely anywhere in the logic of the project.

If the instance data of `Pole_Lights_DB.Pole_Light_1` is accessed, the following will be available:

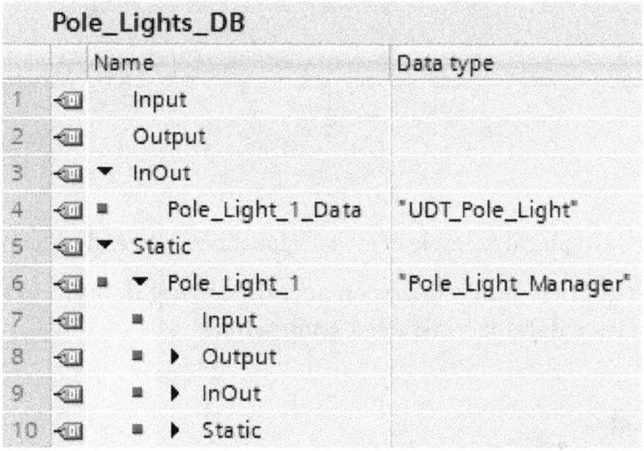

			Name	Data type
			Pole_Lights_DB	
1	◀		Input	
2	◀		Output	
3	◀	▼	InOut	
4	◀		Pole_Light_1_Data	"UDT_Pole_Light"
5	◀	▼	Static	
6	◀	▼	Pole_Light_1	"Pole_Light_Manager"
7	◀		Input	
8	◀	▶	Output	
9	◀	▶	InOut	
10	◀	▶	Static	

Figure 2.18 – The Pole_Lights_DB global instance data block

This is instance data and it's structured with the relative interface requirements, where `Pole_Light_1` is a static instance of `Pole_Light_Manager` within this data. Accessing this path allows you to access the interface of `Pole_Light_1` and any static data within that instance.

Configuration options

Depending on whether a data block contains instance or global data, different options exist in different places so that you can configure the desired behavior.

Start value

The **start value** is the value that a variable will contain before the PLC is scanned for the first time. Start values behave differently, depending on where they are located:

- **Global data**: The start value is set via the global data block and is used when the PLC is started to set the required value.

- **Function block/instance data**: The start value can be set in two places for instance data:

  - **Function block interface**: The interface variable is set to the required start value when the function block is scanned for the first time.

  - **Instance data block**: The interface variable is set to the required start value when the PLC is started.

> **Note**
>
> If the start value is declared in the instance data block, it will override the setting of the function block. This happens when the PLC is downloaded and often causes issues if it's set incorrectly as the data's value will be overwritten with the start value. This only occurs if the download causes reinitialization.

Retain

The **Retain** option is used to store data in the non-volatile section of memory in the PLC. This means that if the PLC were to suddenly suffer from a power loss, the data would be retained.

The options for whether a value should be retained or not are a little more complex as it also depends on if optimized data is in use (by default, all new blocks are optimized). The following options are available:

- **Global data – optimized/non-optimized**:

  - Simply checking the **Retain** option will add the variable to the Retain memory set.

  - If one variable is set to Retain, all the variables must be set to Retain and TIA Portal will automatically set the other variables to be retained.

- **Instance data – optimized**:

  - Retain mode cannot be set in the instance data block, but it can be set in the function block interface.

- A dropdown list is displayed that allows the programmer to define whether **Retain** is **Non-Retain**, **Retain**, or **Set in IDB**. The latter allows the behavior of a non-optimized function block to be issued.

- **Instance data – non-optimized**:

  - `Retain` mode cannot be set in the function block and must be set in the instance data block.

> **Note – Non-Optimized Child in an Optimized Parent**
>
> While having a non-optimized function block instantiated inside an optimized function block should be discouraged, sometimes, it cannot be helped. In these situations, the child function block will state **Set in IDB** as the **Retain** option and it will not be possible to change it. Here, the programmer can set the **Retain** option in the instance data block of the parent function block.

Interfaces and the effects on running PLCs

The different configurations, types, and combinations of interfaces affect not only the behavior of the PLC but its performance. It is important to understand that a poorly configured data management system will result in poor PLC performance.

One of the biggest factors of this in TIA Portal is whether data is optimized or not.

Optimized data/non-optimized data

The concept of a PLC optimizing its data is not a new one. There are many PLCs that do this all the time, but Siemens has allowed programmers to choose if they want an object to be optimized or not.

> **Note**
> By default, all objects that are created in TIA Portal have optimization enabled.

Differences between optimized and non-optimized data

The main difference between optimized and non-optimized is the amount of memory that's used to store information. The following diagram shows the difference between how data is defined in a data block and how it is represented in the PLC:

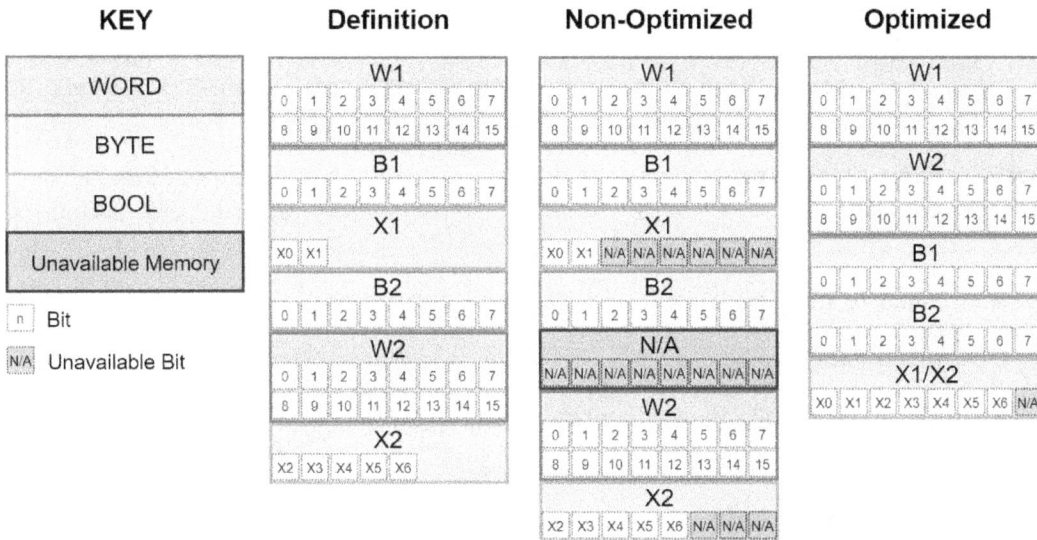

Figure 2.19 – Differences between non-optimized and optimized data

When the preceding data is laid out in a data block, the non-optimized version of the block results in memory gaps, where data cannot be stored but must be reserved by the PLC to retain data consistency.

This can be seen in TIA Portal by viewing the **Offset** column (when non-optimized).

Note that the **Offset** column displays the byte at which the variable address starts:

	Name	Data type	Offset
▼	Static		
▪	W1	Word	0.0
▪	B1	Byte	2.0
▪	X1	Bool	3.0
▪	X2	Bool	3.1
▪	B2	Byte	4.0
▪	W2	Word	6.0
▪	X3	Bool	8.0
▪	X4	Bool	8.1
▪	X5	Bool	8.2
▪	X6	Bool	8.3
▪	X7	Bool	8.4

Figure 2.20 – Non-optimized data block with unavailable reserved data

The preceding screenshot shows that between offset addresses **3.1** and **4.0**, there are no addressed variables. The same happens between **4.0** and **6.0**. This is because a **Byte** is 1 byte long (4.0 to 5.0) but a **Word** (2 bytes long) cannot start on an odd offset, so a **Byte** (5.0 to 6.0) must be reserved to preserve the consistency.

Figure 2.19 shows that optimized data does not have as many reserved or not available areas in the data. Here, the order of the data has been reorganized to optimize the amount of space it takes up in the PLC.

There are benefits to using both optimized and non-optimized data:

- **Optimized data benefits**:

  - A smaller footprint in the PLC's memory.

  - Enhanced logic instructions.

  - Advanced functions are only available to optimized blocks.

- **Non-optimized data benefits**:

  - A known data structure at design time.

  - Different Variant data types are available.

  - Easy to use with open network protocols such as Modbus.

Mixing optimized and non-optimized data

Note that mixing optimized and non-optimized data is not recommended. When data is moved from one memory layout to the other, the CPU needs to work extra hard to ensure the data is processed from the location specified in the program.

The scan time will likely increase while this data is copied and moved to the correct location.

Passing data through interfaces

To use function blocks, interfaces must be created that pass data in and out of the function block (unless you're using global data within the function block). Interfaces are how data that's external to the function block is interconnected to the data within the instance data. There are four interface types:

- **Input**: Data that's copied into the function block's instance data.

- **Output**: Data that's copied out of the function block's instance data.

- **InOut**: Data that's referenced within the function block.

- **Static**: Data that's stored within the function block's instance data.

The following screenshot shows what each interface type looks like:

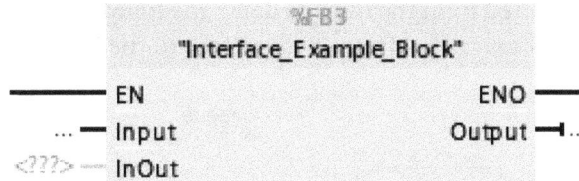

Figure 2.21 – Example of a function block interface

Notice that *Static* is not present in the interface. This is because it is stored in the instance data and not interconnected to any outside data.

The *Input* type is displayed with a straight line and accepts the value of the stated interface data type. This also accepts ladder logic beforehand if the result on the wire is the accepted data type (Boolean, for example). This interface also can't be connected, and the start value will be used from the instance data instead.

The *Output* type is displayed with a straight line and termination point, which only allows a variable to be placed. This interface also can't be connected, and the start value will be used from the instance data instead.

The *InOut* type must be connected; otherwise, the project won't compile or memory access violations will occur when the PLC tries to run. Only the data type that's been defined is accepted.

> **Note**
> The interface pins may be different colors, depending on the data type that is being passed to them.

Copying data to instance data

Depending on the type of interface that's being used, the PLC will behave differently when it comes to memory management and the impact this has on memory reserves. If *Input* or *Output* interfaces are used in a function block, the data is copied. If *InOut* interfaces are used, the data is simply pointed to in the outside data. The following diagram shows an example of how data is interconnected from the interface to the instance data block:

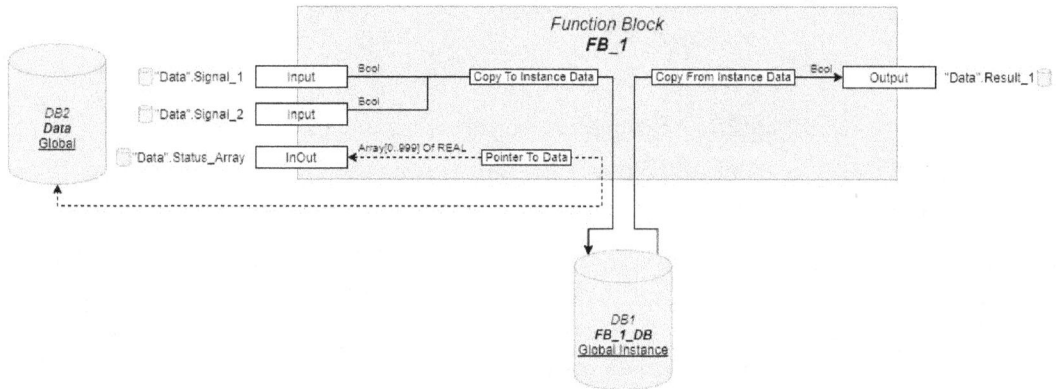

Figure 2.22 – Example of how data is interconnected from the interface to the instance data block

When an interface is processed by the PLC's runtime, the data at an *Input* interface is copied into the global instance at `<InstanceData blockName>.<InputVariableName>`. This means that all the data is essentially global since, eventually, the hierarchy will dictate that a global object will be at the top of all the data. Here, the interface interconnects data from outside the instance data with data inside the instance data.

This means that the instance data holds a containerized copy of the data that was passed to it when the interface was processed. For example, if some input is passed to a function block and the internal logic of the function block changes the value of the input that was passed to it, the variable that's outside of the function block does not change.

Referencing data

The preceding diagram contains an *InOut* interface that behaves differently from that of the *Input* and *Output* interfaces. An *InOut* interface creates a local pointer to the data, a method by which the PLC can *look up* where the data to be used is stored. This becomes increasingly important in large applications with large datasets that are used within function blocks.

The following diagram shows the different data and its sizes in the *Data* data block shown in the preceding diagram:

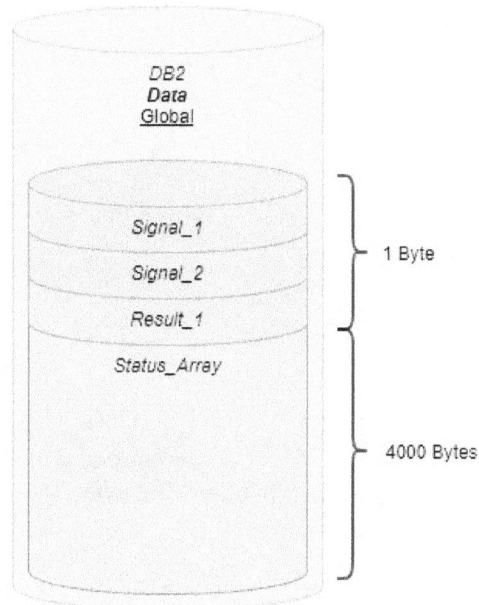

Figure 2.23 – Example of the data's size in bytes

The size of the data that's being copied into a function block costs processing time.

The `Status_Array` value is 4000 bytes long (1 real = 4 bytes, 1,000 elements in the array).

If *InOut* copied the data to the instance dataset, this would mean that the PLC would require 4000 bytes in the *Data* data block to store the data, 4000 bytes in the instance data of the function block, and that 8000 bytes would be copied in *total* – 4000 into the instance data and 4000 out of the instance data. This would be considered a memory-heavy function block.

InOut creates a pointer that's made up of a reference to where the data to be used exists. Although the variables connected to the *Input*, *Output*, and *InOut* interfaces can all be defined in the same way (symbolically by name, as shown in the following screenshot), the way the PLC uses this information is different:

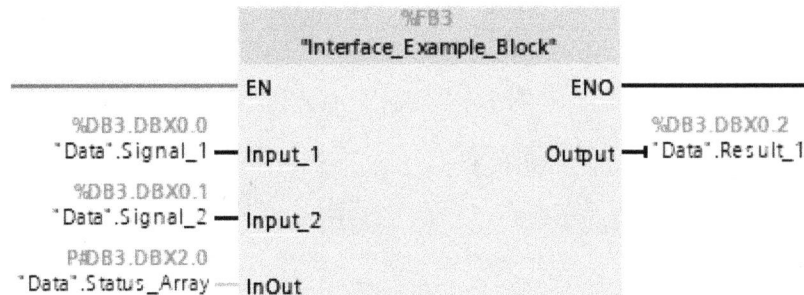

Figure 2.24 – Pointer addressing on the InOut interface

Notice that the `Input_1` and `Input_2` addresses above the variable (`%BD3.DBX0.0`, for example) start with a `%`, whereas `InOut` begins with a `P#`. These symbols identify where and how the data is being used:

- `%`: Direct access – the PLC will directly access this data.

- `P#`: Pointer:

  - The PLC will create a 6-byte structure that contains the following:

    - The length of the data

    - The pointer to the data

  - This data is not accessible and is not represented in the configuration environment other than by `P#`.

The `P#DB3.DBX2.0` address that `"Data".Status_Array` represents is a pointer address and points explicitly to `Data block 3, offset 2.0`.

> **Note**
>
> If *InOut* were to be hardcoded with a pointer, `"Data".Status_Array` would need to be replaced with `P#DB3.DBX2.0 BYTE 1000`. However, pointers do not carry a data type and the compiler would fail, stating that the *InOut* interface has an invalid data type, despite pointing to the same place it does symbolically.

Memory advantages

This approach of using *InOut* interfaces for large data significantly reduces memory impact and reduces the time it takes to process the code. This is because the copy that's required to copy data from outside the function block and into the instance data does not occur, nor does the instance data need to store or reserve any memory for the pointer values.

This can be seen if large data is placed on the interface of a function block in both the *Input* or *Output* interfaces and also the *InOut* interface:

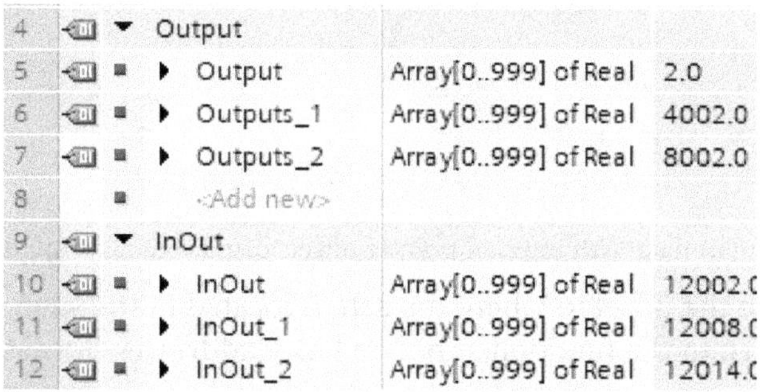

4	▼		Output		
5	■	▶	Output	Array[0..999] of Real	2.0
6	■	▶	Outputs_1	Array[0..999] of Real	4002.0
7	■	▶	Outputs_2	Array[0..999] of Real	8002.0
8	■		‹Add new›		
9	▼		InOut		
10	■	▶	InOut	Array[0..999] of Real	12002.(
11	■	▶	InOut_1	Array[0..999] of Real	12008.(
12	■	▶	InOut_2	Array[0..999] of Real	12014.(

Figure 2.25 – Example of memory requirements between interfaces

Note how 4000 bytes have been reserved for each instance of an Output variable but that for InOut, only 6 bytes have been reserved for the pointer structure.

Drawbacks

Using *InOut* data can have some drawbacks. Remember that any changes that are made to the variables that are passed via a pointer will be immediately applied to the source of the pointer. This can become problematic if asynchronous calls for the data are made by an outside source such as SCADA, or if interrupt OBs access data that is only partially completed.

Using InOut interface variables

InOut variables are *de-referenced* from their pointer automatically in TIA Portal, which means that they can be used in the same manner as normal variables when they're declared in an interface:

InOut			
▼ InOut		Array[0..999] of Real	12002.0
■	InOut[0]	Real	0.0
■	InOut[1]	Real	4.0
■	InOut[2]	Real	8.0
■	InOut[3]	Real	12.0
■	InOut[4]	Real	16.0
■	InOut[5]	Real	20.0

Figure 2.26 – InOut data with internal pointer offsets displayed in an InOut interface

This simply means that the offset to where the data is located in the source data is applied from the start position and the function/function block addresses the data via the pointer and offset.

Summary

This chapter covered the fundamental requirements for structuring and managing the data that's used throughout a project. We have highlighted that TIA Portal is a **hierarchical development environment** and that changes to dependency objects will affect objects that depend on them. This can ripple through a project and affect any object.

We also learned about global/instance data and demonstrated that these data types are at the heart of data management and that we should consider how they are used throughout a project. After that, we learned about optimized and non-optimized data and highlighted how it affects a project and how it may or may not be suited for the application you're designing. Remember that how data is passed through the interfaces of program blocks can make a big difference to the efficiency of the PLC's runtime.

The next chapter builds upon the lessons we've learned regarding structure and hierarchy. We will introduce the concept of **user-defined types** (**UDTs**) to create custom data types that enforce project structures further. This approach, combined with libraries, creates a solid platform where custom software standards can be built.

3
Structures and User-Defined Types

This chapter explores how to create, utilize, and deploy structures as **structs** and **User-Defined Types (UDTs)** effectively in a TIA Portal project. UDTs are the most underutilized tool in PLC programming and are extremely efficient and provide the rigid data structure required for asset-based programming. This chapter also expands on previously learned information about interfaces and how they affect data usage in logic objects.

After reading this chapter, a programmer should feel comfortable with both structs and UDTs in TIA Portal and have enough knowledge to do the following:

- Add structs and UDTs to a project.
- Know where UDTs are stored in a project.
- Know the differences between structs and UDTs.
- Know how to simplify interfaces to **program blocks**.
- Know of potential drawbacks of using structs/UDTs and how to mitigate or overcome the issues that arise.

The following topics will be covered in this chapter:

- What are structs and UDTs?
- Creating structs/UDTs – best practices
- Simplifying interfaces with structs/UDTs
- Drawbacks of structs and UDTs

What are structs and UDTs?

Essentially, structs and UDTs can be described as performing the same function: to hold data that has different types under a parent variable. While they perform the same fundamental function, a UDT is much more powerful than a struct. This is because a struct is a singular instance of a group of **variables** and a UDT is an actual data type.

Both of these datatypes can be described as a *structure*.

Structs

Structs are groups of variables defined under a parent variable with the datatype of struct. Siemens regards the use of structs as *use only when necessary* as they can add considerable overhead to a PLC's performance if used too often. *Figure 3.1* shows two sample pumps' **global data blocks** for storing information associated with the asset:

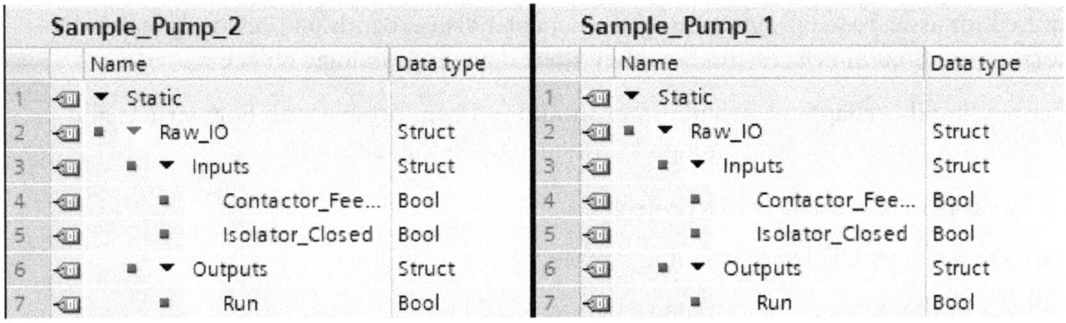

Sample_Pump_2			Sample_Pump_1		
	Name	Data type		Name	Data type
1	▼ Static		1	▼ Static	
2	▼ Raw_IO	Struct	2	▼ Raw_IO	Struct
3	▼ Inputs	Struct	3	▼ Inputs	Struct
4	Contactor_Fee...	Bool	4	Contactor_Fee...	Bool
5	Isolator_Closed	Bool	5	Isolator_Closed	Bool
6	▼ Outputs	Struct	6	▼ Outputs	Struct
7	Run	Bool	7	Run	Bool

Figure 3.1 – Example of structure definition in global data blocks

Both of these pumps use exactly the same struct, Raw_IO; however, they are not tied together in any way. Each instance of the Raw_IO struct exists in its own instance; modifying one will not modify the other.

> **Siemens Terminology**
>
> Siemens refers to the struct data type as an **anonymous structure**. This is because it has no hard definition (like a UDT does). A hard definition means that the project contains an object that defines how the structure is constructed. These are found in the `PLC Datatypes` folder in the Project tree.

Composure

Due to its composition of different types, a struct tag behaves very differently in **optimized** and **non-optimized** blocks:

- Optimized:

  - The first word starts on a non-optimized memory address (byte with an even address).

  - Subsequent variables are then ordered by size and do not reflect the structure in the editor.

- Non-optimized:

  - The first word starts on a non-optimized memory address (byte with an even address).

  - Variables are ordered as they appear in the editor.

  - Packing bytes (unavailable memory addresses) are used to ensure memory does not start on uneven byte values.

Accessing variables

A struct's internal variables can be accessed by using the entire call path with dots (.) as separators. An example is shown here:

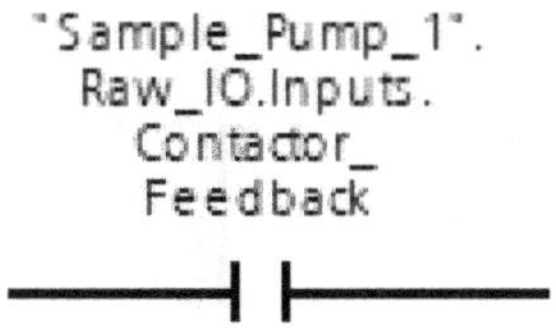

```
"Sample_Pump_1".
  Raw_IO.Inputs.
     Contactor_
      Feedback

——————| |——————
```

Figure 3.2 – Accessing inner variables in a structure

Each dot represents a layer in the structure. Each layer can be passed around the project through interfaces or (if globally accessible) accessed directly.

In structs, the structure composure must be known if it is to be copied (moved) to another structure that is the same. If the two structures do not match up in size and composure, the compiler will throw a warning.

> **Note**
> Two structures with different internal variables can be assigned to each other as long as the internal variable datatypes match. *The names of the variables do not need to be the same.*

Nesting

Note that in *Figure 3.1*, the `Raw_IO` struct contains another two struct definitions with the names of `Inputs` and `Outputs`. Both of these are also structs that contain variables for use within the project logic.

Structs can contain other structs or even UDTs; however, not all data types are available in structures and a maximum of eight nested depths is possible (unless using `InOut` interfaces, where nine is then possible).

UDTs

UDTs are essentially a structure, however, they are defined outside of a program block. This means that they are accessible *anywhere* in child objects of the CPU, including tags. *Figure 3.3* demonstrates the declaration of a UDT (on the right) and the usage of the UDT as a data type on the left:

	Sample_Pump_2				UDT_Sample_Pump_Raw_IO	
	Name	Data type	...		Name	Data type
1	▼ Static			1	▼ Inputs	Struct
2	▼ Raw_IO	"UDT_Sample_Pump_Raw_IO"		2	Contactor_Feedback	Bool
3	▼ Inputs	Struct		3	Isolator_Closed	Bool
4	Contactor_Fee...	Bool	fal	4	▼ Outputs	Struct
5	Isolator_Closed	Bool	fal	5	Run	Bool
6	▼ Outputs	Struct		6	<Add new>	
7	Run	Bool	fal	7	<Add new>	

Figure 3.3 – Example of UDT definition and usage in a Global Data block

Note that the UDT_Sample_Pump_Raw_IO instance in Sample_Pump_2 has been defined with a Raw_IO name. This means that the two data blocks shown in *Figure 3.3* result in exactly the same data structure, but by two different means.

The advantage of the UDT method is that if 10 instances were defined and the UDT is updated to a newer version, all 10 instances receive the update, as opposed to structures, where each instance would have to be updated manually.

Creating a UDT

Unlike structures, UDTs are not defined in the declaration space of program blocks, they are defined in the **Project tree**, in the PLC data types folder:

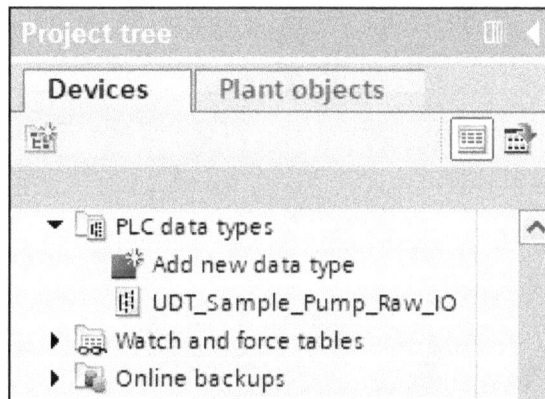

Figure 3.4 – PLC data types folder where new UDTs can be created

By double-clicking on **Add new data type**, a new UDT is automatically created and opened. No dialog is displayed as UDTs have very limited options available and are mostly to do with commenting.

Composure

Just like a struct, a UDT behaves differently when in optimized or non-optimized memory. However, UDTs cannot be set to optimized/non-optimized themselves; it's the location in which they are instantiated that is selected as optimized or non-optimized.

This means that it's possible to have two instances of the same UDT in two different data blocks or function blocks, one of which is optimized and the other is not. This isn't immediately apparent, and passing data between the two instances can cause an undesired overhead if they are not of the same memory layout.

Optimized/non-optimized

Remember that a block can be set to **optimized** or **non-optimized** by right-clicking the object in the Project tree and selecting **Properties**, then in the **Attributes** section, checking or unchecking the **Optimized block access** checkbox:

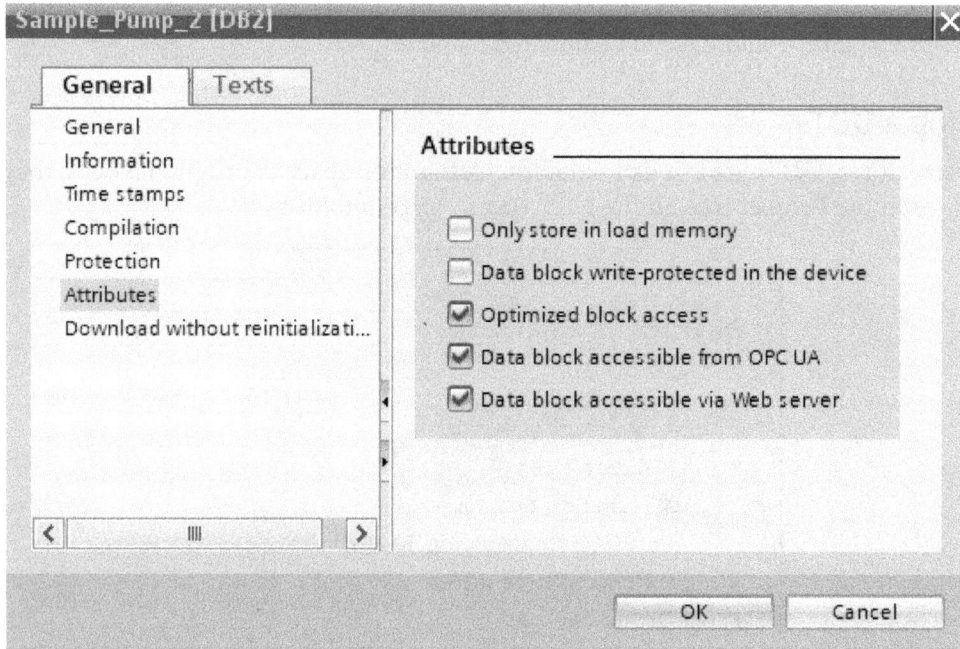

Figure 3.5 – Dialogue for setting Optimized block access

Naming conventions – good practice

When a new UDT is created, it will by default be assigned a `User_data_type_1` name. It is strongly recommended that this be changed to something that represents the usage. It is a good idea to prefix the data type with `UDT_` as all UDTs are easy to find when declaring data types by typing `UDT`. The **IntelliSence** prompt will then display a list of possible UDTs:

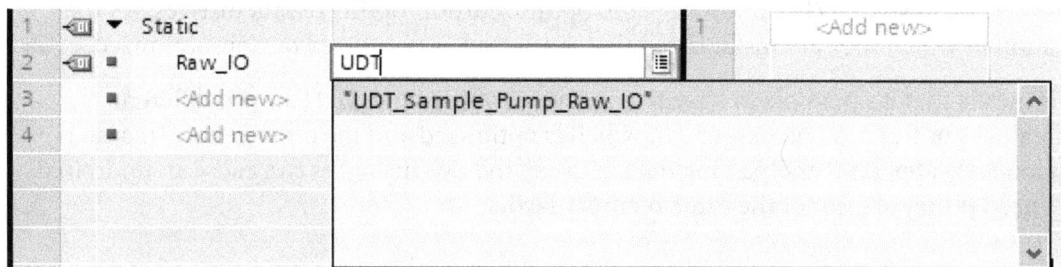

Figure 3.6 – IntelliSense displaying available UDTs

By implementing a proper naming convention, as *Figure 3.6* demonstrates, programming using UDTs is made simpler for programmers, especially those that may not have used the UDTs before and do not know them by specific names.

Nesting

As with structs, UDTs can also contain other UDTs (as well as structs) as shown here:

Sample_Pump_2		
	Name	Data type
1	▼ Static	
2	▼ Raw_IO	"UDT_Sample_Pump_Raw_IO"
3	▼ Inputs	"UDT_Basic_Pump_Raw_Inputs"
4	Contactor_Feedback	Bool
5	Isolator_Closed	Bool
6	Local_Emergency_Stop_Healthy	Bool
7	External_Fault_Healthy	Bool
8	▼ Outputs	Struct
9	▼ Pump	"UDT_Basic_Pump_Raw_Outputs"
10	Run	Bool
11	▼ Panel_Indicators	"UDT_Basic_Asset_Panel_Indicators"
12	Healthy	Bool
13	Fault	Bool
14	Manual	Bool
15	Auto	Bool
16	Off	Bool
17	Running/Open	Bool
18	Stopped/Closed	Bool
19	Starting/Stopping/Travelling	Bool

UDT_Sample_Pump_Raw_IO		
	Name	Data type
1	▼ Inputs	"UDT_Basic_Pump_Raw_Inputs"
2	Contactor_Feedback	Bool
3	Isolator_Closed	Bool
4	Local_Emergency_Stop_Healthy	Bool
5	External_Fault_Healthy	Bool
6	▼ Outputs	Struct
7	▼ Pump	"UDT_Basic_Pump_Raw_Outputs"
8	Run	Bool
9	▼ Panel_Indicators	"UDT_Basic_Asset_Panel_Indicators"
10	Healthy	Bool
11	Fault	Bool
12	Manual	Bool
13	Auto	Bool
14	Off	Bool
15	Running/Open	Bool
16	Stopped/Closed	Bool
17	Starting/Stopping/Travelling	Bool
18	<Add new>	
19	<Add new>	

Figure 3.7 – Example of nested UDTs and structures

Figure 3.7 shows an example of nesting with UDTs. The UDT_Sample_Pump_Raw_IO data type now contains instances of UDT_Basic_Pump_Raw_Inputs and a struct that contains a further two UDTs for output management.

This demonstrates how UDTs can be used within each other to build large data structures that are defined as an explicit data type for use within the project.

Creating struct/UDTs – best practices

When creating a struct or UDT, it is always best to think of the *structure* of the data. Remembering that other programmers need to easily pick up and use the data is the best way to ensure it is structured efficiently.

Understanding what is required

Choosing which type of structure is best for the application that is being built is fundamentally down to how it will be used.

Anonymous structures (structs) are best for grouping variables or other structures together, but not for use in interfaces.

The reason for this is that structs are not linked to each other. If 10 FBs make use of the same struct layout in interfaces and a new requirement is added to the struct, all 10 FBs require editing. UDTs are instantiated as a data type, so if a UDT is updated, all instances of the UDT update too. This is what makes UDTs more suited for creating standard code.

Defining structure variables

Before creating a structure, define exactly what needs to be stored or used within the structure by looking at the asset or object the UDT is being designed to interact with, as shown here:

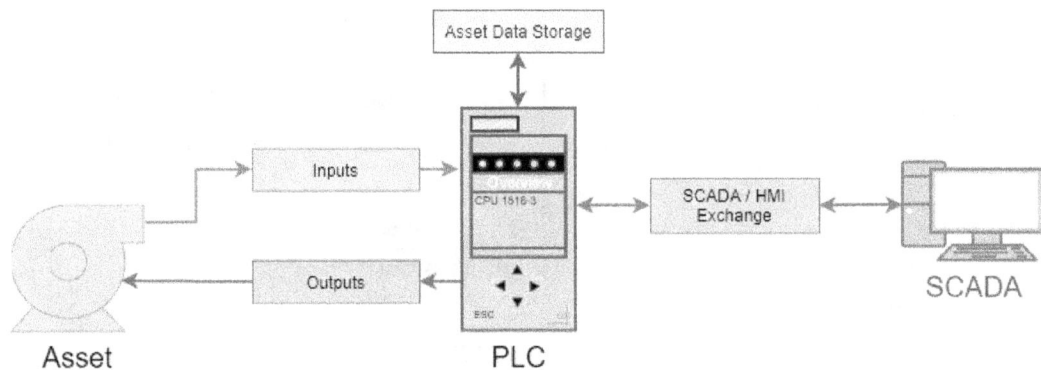

Figure 3.8 – Example of an asset-based structure design

The process of defining everything required can appear daunting at first; producing something such as *Figure 3.8* can help organize data into the required areas.

For example, an asset that is a *pump* may require the following areas:

- **Inputs**: An area for inputs from the pump to be written to from the I/O abstraction layer.

- **Outputs**: An area for outputs to the pump to be written from the I/O output layer.

- **Asset Data Storage**: The data that needs to be stored between scans and between power cycles of the CPU. In the case of a pump asset, this may include items such as total running hours or number of starts.

- **SCADA/HMI Exchange**: An area designed to be exchanged between the PLC and SCADA. This will typically consist of variables that are written from SCADA and variables that are read by SCADA.

By defining each of these areas as a UDT (or at least a structure that is part of a parent UDT), all assets that make use of the UDT can be updated in one go by updating the UDT used by them.

Once the defining of the variables has been decided in the relevant areas, it can be constructed as a series of nested UDTs and structures in TIA Portal, as shown here:

		Name	Data type
1	▼	IO_Layer	Struct
2	■ ▶	Pump_Inputs	"UDT_Basic_Pump_Raw_Inputs"
3	■ ▶	Pump_Outputs	"UDT_Basic_Pump_Raw_Outputs"
4	■ ▶	Panel_Outputs	"UDT_Basic_Asset_Panel_Indicators"
5	▼	Asset_Data	Struct
6	■	Hours_Run	Real
7	■	Number_Of_Starts	DInt
8	■	Number_Of_Failures	DInt
9	▼	SCADA_Data	"UDT_Basic_Pump_SCADA_Data"
10	■ ▼	Read	Struct
11	■ ▶	HOA	Struct
12	■ ▶	Asset_Status	Struct
13	■ ▶	Control_Status	Struct
14	■ ▶	SCADA_Control	Struct
15	■ ▼	Write	Struct
16	■ ▶	HOA	Struct
17	■ ▶	Control_Commands	Struct

UDT_Direct_Online_Pump

Figure 3.9 – Example of an asset UDT

This is then a UDT that relates only to a specific type of asset that the project uses; in the case of *Figure 3.9,* this is a *Direct Online Pump*. As shown, this consists of many different UDTs, structs, and base type variables (such as **Real** or **Dint**).

Remember the Dependencies

The reason why nested UDTs are used is so that if a dependent type is updated, the main UDT will update too. In this example, if the requirements for SCADA were to change and SCADA required an additional button for all *Direct Online Pumps* to perform a rotation check on the pump, this could be added to the write struct in UDT_Basic_Pump_SCADA_Data. This would mean *any* type of pump that uses UDT_Basic_Pump_SCADA_Data would update with the new functionality, including UDT_Direct_Online_Pump.

Finding commonalities between assets

This is a step that is often overlooked during development phases, yet it is an extremely useful exercise to go through. Finding commonalities between assets is the difference between having to maintain 100 UDTs or 50 UDTs that are shared between different assets, as shown in the following diagram:

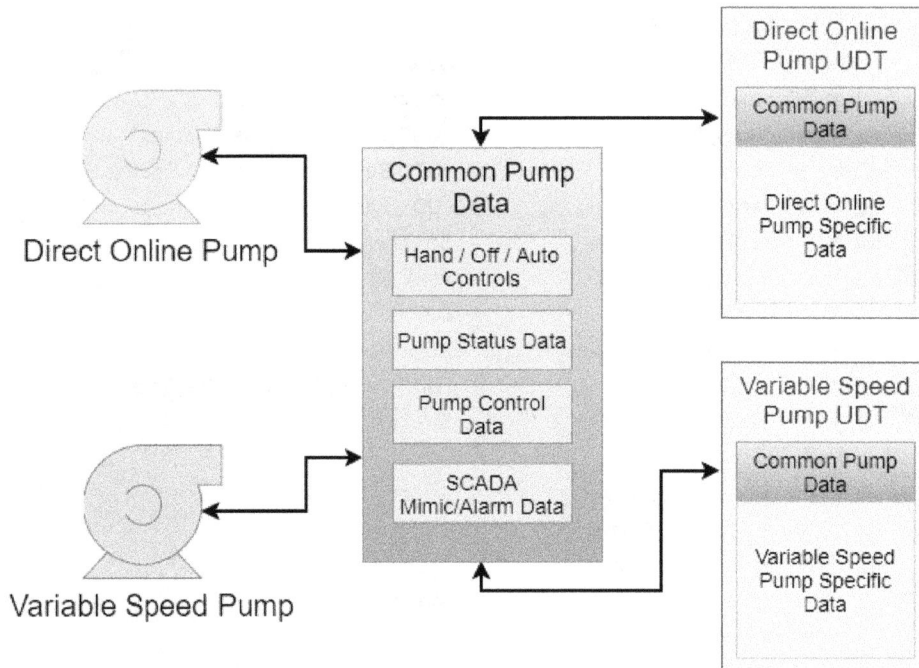

Figure 3.10 – Example of finding commonalities between devices

The example shown in *Figure 3.10* demonstrates the importance of creating common UDTs between assets. If the four UDTs were to be declared as structs inside the relative main UDTs and the requirements were changed, both assets would need updating. The same can be said if each asset had its own UDT for any of the four common data areas.

By finding these commonalities and declaring them as their own UDTs, `Direct Online Pump` and `Variable Speed Pump` can both depend upon them as nested UDTs in their own UDTs. This means if any of the common UDTs are updated, *both* assets update.

> **Considerations**
>
> It's easy to take this concept too far and find that later on (when a particular asset requires a modification to the data) it's not possible to make the desired changes without affecting many assets. Be sure to weigh up the factors of modification before deciding to make something common to other assets.
>
> If in doubt, make it specific to the asset being developed; it can always be made common to other assets with relative ease later.

Naming conventions

By using UDTs/structs, long variable names can be created, which many programmers are used to abbreviating. In TIA Portal, variable names are restricted to 128 characters, which is far higher than the average development environment. In addition, nested UDTs/structures have a 128-character limit at each nest level. This means that a three-layer deep variable could end up having a name that is 384 characters long!

While this sounds like it should be discouraged, it should not. The length of a name is insignificant against the time required to abbreviate and then comment on what the abbreviation means, as shown here:

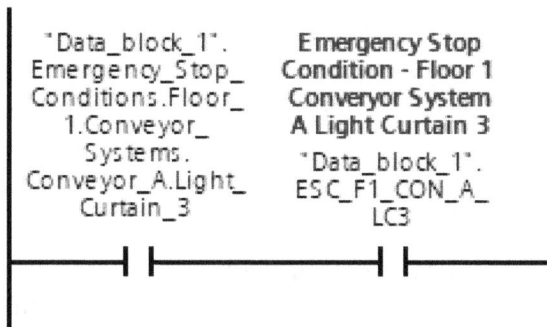

```
    "Data_block_1".          Emergency Stop
    Emergency_Stop_        Condition - Floor 1
    Conditions.Floor_       Conveyor System
       1.Conveyor_          A Light Curtain 3
         Systems.
    Conveyor_A.Light_         "Data_block_1".
       Curtain_3             ESC_F1_CON_A_
                                  LC3

     ─────┤ ├──────────────────┤ ├──────
    │
```

Figure 3.11 – Example of two different naming conventions

Figure 3.11 is an example of a structured naming convention that uses structs and a simple variable that still identifies the asset location, but without using structs. Both of these approaches allow a programmer to identify that the variable sits in the following environment:

Emergency Stop Condition → Floor 1 → Conveyor Systems → Conveyor A → Light Curtain 3

The use of the struct, however, gives the opportunity to pass all data at any of the given layers. For example, a function block may accept the struct for *Conveyor Systems* so that all conveyors can be checked. This is much harder to achieve and uses the second approach.

It is also worth noting that, without a comment, the structured approach is still easy to read. The abbreviated variable requires a comment or an understanding of the application to work out the abbreviations.

This example highlights the important understanding that a *naming convention* is more than just the name of a variable when structs/UDTs are in use. They define the data path to a variable.

Simplifying interfaces with structs/UDTs

Structures (structs/UDTs) can be declared in the following interface scopes:

- Input
- Output
- InOut
- Temp
- Static

Structures cannot be declared as either of the following:

- Constant
- Return

With this in mind, structures can be used to help simplify interfaces with program blocks. They can also help multiple blocks come together to access different parts of common data.

Passing inputs as a single struct

Passing data in and out of function blocks can help keep the complexity of the program object to a minimum, as shown here:

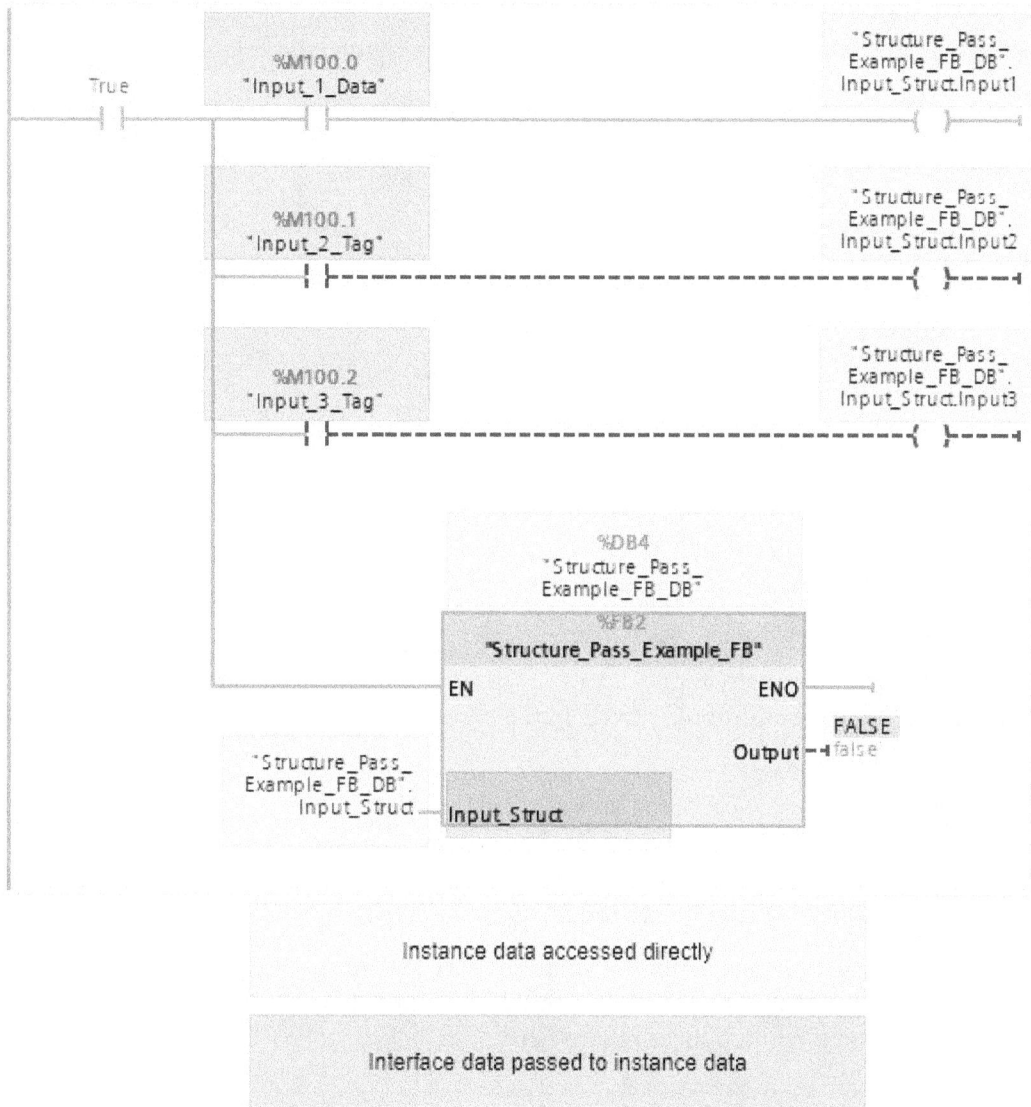

Figure 3.12 – Example of grouping inputs into a struct data type

Figure 3.12 demonstrates how using a struct could minimize the size of the interface used with the Structure_Pass_Example_FB function block. By accessing the data block that is holding the **instance data** for the function block, the Input interface data can be accessed and written to before the block is processed.

By passing `Input_Struct` from the instance data to the `Input_Struct` interface pin, the already populated instance data is used in the logic, as shown here:

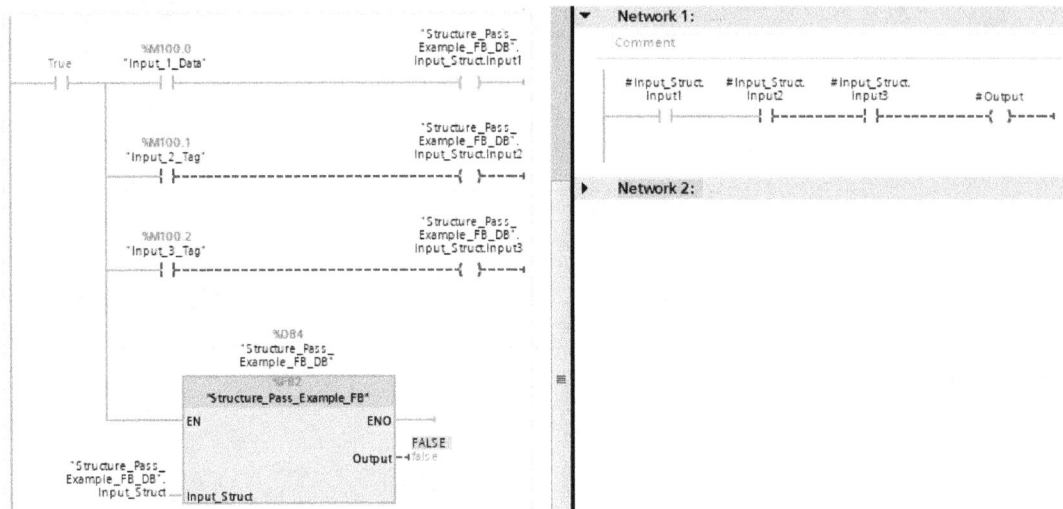

Figure 3.13 – Example of the passed struct variable being used inside the function block

This approach allows for many inputs into a function block to be rolled into a single input without having to create additional structs in the parent program block.

Functions

The same approach cannot be used for functions, as functions have no instance data. Because there is no instance data, the interface of the function cannot be accessed outside of the function to pre-load data before the function is called.

To utilize a similar approach and allow structures to be passed as interface inputs, a struct in the **Temp** interface of the parent object should be created that matches in the `Input` struct that is used in the function, as shown here:

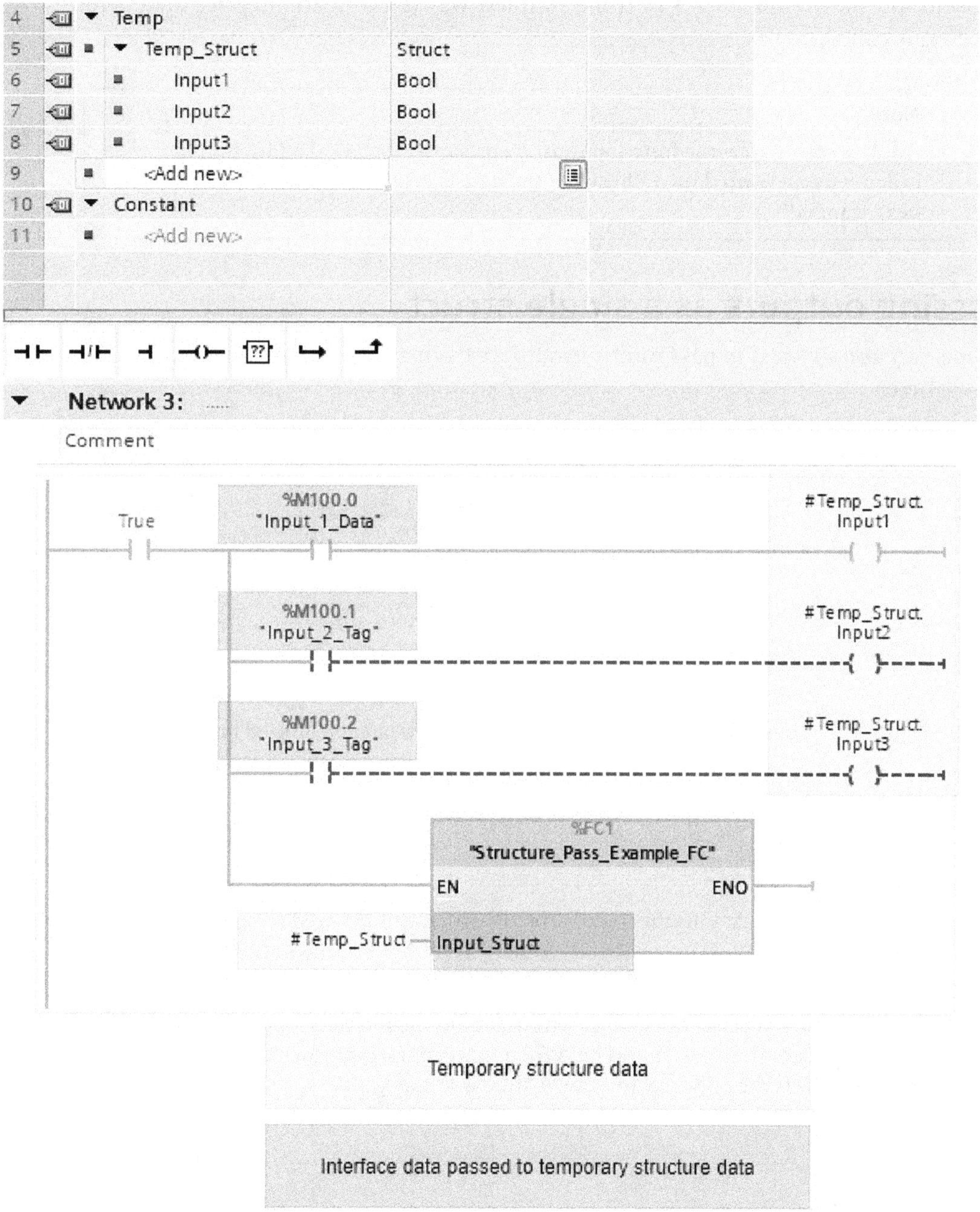
4 Temp
5 Temp_Struct Struct
6 Input1 Bool
7 Input2 Bool
8 Input3 Bool
9 <Add new>
10 Constant
11 <Add new>
Network 3:
Comment
True
%M100.0
"Input_1_Data"
#Temp_Struct.
Input1
%M100.1
"Input_2_Tag"
#Temp_Struct.
Input2
%M100.2
"Input_3_Tag"
#Temp_Struct.
Input3
%FC1
"Structure_Pass_Example_FC"
EN
ENO
#Temp_Struct
Input_Struct
Temporary structure data
Interface data passed to temporary structure data

Figure 3.14 – Solution for using functions with Temp struct data

The solution shown in *Figure 3.14* is the simplest way to pre-load interface data before a function is called.

> **Note**
>
> All data required by the function must be processed before the function is called if using **temp data**. Otherwise, the data is lost by the time the function is next scanned!

Passing outputs as a single struct

Structs can also be used to pass multiple outputs from a single program block, as shown here:

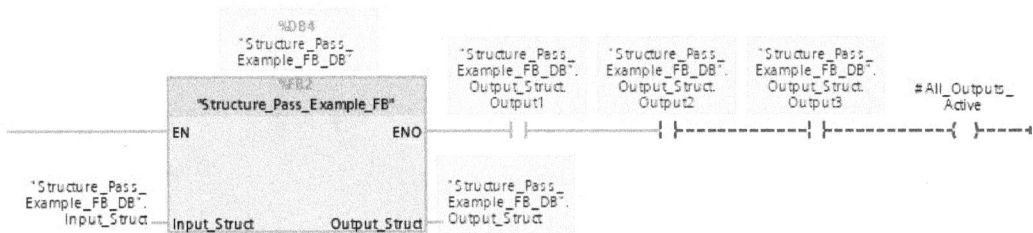

Figure 3.15 – Example of a struct being passed as an output and utilized in further logic

Figure 3.15 is an example of a function block passing output data to its own instance, which is then referenced directly in further logic.

> **Note**
>
> The function block will still work correctly with nothing wired to `Input_Struct` or `Output_Struct`, as the logic accesses DB4 for the instance data anyway.
>
> In some cases, it is not possible to wire values to the interface pins when the data originates from instance data. When this occurs, omitting the interface connections is the only way to achieve the same result.

Functions

The same can also be achieved for functions but will require another structure in which to move the output data, as shown here:

Figure 3.16 – Example of a function passing multiple elements as a single Output Struct

In *Figure 3.16*, Output_Struct is a variable of the struct type (that matches the function's Output_Struct interface) declared in the *parent object's interface*.

Passing InOut data as a single struct

Just like both input and output data, structs can be passed through the **InOut interface**. The data that is passed through the interface must be from an external source (such as a global data block or instance data) even if the InOut interface is part of a function block. While there are some exceptions to this rule, it is a good idea to use external data with InOut interfaces to help with the containerization of data. *Figure 3.17* shows an example of an InOut interface with associated data being passed:

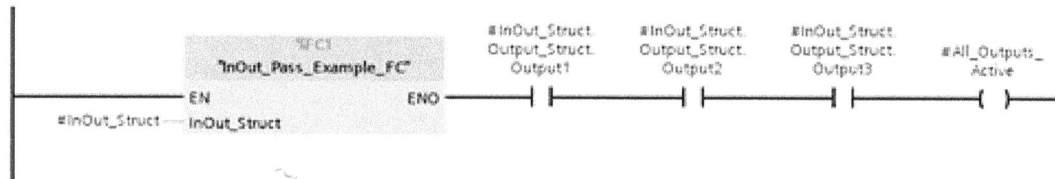

Figure 3.17 – Example of a function passing an InOut variable for both reading and writing access

This approach means that all data for reading and writing must be contained within the variable associated with the InOut interface.

Structures in static and temporary memory

Structs and UDTs can be declared as part of static/temporary memory in a function/function block. This has some useful use cases as it allows for structures to be created as instance data inside a function block.

> **Note on UDTs**
>
> Structs and UDTs can be used in the same way, but remember that UDTs carry an elevated status of being a datatype. This means that if the UDT is updated, all interfaces of that datatype also update.
>
> If a TIA Portal project makes heavy use of structures in interfaces, it can quickly become overwhelming if structs are used instead of UDTs as each interface that uses a struct needs to be updated manually.

Creating static declarations of UDTs or structs

Structures or UDTs defined in static data are created in the same manner as a normal variable:

6		▼	Static			
7		▼	Diagnostics	"UDT_Block_Diagnostics"		
8			Last_Called	LDT	LDT#1970-01-01	LDT#2021-08-05-17:56:27.167442490
9			Call_Count	DInt	0	330347
10			Runtime	LReal	0.0	1.35910541332928E-05
11			Runtime_Memory	LReal	0.0	16#0000_014A_8810_FB16

Figure 3.18 – Example of a UDT used in the static declaration of a function block

Figure 3.18 is an example of a common UDT that can be used in all program blocks – UDT_Block_Diagnostics. It is declared in the static declaration scope and allows a function or function block to record information such as the last time the block was called, how many times the block has been called, and the length of time the block takes to execute.

This information is accessible via the instance data and is not used in the program itself but is used to help the programmer debug their program and provide information to help diagnose faults and issues.

Creating temporary instances of UDTs or structs

In the same manner as static declarations, UDTs or structs can be created in the temporary declaration scope too:

10		▼	Temp			
11		■	Status_Word		Word	0.0
12		▼	Statuses	AT"Status_Word"	Struct	0.0
13		■	Healthy		Bool	0.0
14		■	Ready_To_Run		Bool	0.1
15		■	Request		Bool	0.2
16		■	Torque_OK		Bool	0.3
17		■	Temp_OK		Bool	0.4
18		■	Voltage_OK		Bool	0.5
19		■	Break_Off		Bool	0.6
20		■	Forward_Direction		Bool	0.7
21		■	Reverse_Direction		Bool	1.0

Figure 3.19 – Example of a struct declared in the temporary scope

Figure 3.19 is an example of a struct declared in the temporary scope. This example shows a status word being overlayed by a structure so that the logic has access to individual Boolean variables via a symbolic name.

Drawbacks of structs and UDTs

Despite Siemens' excellent approach toward UDTs and structs, there are some drawbacks to using them.

Most of the issues are small and are easily managed, however, a potentially confusing issue (especially to programmers that are new to using libraries and UDTs) occurs when a UDT and a block dependent upon the UDT both appear in a **Project library**.

Libraries

UDTs that are stored in a library and are utilized in many places can cause a large number of objects to upgrade to a newer version if modified.

Remember
A program block is dependent upon a UDT if its interface contains a declaration of the UDT.

When a project library has no objects in test, it would look something like the following:

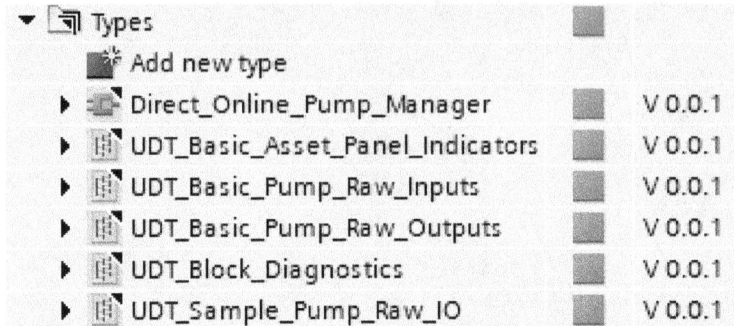

Figure 3.20 – Project library containing UDTs and function blocks

If UDT_Sample_Pump_Raw_IO is modified and a new version is created, the project library indicates that the UDT is in test:

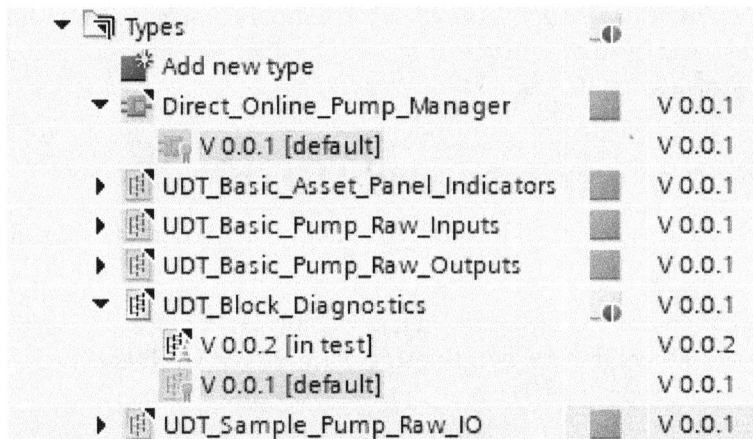

Figure 3.21 – Project library displaying in test UDTs

However, Direct_Online_Pump_Manager is dependent upon UDT_Block_ Diagnostics and this is not detailed anywhere in the project library at this moment in time.

If UDT_Block_Diagnostics is modified (*v0.0.2*) and new items are added, the change in the UDT ripples through the **hierarchy** and Direct_Online_Pump_Manager becomes inconsistent as the UDT no longer matches the interface of the function block.

If the `Direct_Online_Pump_Manager` function block was not tied to a library, it would be a simple case of updating the interface:

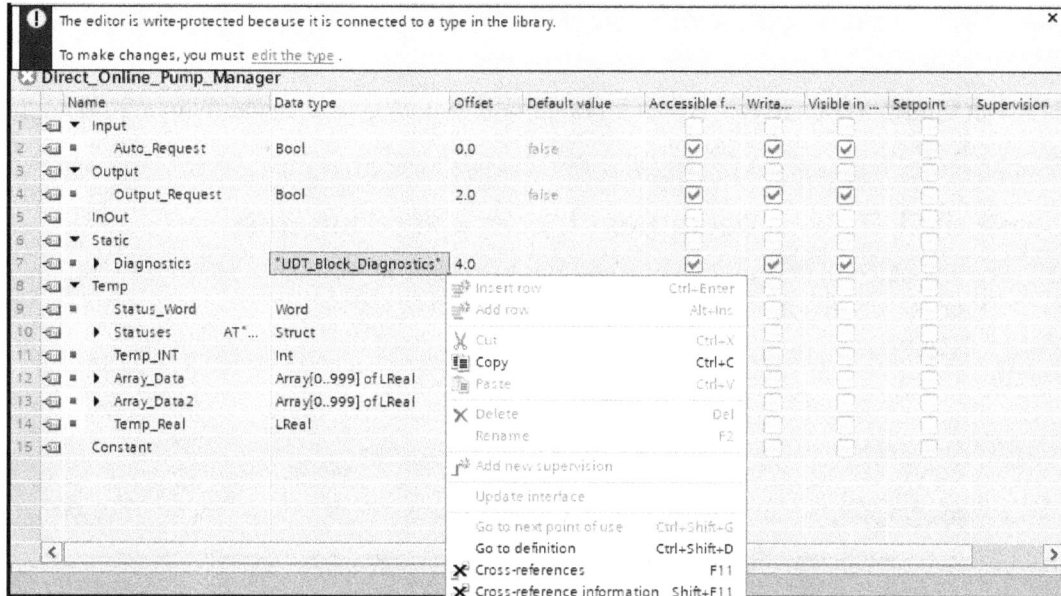

Figure 3.22 – The Update interface option is not available because the function block is still write protected

This can lead to some confusion because it is not possible to update the instance of the UDT in the interface until the function block is edited. Compiling at this point also throws errors:

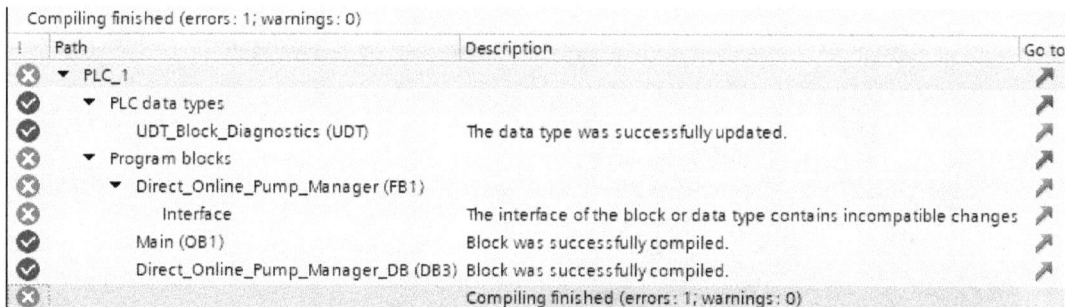

Figure 3.23 – Compiling error as a result of incompatible interface types

From this point, there are two solutions available:

- *Manually* edit each instance of an object that is dependent upon the UDT.

- *Release* the UDT and select TIA Portal to place objects that are dependent on the old version into test.

Manually updating dependent types

This method requires the programmers to keep opening all of the blocks that use the modified UDT and editing them to use the new version of the UDT until they no longer need to be edited.

This is a straightforward approach but can be time-consuming if there are many blocks to be changed.

Release the UDT with dependent blocks placed into test

TIA Portal can place all dependents into edit on the release of the new version of a UDT (or any other object) that it is dependent upon, as shown here:

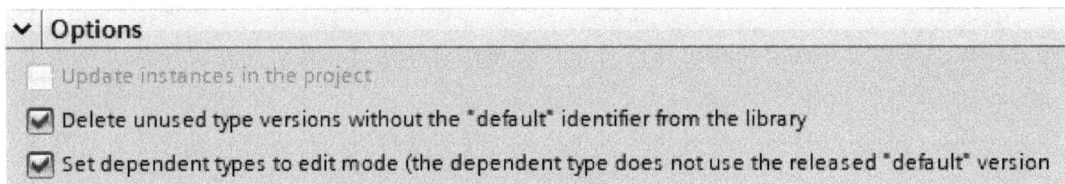

Figure 3.24 – Options when releasing a new version

Figure 3.24 shows the options available when releasing an object that has other objects dependent upon it. The last option, **Set dependent types to edit mode,** will (on release) set all objects that were dependent on the previous version to test with a new version:

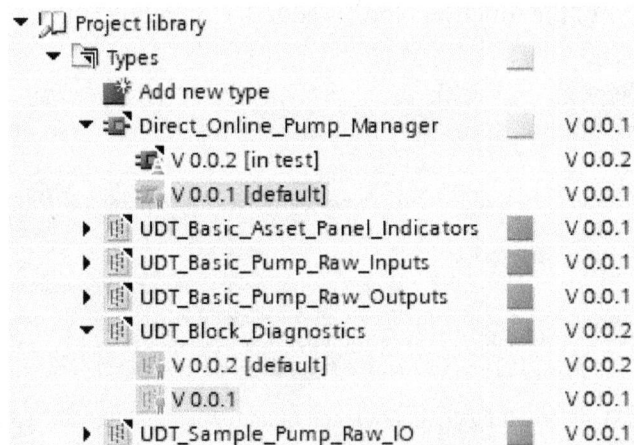

Figure 3.25 – UDT released, dependent objects now in test

If a compile is now performed on the entire project, TIA Portal will update the *in test* versions with the new instance of the UDT and compile successfully without any further input from the programmer.

The objects in test can then be released back to the project library with a new version.

> **Remember**
>
> The project library can release multiple objects at once by selecting a parent object, right-clicking, and choosing **Release all**.
>
> If the **Delete unused type versions without the "default" identifier from the library** option is checked, any unused library elements will be deleted, including the UDT version that is no longer in use.

Once these steps have been completed, the project library will have the healthy status again:

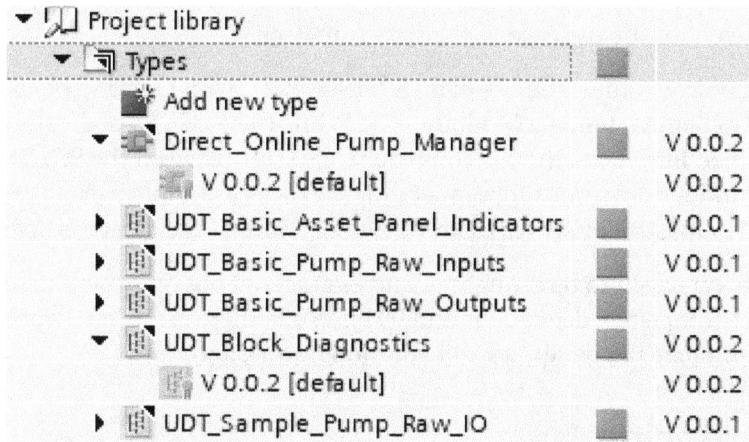

Figure 3.26 – Project library with the healthy status

Note now that the UDT and the function block that is dependent upon it are both *v0.0.2*, and *v 0.0.1* no longer exists.

Considerations

This may seem like it's complicated at first, but there are very good reasons why the project library behaves this way. When modifying a UDT that is used in many other functions/function blocks, it can become messy and overwhelming with potentially hundreds of objects all now in test.

By allowing the programmer to focus solely on the UDT without placing other items in test, it allows the programmer to focus on data and then logic separately.

Once the UDT has been released, all objects that are dependent upon it then need to be re-compiled to have access to the new data, and any logic changes need to be made.

Lack of open protocol support

Remember that structs and UDTs are ultimately data types that a programmer had created. There is very little possibility that outside of the application it is being used in, support can be given for the use case or the structure itself.

A real-world example would be the use of an **S7-1500** PLC, with UDTs to be interfaced with a third-party client system via Modbus TCP. While the PLC can make full use of the UDTs and simplified interfaces, the Modbus exchange layer between the SCADA and the PLC cannot, as Modbus does not support sending UDTs as a complete data type. It is not possible for the Modbus client/server to know the structure of the datatype beforehand.

These kinds of caveats need to be thought of and engineered out or around before building a project completely dependent on UDTs. While there are always workarounds and methods to mitigate the issues, it's still additional work and time.

Cross-referencing

Cross-referencing effectively becomes broken by using UDTs/structs with interfaces and libraries. This is because Siemens implements a *best practice* that project/global library types should not contain references to globally accessible data, as shown here:

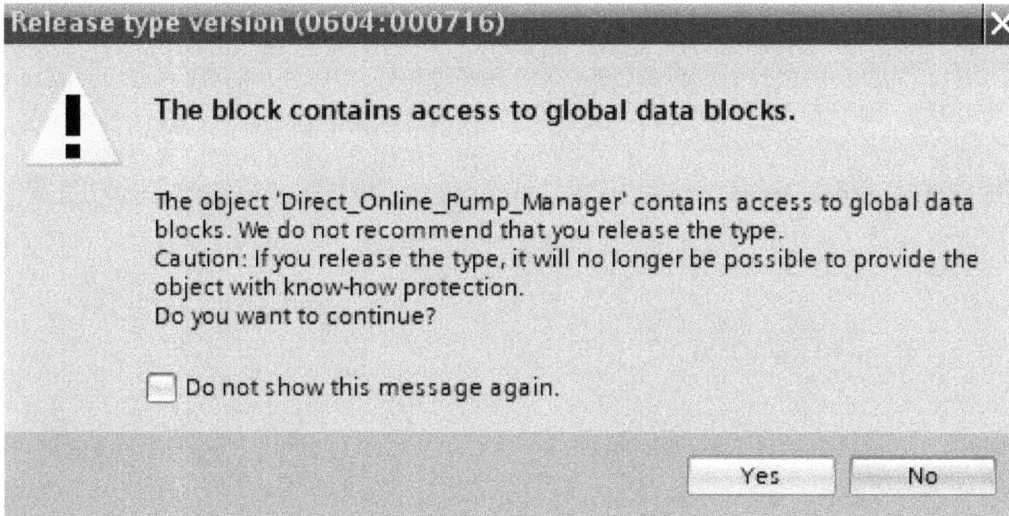

Figure 3.27 – Notice in TIA Portal that discourages the use of globally accessible data in a typed block

In order to satisfy Siemens' approach, the data should be passed to the block via an interface. This must then be `true` for *all* child instances of this data:

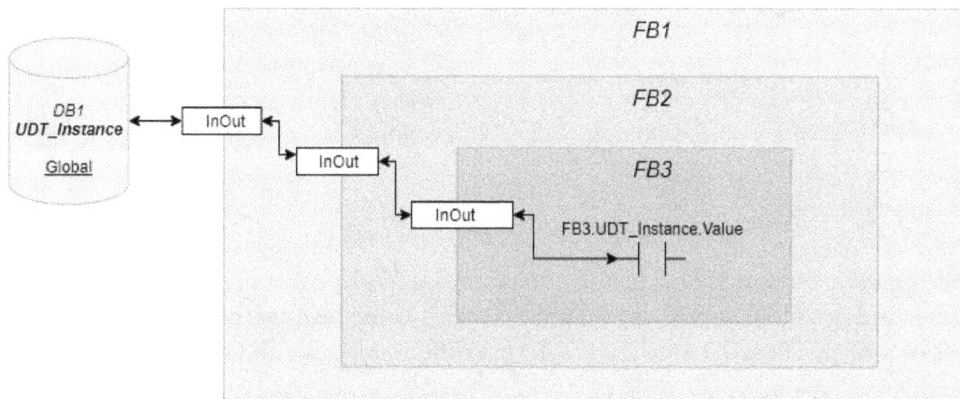

Figure 3.28 – Example of all nested child objects also passing data via interfaces

While the example in *Figure 3.28* does not present any logical issues (in fact there are large memory savings to be made with the approach shown), attempting to perform a cross-reference on `DB1.UDT_Instance.Value` would result in zero results, as **DB1** is not being directly referenced. Instead, the value is referenced via the interface of **FB1**, **FB2**, and **FB3**, which all could have different symbolic names.

Solution

TIA Portal does have a very useful project-wide search that helps mitigate this issue. By pressing *Ctrl + F* twice, or by searching in a location in the Project tree that is not a block object, the project search window will open:

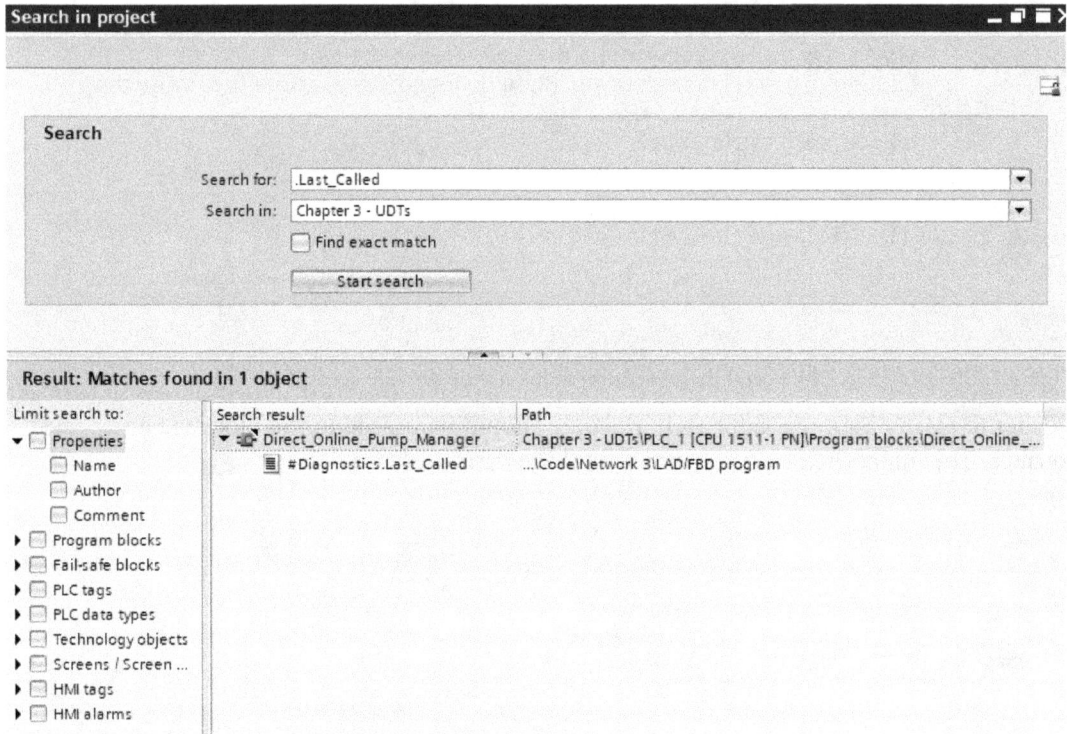

Figure 3.29 – Search in project

By searching for a variable that is inside a struct or UDT prefixed with . , TIA Portal will return interfaces and logic networks that match. This is the best method to return all instances of a variable used inside a UDT across different blocks and locations.

If the variable that is being searched has a common declaration name, such as . Fault, it may be advisable to make the search term more explicit by moving a level up the structure if possible. For example, a structure that contains Pumps . Fault and Valves . Fault would both return results if . Fault was searched for, but searching for . Pumps . Fault would only return instances of the structure where . Pumps . Fault is used.

Overusing UDTs/structs

Remember that the objective for using structures is to create standardized and reusable data that can be used alongside standardized and reusable code. Lumping large datasets together under huge structures can be very memory intensive and extremely wasteful if all of the variables are not required.

Don't forget that UDTs common to more than one logic set may carry variables that are not used in some function blocks. While this is okay, it should be carefully considered while expanding UDTs that the UDT is not reaching a point where it contains too many variables that may or may not be used. When a UDT or struct contains too many *optional* variables, it becomes difficult to maintain and affects many different areas of the project.

UDTs should be precise and explicit in their use.

Summary

This chapter has introduced concepts behind using structured data in the form of structs and UDTs. It has highlighted the benefits of using structures, and also some of the pitfalls and issues that can be faced in large projects by using them. Remember that a structured project is far easier to navigate through, standardize, and work with in the future.

After reading this chapter, the *asset-oriented approach* should make sense from a structural perspective: containerizing data so that data that is only relative to a particular asset lends itself to structures (structs or UDTs).

Identifying what is needed in structures is always the hardest part, as structures can be difficult to keep updating in a system that is already operational, so it's best to try and plan for common use cases first.

TIA Portal offers a robust and highly flexible solution to managing data using structures. With two different types of structured data (struct and UDT), programmers can simplify project structure greatly, without compromising logic.

The next chapter introduces programming languages where these structures can actually be called and utilized in development languages. This chapter will give a basic overview of all of the available languages in the standard TIA Portal V17 environment and also how to choose the best programming language for the task.

Section 2 – TIA Portal – Languages, Structures, and Configurations

Learn how to program in the different languages that TIA Portal offers, including utilizing previous learning about structures and different configuration options.

This part of the book comprises the following chapters:

- *Chapter 4, PLC Programming and Languages*
- *Chapter 5, Working with Languages in TIA Portal*
- *Chapter 6, Creating Standard Control Objects*
- *Chapter 7, Simulating Signals in the PLC*
- *Chapter 8, Options to Consider When Creating PLC Blocks*

4
PLC Programming and Languages

While the previous chapters have explained how **structure** helps build a strong foundation where data can be managed effectively, the next few chapters will focus on PLC programming and the various languages that Siemens offer in TIA Portal.

Without understanding at least one of the many available languages, a programmer will find it difficult to create an executable program. This chapter explores the different languages available in TIA Portal, including the new Cause and Effect language that was introduced in TIA Portal version 17.

After reading this chapter, you should know the following about each available language type:

- The language's basic composition (graphical, text-based, and so on)
- Common use cases

The following topics will be covered in this chapter:

- Getting started with languages
- Selecting the best language for the task
- Differences between Siemens and other PLC vendors

Getting started with languages

PLCs can be programmed in a variety of different **languages** and different PLCs from different manufacturers may implement slightly different variations of those languages.

Not all PLCs support multiple languages, and the ones that do support more than one language do not necessarily allow for that language to be fully implemented. The general correlation between PLCs and language support comes down to cost – the more expensive a PLC is, the more likely it is to support more than one language type.

> **Note**
>
> Siemens PLCs, when programmed with TIA Portal, support multiple languages, even at the lower ranges.

Available languages

Before you select a language for the task at hand, it's important to know what is available and what that may mean for the project that's being developed.

LAD – ladder logic

Ladder logic is a graphical language based on **contacts**, **coils**, and **instruction blocks**.

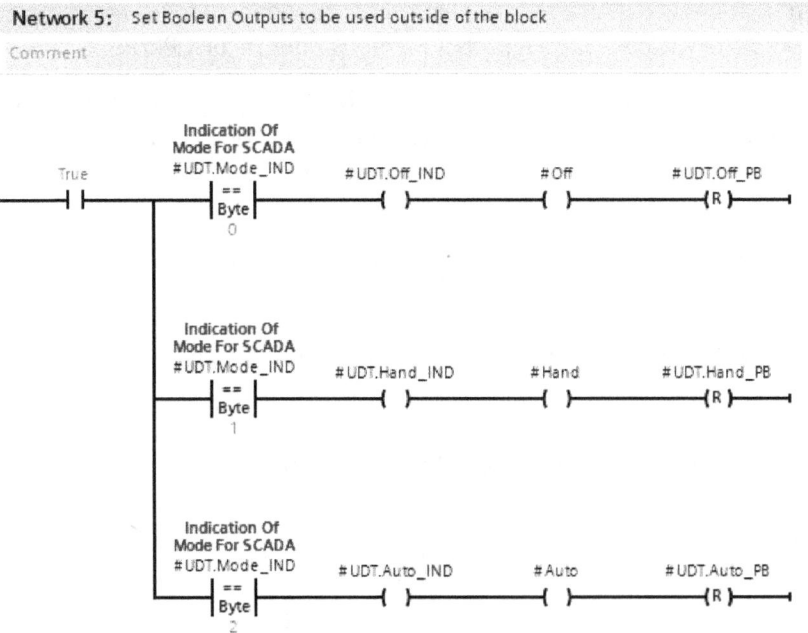

Figure 4.1 – Example of ladder logic

There are multiple different types of contacts and different types of coils. Contacts read data, while coils write data. These changes are based on how the data that's associated with them behaves.

Ladder logic is still the most used programming language in PLCs; however, this may not always be the case. Ladder logic is preferred because of its likeness in reading to electrical drawings. The idea of contacts being open or closed to allow *logic to flow through* to coils that energize outputs is very similar to relay panel wiring.

Ladder logic is one of the easier languages to learn and probably the most well documented. It's best suited for Boolean logic, but it's not uncommon for entire systems to be programmed in ladder logic simply because it's easier to maintain a single language.

Function block diagram (FBD)

FBDs are often overlooked in favor of ladder logic. It's similar to ladder logic as it is managed in networks and has the same left-to-right flow. However, the entire language consists of **blocks**, where the function that the block represents is indicated at the top of the block.

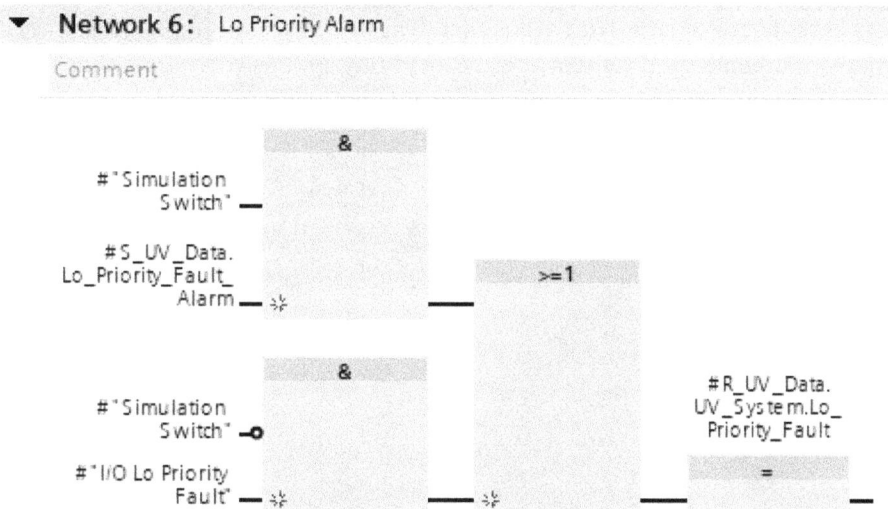

Figure 4.2 – Example of an FBD

While being a graphical language, the same as ladder logic, the logic's layout is quite different. Each block represents a function and operates similarly to a **program block**, with inputs on the left and outputs on the right. **Assignments** are used to set values and AND and OR blocks are used to evaluate the status of signals.

Structured control language (SCL, structured text)

SCL in Siemens is more commonly referred to as **structured text**. It is a textual language that is considered the most complex of the programming languages. SCL is one of the harder languages to learn due to it having a larger instruction set and syntaxes that need to be learned.

```
IF #Node_Index <> 0 THEN
    //Check if the mask for the alarm contains a 1, if so, process the alarm check
    "Global_Alarm_Manager"(Condition := #Temp_Diag_Data[#Master_Index].M1_LCE_Slave_A[#Node_Index] AND
                           #Alarm_Masking_Array[#Node_Index],
                   Index := ((#Master_Index + 225 + (#Master_Index * 64)) + #Node_Index) - (1 * #Master_Index),
                   Alarm_Data := "H1_Alarms".SYS,
                   AB_Data:= "H1_Alarms".AB,
                   ARM:= "H1_System".ARM,
                   Alarm_Active => #Alarm_Active,
                   Alarm_Status => #Alarm_Status);

END_IF;
```

Figure 4.3 – Example of SCL

SCL makes extensive use of instructions that are not available in other languages, such as IF and FOR. These additional instructions allow the language to perform repetitive or complex logic with much more ease than that of graphical languages.

While SCL is best suited for small functions that deal with math or looping logic, it's not uncommon to see it being used for just about everything. SCL is very popular as it is the easiest language to port between different environments since it is just text.

GRAPH

Similar to a sequential function chart, **GRAPH** is a graphical language that is best suited for managing *sequences* and *parallel* processes.

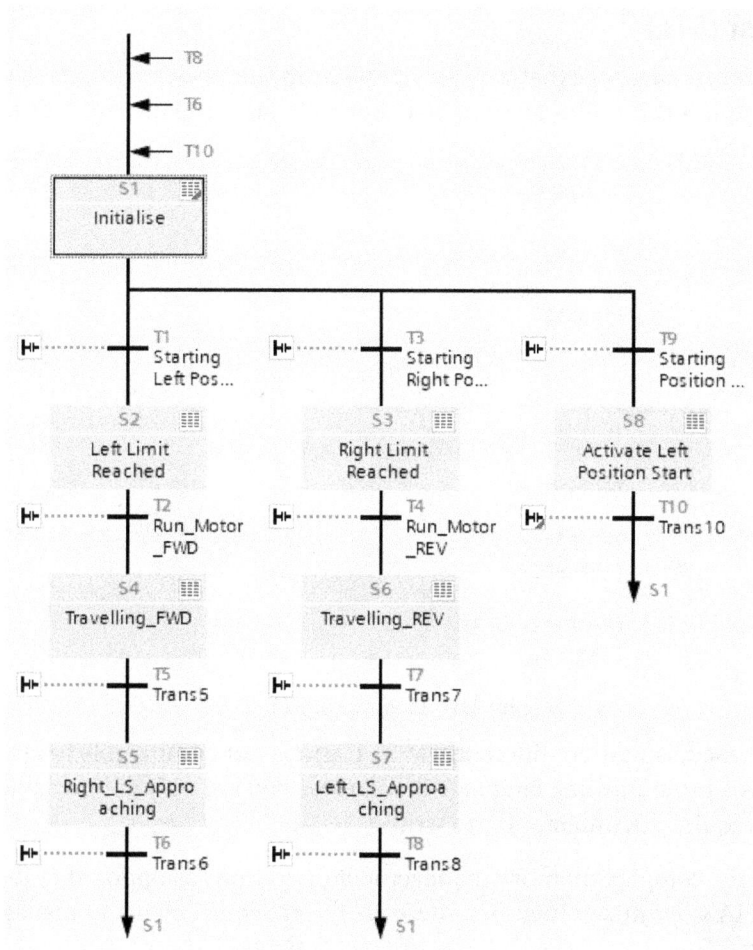

Figure 4.4 – Example of GRAPH

GRAPH comes with additional features such as interlock monitoring for each step (the interlock must be healthy for the step to take place) and supervisions that fail a step and stop the sequence from continuing. TIA Portal does a good job at creating an environment that makes managing a sequence much easier, with all the relative building blocks in place already.

> **Note**
>
> GRAPH is not strictly a programming language in its own right. It relies upon ladder logic to set transitions, interlocks, and supervisions.

Statement list (STL)

STL is another textual language that was very popular in the earlier Siemens days. STL is becoming less and less utilized in favor of SCL, but STL has some instructions that are not possible to use in SCL.

```
Network 1:
Comment
 1   //Amplify signal
 2        L     #Signal
 3        L     20                                          20
 4        *I
 5        T     "Amplified_Signal"                          %MW10
 6
 7   //Check Enable Signals And Enable Output Signal
 8        A     #Enable_1
 9        A(
10        O     #Enable_2
11        ON    #Enable_3
12        )
13        =     #Enable_Signal
14
```

Figure 4.5 – Example of STL

STL, while text-based, is written differently to SCL and is not comparable to any other language set. It is a programming language that, unless you are very comfortable with it, will most likely require documentation to write.

STL is best used for complex memory management operations as opposed to logic. However, with TIA's optimized memory, there are fewer requirements to manage data in memory directly.

Cause and effect matrix (CEM)

The CEM language is a newly introduced language since TIA Portal version 17. It allows logic to be connected in a matrix of columns and rows, where rows make up causes and columns make up effects.

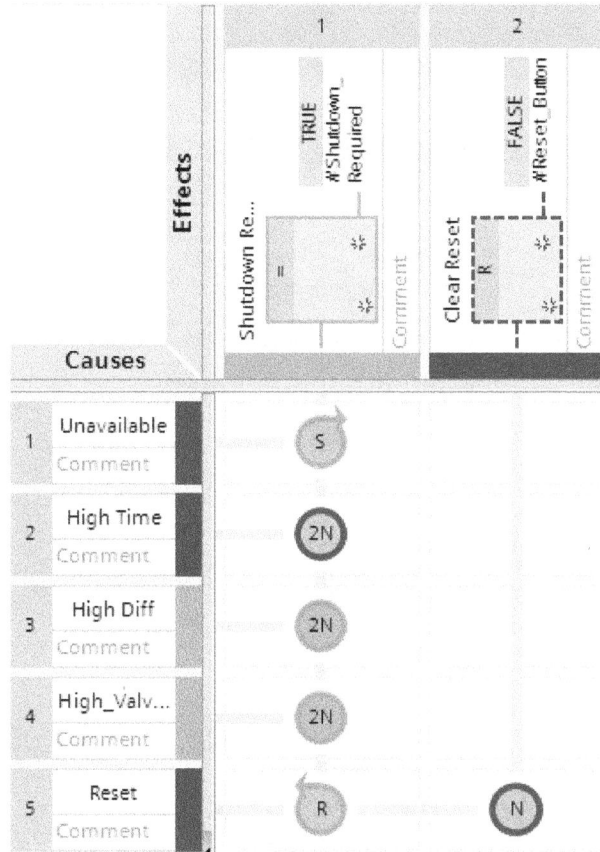

Figure 4.6 – Example of CEM

CEM is a highly visual language that exists simply to map several inputs to their eventual outputs. The system allows you to group the causes so that more than one cause may need to be active for the effect to be actioned.

This language is particularly useful in complex interlock logic or safety logic. Documentation may refer to risk matrixes or cause and effect matrixes that explain what needs to happen in certain scenarios. This language helps translate those documents into logic with ease.

Languages in program blocks

The type of object that's being used restricts the usage of certain languages. This is because functions have no instance data, so languages such as GRAPH and CEM are unavailable since these languages require static data to hold information relative to the way GRAPH and CEM construct logic.

Function blocks

In TIA Portal, **function blocks** can be written in one of six languages, as shown in the following screenshot:

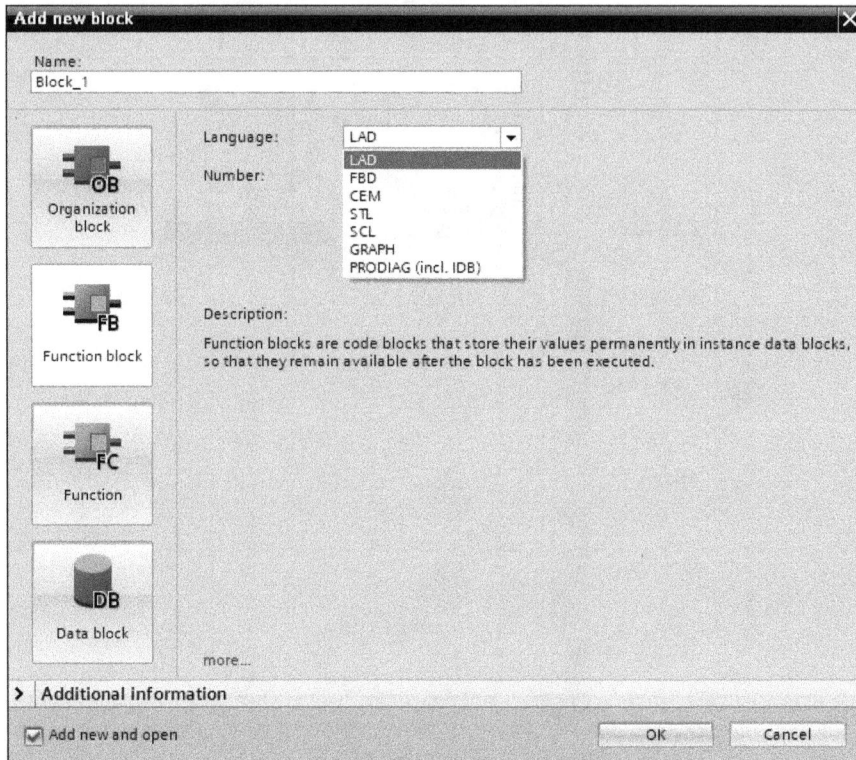

Figure 4.7 – Adding a function block and the available languages

A function block can be written in one of the following languages (as shown in the preceding screenshot):

- **LAD**
- **FBD**
- **CEM**
- **STL**
- **SCL**
- **GRAPH**

> **Note**
>
> CEM is a new language that's been implemented in TIA Portal since version 17.
>
> PRODIAG is not a programming language and has been excluded from the preceding list.

Functions

A **function** can be written in one of four languages, as shown in the following screenshot:

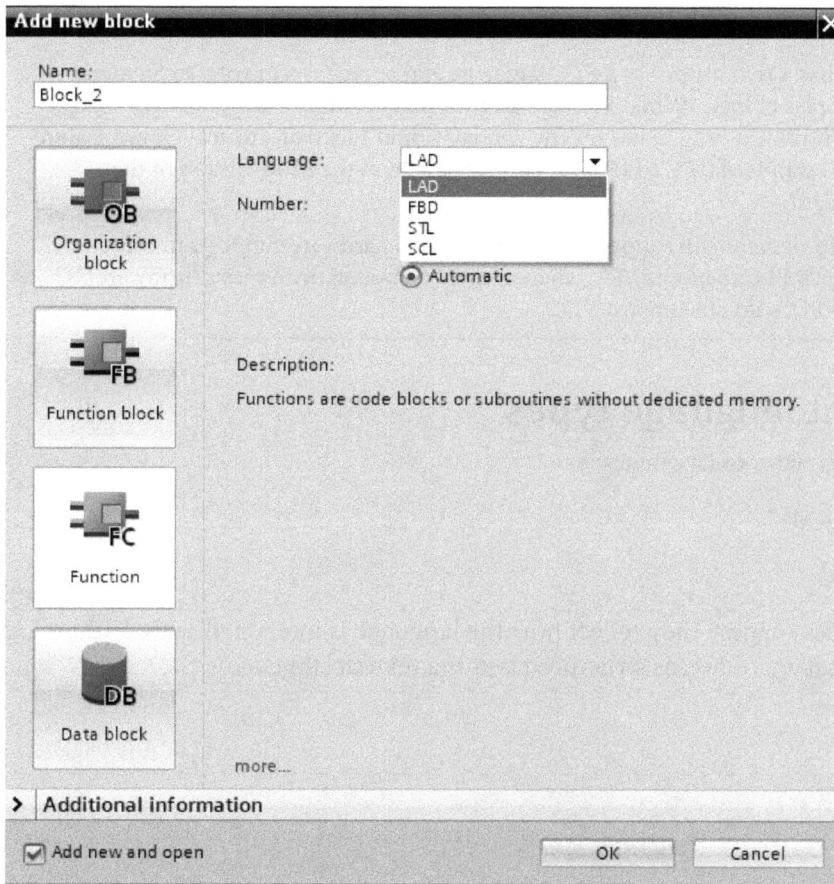

Figure 4.8 – Adding a function and the available languages

A function can be written in one of the following languages (as shown in the preceding screenshot):

- **LAD**
- **FBD**
- **STL**
- **SCL**

> **Note**
>
> STL has a rich history as a PLC language and appears frequently in Siemens' examples of logic. While STL was an extremely common language years ago, it is now diminishing in usage. The **International Electrotechnical Commission (IEC)** standard (IEC 61131-3) deprecated STL in the third edition of the standard.
>
> While Siemens still supports STL, not all PLC hardware supports it. Most S7-1500 PLCs support STL, whereas most (if not all, with recent firmware) S7-1200 PLCs do not support STL.

Different language types

There are two types of languages:

- Graphical
- Textual

As their names suggest, they reflect how the language is interacted with. Ladder logic is a graphical language, whereas structured text is a textual language.

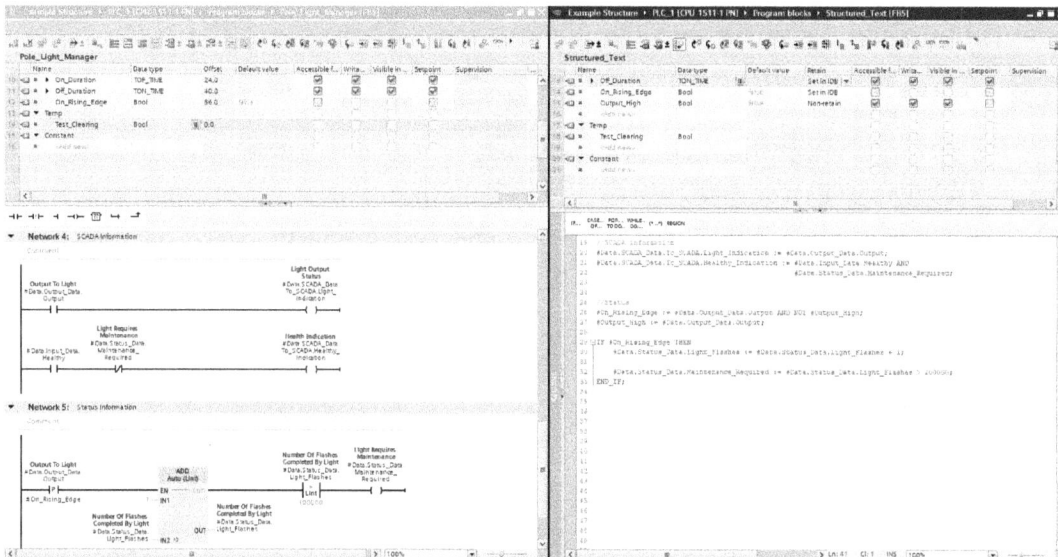

Figure 4.9 – Example of graphical versus textual language

There are strengths and weaknesses between graphical languages and textual languages that vary depending on the programmer, but they can be summarized. Let's take a look.

For graphical languages, the strengths and weaknesses are as follows:

- **Strengths**:
  - Generally easier to read.
  - People with different skill sets, such as electricians, tend to find it easier to work with and relate to other graphical tools such as wiring diagrams or schedules.
  - Easy to use due to its drag-and-drop style programming.
  - Easy to monitor with a graphical representation of `true`/`false` logic.

- **Weaknesses**:
  - Generally less compact than textual languages.
  - Its less structured approach allows for programmers to have variations in how code is laid out.
  - Comments are bound to very specific areas – at the top of a network, for example.

For textual languages, the strengths and weaknesses are as follows:

- **Strengths**:

  - Portability is greatly increased with textual languages as nearly all development platforms accept text from the OS clipboard.

  - Can be programmed outside of the development environment and copied into the development environment (in a function block in SCL, for example).

  - Comments can appear anywhere, and in any format required (multiline, for example).

  - Areas of code can be commented out for quick testing/commissioning.

- **Weaknesses**:

  - Generally harder to read for those of other skill sets as the structure is not immediately obvious.

  - Most development environments do not have a *drag-and-drop* style approach for textual environments, making the language harder to use.

  - Online monitoring is weak compared to graphical languages.

  - More prone to errors as more user entry (via keyboard) is provided than graphical languages.

Variations in languages

There are fundamental differences between all languages, but there are some particular differences between textual languages and graphical languages that can catch programmers out, especially if they are new to a language.

For example, in graphical languages, the condition to be set is on the *right*, whereas in textual languages, the condition to be set is on the *left*, before the arguments that set it.

Another one that can catch programmers out is that generally, in textual languages, when a `boolean` variable is set to a value (and is not a `temp`), it remains at that value until it is set to a different value. In graphical languages, the instruction must be *set* to retain a value if the condition is no longer `True`. This is because, in graphical languages, the logic is scanned unless a `JMP` (jump) instruction is used, but in textual languages, `JMP` instructions are used behind the scenes if statements such as `IF` statements are not true. Code that is jumped over does not change value.

```
IF #On_Rising_Edge THEN
    #Data.Status_Data.Light_Flashes := #Data.Status_Data.Light_Flashes + 1;

    #Data.Status_Data.Maintenance_Required := #Data.Status_Data.Light_Flashes > 100000;
END_IF;
```

Figure 4.10 – Example of an IF statement

The preceding screenshot shows an example of this behavior. If `On_Rising_Edge` is `True` and `Data.Status_Data.Light_Flashes` is over `100000`, then `Data.Status_Data.Maintenance_Required` will be `True`. But if `On_Rising_Edge` is `False`, then `Data.Status_Data.Maintenance_Required` will not be updated and will remain in its last known state as it does not get processed due to jumping over the false `IF` statement.

Selecting the best language

It is important to understand that there is no right or wrong language when it comes to programming a PLC. However, there are strengths and weaknesses between different languages. There are times when choosing a particular language over another has its advantages.

The best approach is to look at what is trying to be achieved, both now and in the future, and create a block in a language that fits those needs. A PLC project will generally fair better with mixed languages that fit the needs of the project rather than sticking with a sole language and struggling through areas that are difficult to program.

Understanding the use case

Every block in a project has a *use case*. This relates to the following:

- What the block has been developed to control/manage
- How the block fits in with other blocks
- Who is using/maintaining the block

These simple statements help us understand the best language to implement when designing a block.

Example 1

Control scenario: *A block is required to summate an array of pressure instruments across a manifold and output the average.*

To satisfy this scenario, a programmer will need to consider the following factors:

- The input is an array of values
- Math is involved

Typically, when math is involved, graphical languages are at a disadvantage. This is because either many blocks are required to achieve the desired result, or additional interfacing/variables are required.

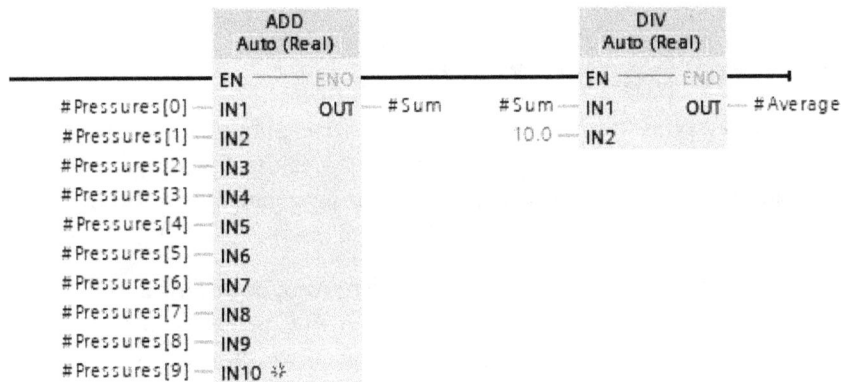

Figure 4.11 – Adding values from an array in LAD

The preceding screenshot shows how the statement would be satisfied in LAD. The following screenshot shows a simplified version in SCL:

```
FOR #i := 0 TO 9 BY 1 DO
      #Sum += #Pressures[#i];
END_FOR;

#Average := #Sum / 10;
```

Figure 4.12 – Adding values from an array in structured text (SCL)

> **Note**
>
> The example shown in *Figure 4.12* may seem small and simple enough that it would not matter as to which language was in use. However, if the example was changed slightly so that the summation happened to a 200-element array, the required changes between the two languages would be extremely different.
>
> The ladder logic would require an additional 190 inputs to the ADD block (which would not be possible as 100 is the maximum, so two ADD blocks would be required) and then have to be divided by 200.
>
> The structured text would require the 9 in the FOR instruction to be changed to 199 to loop through the array and then divide by 200 instead of 10.

Example 2

Control scenario: *A chemical mixing tank requires basic tank-level control and mixing control. The sequence of events starts with a filling valve being opened until the level in the tank is at a "Stop Fill Level." At this point, the filling valve should be closed, and the mixer should be started. The mixer should run for 20 seconds. For the first 10 seconds, the chemical should be screw-fed into the tank via the screw feed motor. Once the mixer has stopped, the transfer valve should open until the "Low Mixing Level Switch" becomes unhealthy.*

If the Low Mixing Level Switch becomes unhealthy at any point while the mixer is running, the system should fail and require the operator input to be reset.

The preceding scenario is simple but complex when it comes to picking a particular language. The key in selecting a language for this use case is that the following details are present:

- The control requirements are a sequence.

- An operator is required to interact with the control.

When sequences are involved, it's best to stick with a graphical-based language as these have much better monitoring facilities, so diagnosing issues will usually be faster.

Let's consider TIA Portal's GRAPH language, which is the best fit for sequences.

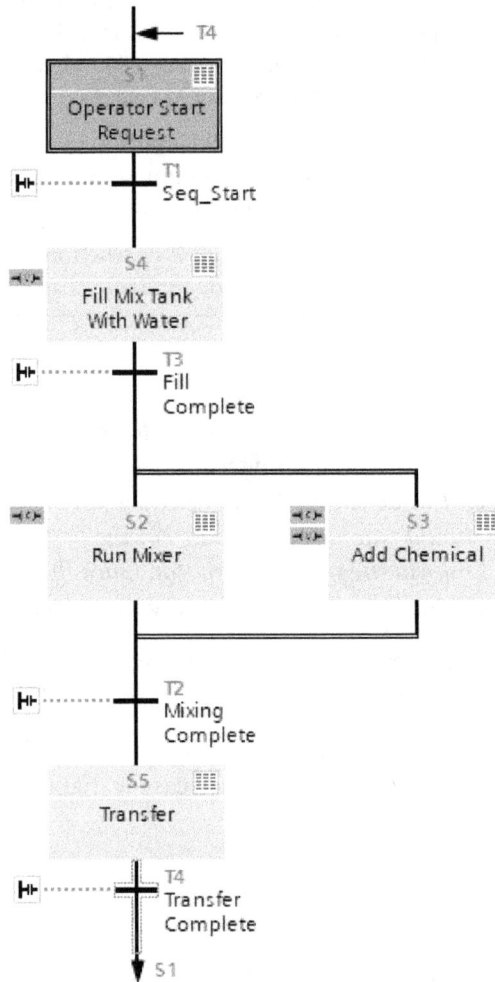

Figure 4.13 – TIA Portal's GRAPH language

GRAPH is a language that is perfectly suited for sequences and displays a sequence overview that gives immediate insight into which step the sequence is currently executing, as well as the conditions/transitions that are required to move to the next step.

This type of graphical approach makes it very easy and simple to follow, without having to look at the logic. GRAPH (or SFC in many other editors) is great for those who are trying to find a fault in a sequence because of this compartmentalized approach.

Compared to other graphical languages, such as FBD, it's easy to see how GRAPH is simpler at first glance.

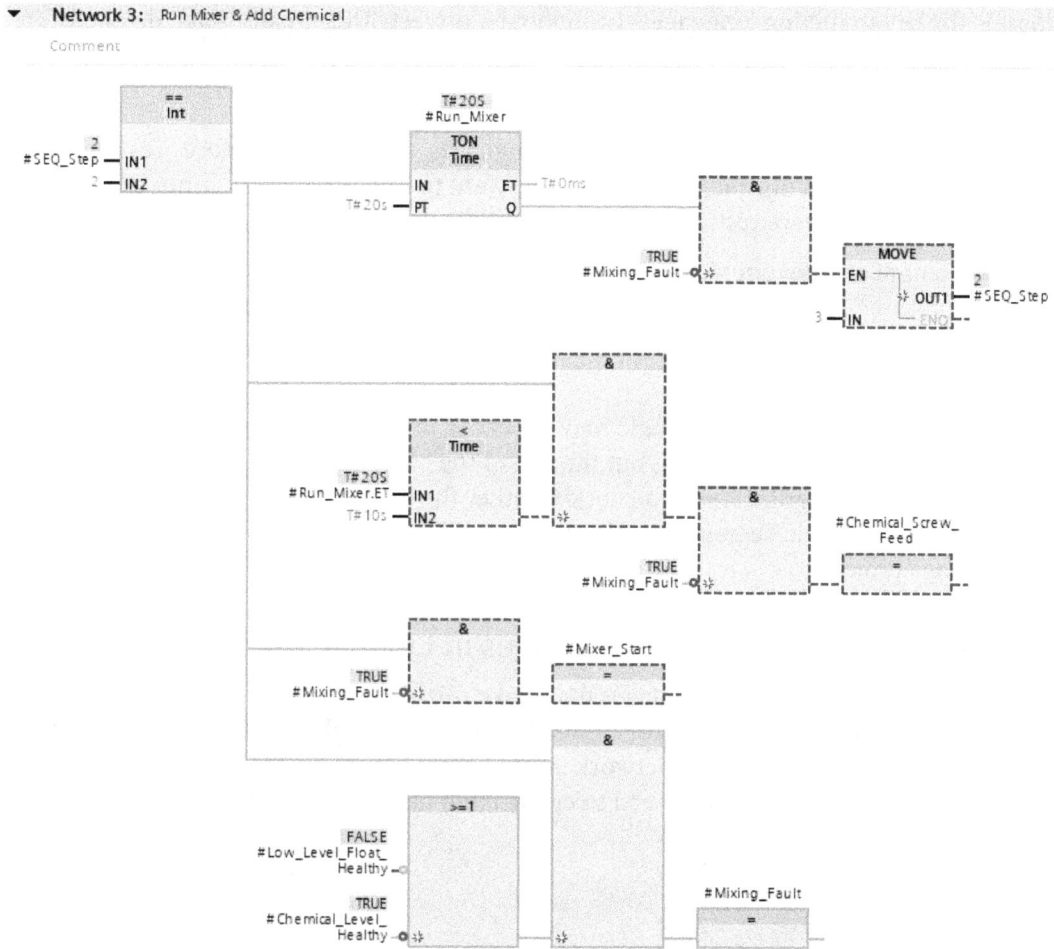

Figure 4.14 – FBD with logic for mixing the chemical tank

The preceding diagram shows that FBD cannot package up code into nicely laid out steps. Instead, the logic is also shown at the sequence step. While this is not a problem, it can make things more tedious when you're trying to locate an issue or even find what step the sequence is currently executing.

Both GRAPH and FBD are graphical languages, but they have opposing strengths and weaknesses for this particular use case. Both logical outputs are the same; the scenario is controlled as per the instructions, but the methods that are used to deal with the software can be modified and maintained in different manners.

Why is understanding a use case important?

Before any logic can be written in any language, understanding why a particular language fits best is the key to mixing languages throughout a project. This means that a detailed understanding of the end goal of the project is required. You should think and plan around the following:

- **Hardware**: Not all hardware can run all languages, so check this before creating standard blocks in one language. You don't want to find out that it cannot be used later due to hardware restrictions (STL, for example).

- **Maintenance/colleagues/customers**: Who may interact with this object or update it in the future? There's no advantage in writing a block in a language that nobody in a team can maintain when it can be written in a language people can understand just as easily.

- **Fit for purpose**: Sometimes, logic may appear to fit a particular language because of one complex or large aspect, but the rest of the block fits a different language. It's important to recognize this and check whether the logic can be reduced to a child block. This will help segregate the languages while still taking advantage of a mixed language project.

Mixing graphical and textual languages in LAD/FBD

Ladder logic and FBD are both languages that make use of networks. These networks can be used to hold textual languages at the network level too. By right-clicking on a network, the option to create an SCL or STL network is available. This allows graphical and textual languages to be mixed, without the need to create child blocks to change languages.

Network 4: Transfer

Comment

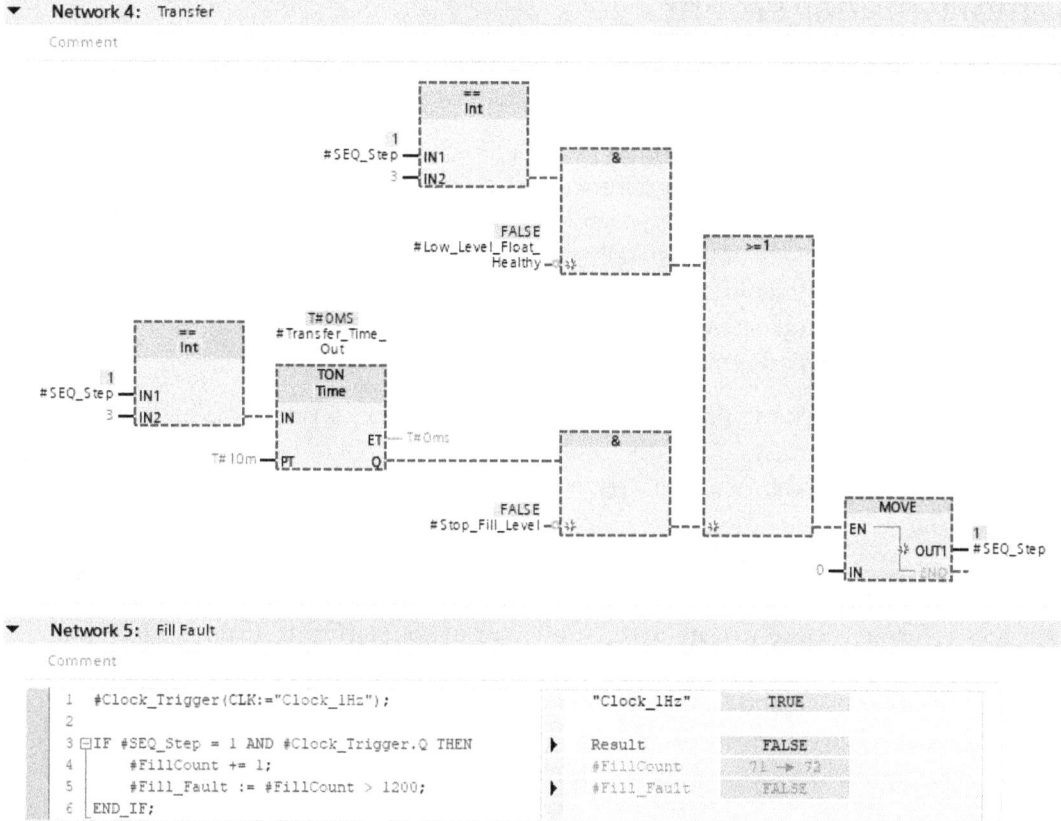

Network 5: Fill Fault

Comment

```
1   #Clock_Trigger(CLK:="Clock_1Hz");              "Clock_1Hz"      TRUE
2
3  ⊟IF #SEQ_Step = 1 AND #Clock_Trigger.Q THEN  ▶  Result          FALSE
4        #FillCount += 1;                           #FillCount      71 → 72
5        #Fill_Fault := #FillCount > 1200;       ▶  #Fill_Fault     FALSE
6  END_IF;
```

Figure 4.15 – Example of a mixed language block. Both FBD and SCL are in use between two different networks

The approach of mixing languages inside LAD/FBD blocks helps reduce the complexity of dealing with large mathematical equations or numerically heavy logic. SCL also offers an extended instruction set that LAD/FBD can leverage by using an SCL network.

Memory management

Depending on the language that is chosen for a block, this can affect the amount of memory the logic consumes in the PLC to execute the logic.

For example, consider a sequence written in ladder logic compared to one written in GRAPH. With ladder logic, a programmer would need to create the sequencing mechanism themselves using equal instructions and a sequence step variable. This is something that is provided automatically in GRAPH. However, GRAPH also automatically provides many variables, per step, that may not be required, such as the amount of time spent in an active step, the previous step number, and the next step number. The list is large and is repeated for each sequence or step.

Similarly, a GRAPH block can only have an optimized memory layout. This means passing structures via InOut interfaces would still require them to be copied, which may introduce memory and scan issues to a project.

Differences between Siemens and other PLCs

Languages in Siemens closely relate to those of other major platforms. Ladder, FBD, SCL (or ST in most other environments), and STL are all ultimately the same. The other languages are either unique to Siemens or have noticeable variations compared to other environments. Despite the most common languages being the same for the most part, there are a few things to watch out for that are subtle and not immediately obvious.

Timers

In TIA Portal, when an IEC timer is used (TON, for example), everything appears normal compared to other IEC environments.

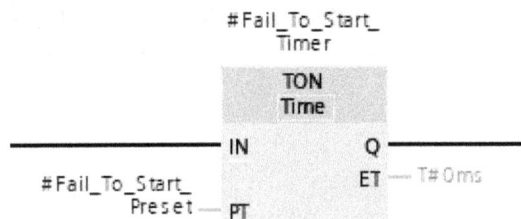

Figure 4.16 – TON timer in TIA Portal

However, if the PT value is changed while the IN input is True, Timer does not respond to the change until the IN input has become False and then True again. This is not immediately obvious and is not the expected behavior compared to other environments.

TIA Portal has a solution to this, which is to place a `PT Coil` before `Timer`. This updates the `PT` value while the timer is running:

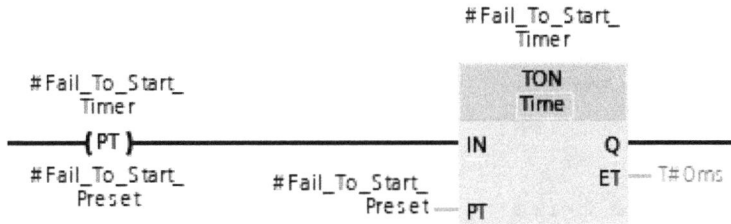

Figure 4.17 – TON timer with a PT Coil

Each language has a variation of how this is performed. For example, in SCL, the `PRESET_TIMER` instruction needs to be used.

> **Note**
>
> `PT Coil` can be used with the TON, TOF, TONR, and TP timers.

Valid networks in ladder logic

In the LAD language, TIA Portal has some uncommon rules regarding what is allowed and not allowed when it comes to positioning logic. For example, it is not possible to bypass a box instruction with a branch that originates from the same wire that the box is situated on.

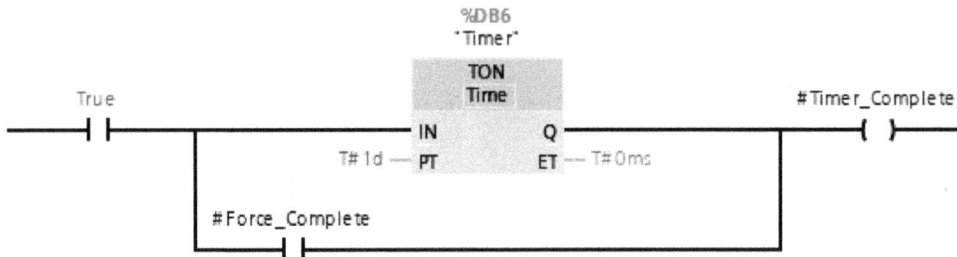

Figure 4.18 – Example of an invalid network

However, removing the first open branch is acceptable, even if the logical outcome is the same as it was previously.

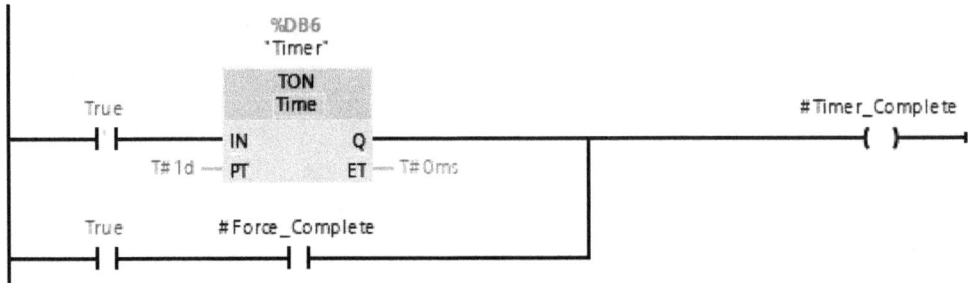

Figure 4.19 – Example of a valid network

This behavior may be unexpected to programmers who have come from other platforms, where *Figure 4.18* would be a valid logic network.

GRAPH is not SFC

Programmers moving from a platform that implements SFC may feel like GRAPH is Siemen's answer to SFC. While they are very similar, SFC is not embedded with interlock and supervision coils like GRAPH is.

Figure 4.20 – Example of a GRAPH interface with interlock and supervision coils

This can confuse programmers coming from SFC. It is also worth noting that some environments that implement SFC are capable of calling the step as a function, thus implementing a new language of choice at each step. GRAPH offers a traditional qualified action approach, with the other areas of logic being written in ladder only.

Bit access in the byte, Word, and Dword variables

Most modern PLCs are capable of accessing bits inside other variables without too much effort on the programmer's part. Most environments opt for the following approach:

```
#Filter_Alarm_Active := #Alarm_Word.13;
```

Figure 4.21 – Accessing a bit in a Word in most environments

In TIA Portal, however, this does not compile as . denotes accessing a child object in a Struct. To access bits in TIA Portal, the following should be used:

```
#Filter_Alarm_Active := #Alarm_Word.%X13;
```

Figure 4.22 – Accessing a bit in a word in TIA Portal

. %X tells TIA Portal to return the bit that's been specified in the variable as a bool data type. This method saves the programmer from having to convert words into *Boolean arrays*.

Summary

You should consider different languages and use cases when you're developing logic blocks. This chapter has introduced all of the available programming languages that TIA Portal offers programmers and explained where they can be used. It also provided an overview of what they look like and how they are used.

This chapter should have helped you appreciate the different languages and what they can offer an overall project.

The next chapter will expand on this by providing more detail on the languages mentioned here and how to program them. A basic sand filter cleaning sequence will be programmed in each of the languages (where possible) to demonstrate how to construct a block in the respective language and what issues will be encountered.

5
Working with Languages in TIA Portal

This chapter uses a common scenario to explore how different languages would approach the logic and control of the scenario.

Working with different languages presents different challenges at different points of programming. Understanding where languages have strengths and weaknesses with a comparable scenario will help programmers learn quickly when to mix and match code.

In this chapter, we'll cover the following main topics:

- The control scenario
- Languages used in TIA Portal

After reading through the chapter, the following languages will have been used in programming for the provided scenario:

- **Ladder Logic (LAD)**
- **Function Block Diagram (FBD)**
- **Structured Control Language (SCL)**

- GRAPH
- **Cause and Effect Matrix (CEM)**

Each of the preceding topic areas will be examined against a control scenario. An overview of the basics of the languages and instructions are provided before a walk-through of the control scenario.

The control scenario

This chapter focuses on a *control scenario* for a basic sand filter. The scenario is simplified enough that the control aspects are simple but enough of the languages is explored to understand how to use them and demonstrate strengths and weaknesses.

The languages explored in this chapter will be used for the following *control scenario*:

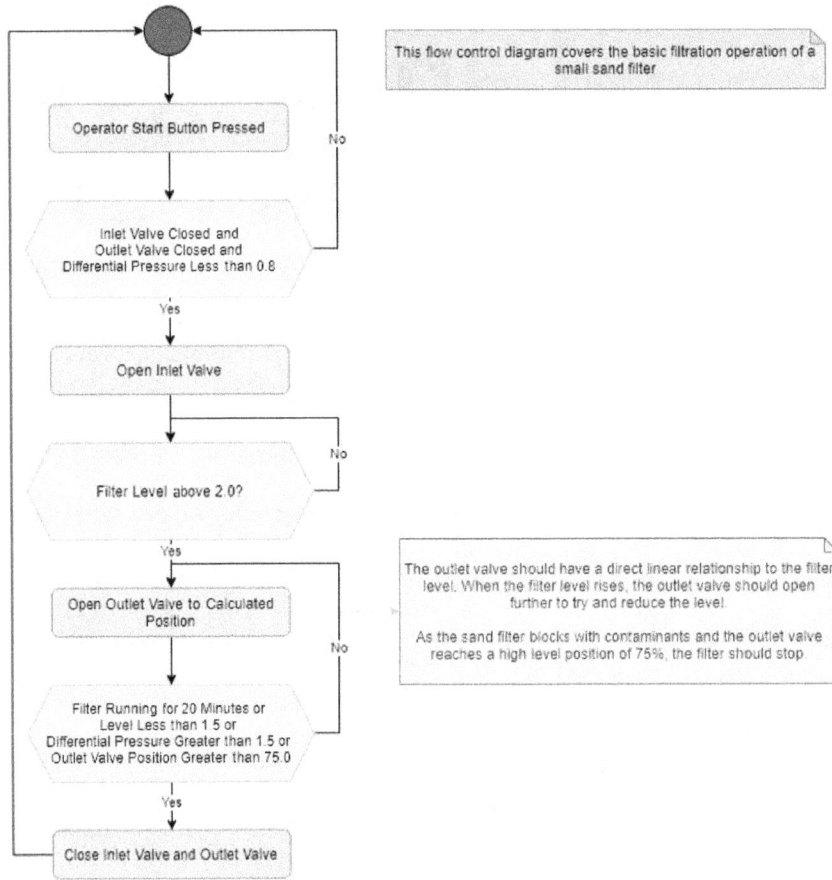

Figure 5.1 – Chapter control scenario

The flow control diagram in *Figure 5.1* demonstrates a simple method to relay how the controls are expected to operate for each of the languages used in this chapter.

A graphical representation of the process is available via a Comfort Panel **Human Machine Interface (HMI)**.

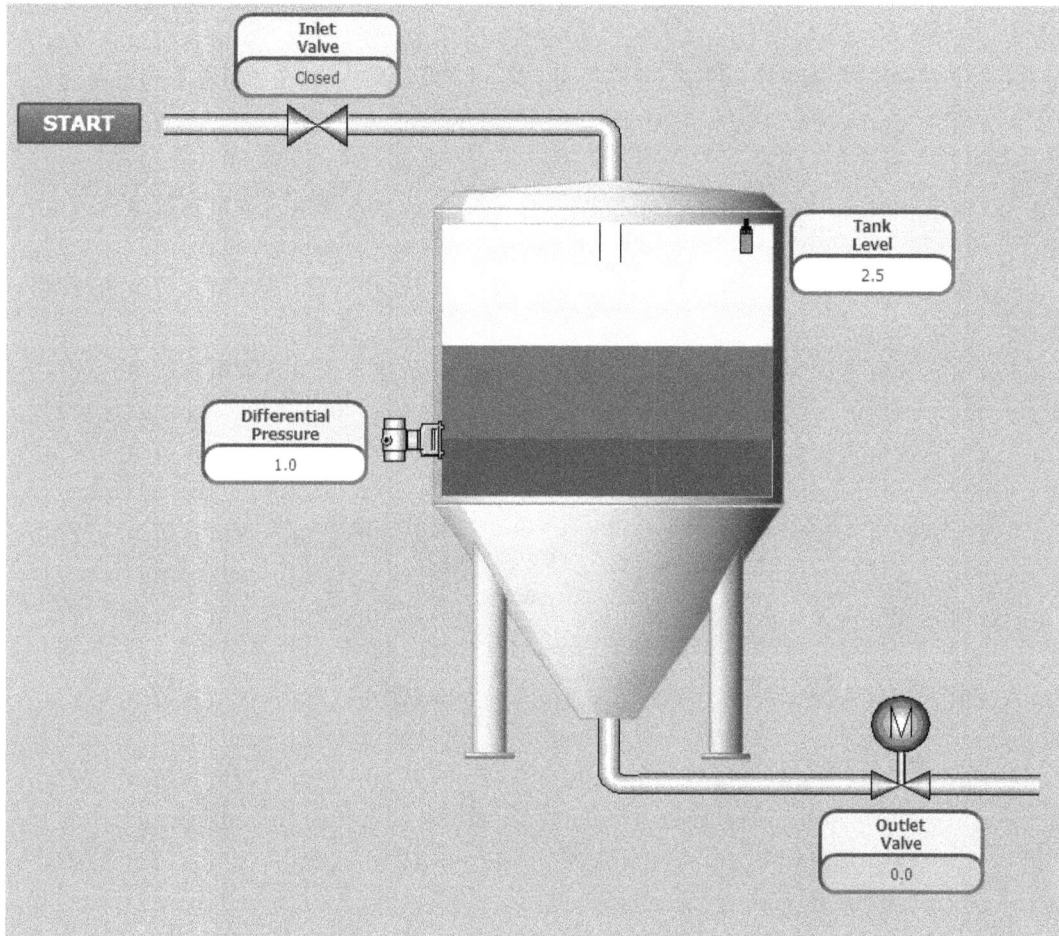

Figure 5.2 – Graphical representation of control system

The liquid to be filtered enters the tank from the left, through the **Inlet Valve** and exits the tank after passing through the filtration media via the **Outlet Valve**. The instrumentation consists of a **Tank Level** and **Differential Pressure**. These are used in the *control scenario* to provide control of the inlet and outlet valves.

The operator will also use this HMI to begin and manually stop the process.

Control overview

In accordance with the *control scenario (Figure 5.1)*, the system should go through the following phases whereby the graphical representation should change to indicate the control steps.

Opening the inlet valve

When the operator presses the **Start** button, the **Inlet Valve** will open, and the tank begins to fill.

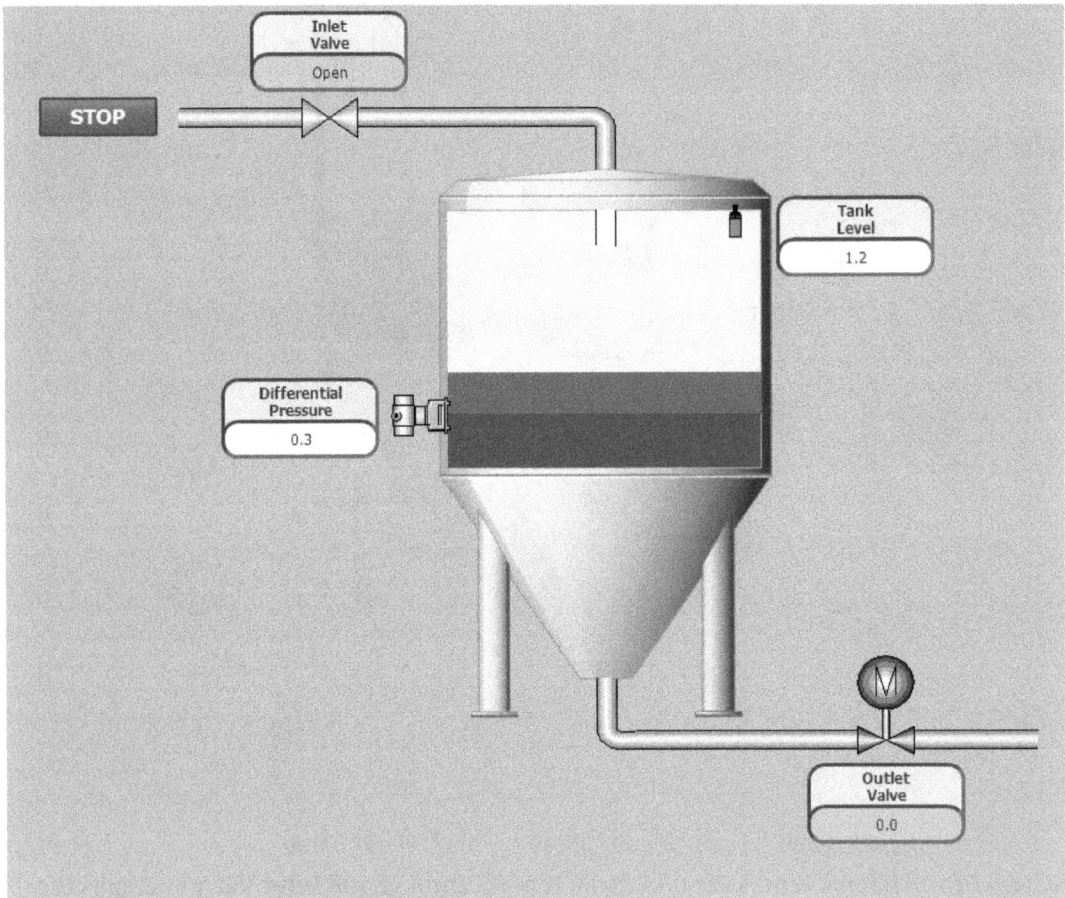

Figure 5.3 – Inlet Valve open and tank filling

At this point, the **Outlet Valve** is supposed to be closed to allow the tank level to rise above 2.0.

Opening the outlet valve

Once the **Tank Level** has risen above 2.0, the **Outlet Valve** should open to a calculated value. At this point, the liquid in the tank will drain through the *filtration media*. The level will begin to drop, and the **Outlet Valve** responds accordingly. Similarly, if the level should rise, the **Outlet Valve** will respond by increasing the position of the valve.

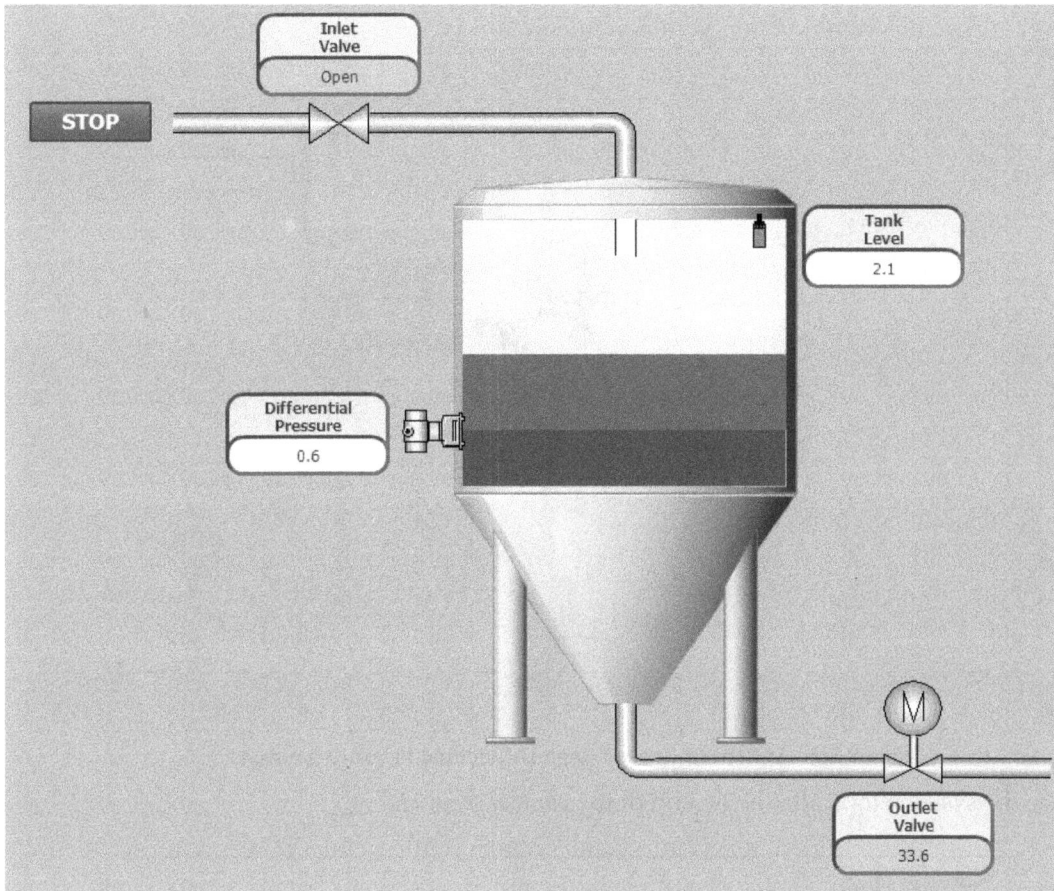

Figure 5.4 – Outlet Valve open to calculated value as Tank Level has reached 2.0

The system continues to operate in this mode, controlling the **Outlet Valve** against the **Tank Level** until a stop condition is True.

Stop conditions

The *control scenario* lists different causes for the system stopping; however, all outcomes are the same. The system should stop, and all valves should close.

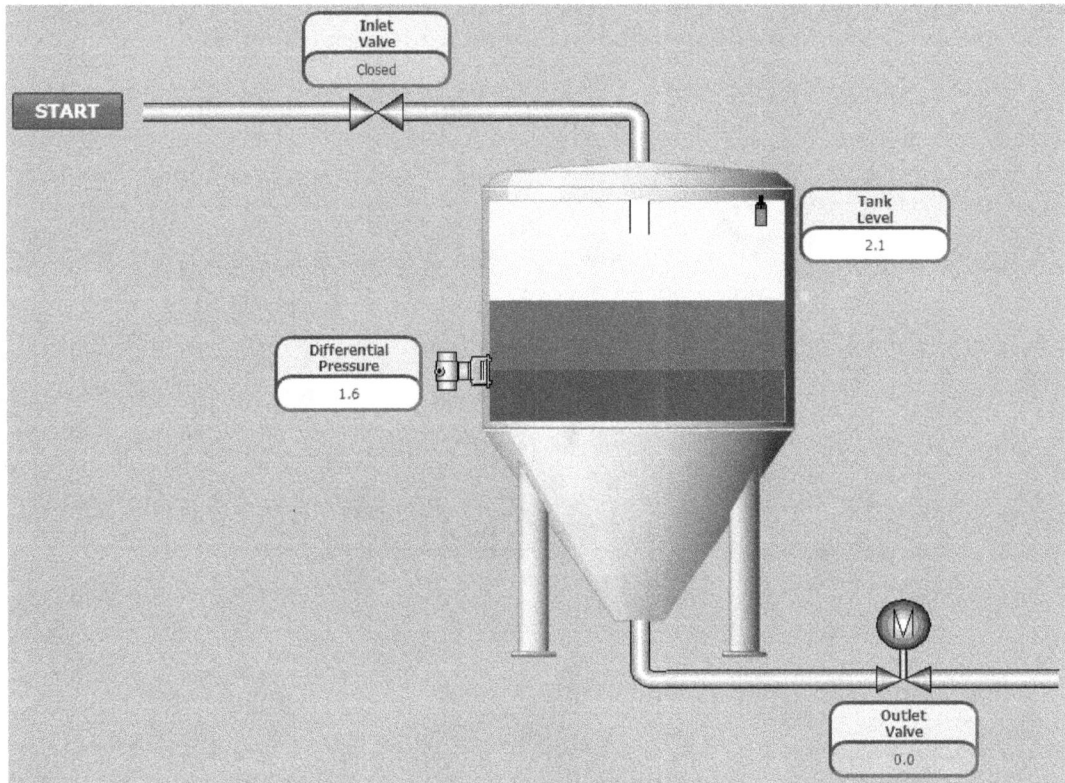

Figure 5.5 – System stopped on high Differential Pressure (above 1.5)

Once the system stops, that is the end of the *control scenario*.

Note

If this were a real system, the filter would require cleaning or a form of manual intervention from the operator of the system. In this case, the system can simply be reset back to the starting values and run again.

Using the HMI

Clicking the **Start** button will begin the process as long as the control logic in use in the PLC meets the *control scenario* conditions.

The **Tank Level** and **Differential Pressure** sensor values can be modified by clicking in the white area that contains their respective values.

Languages used in TIA Portal

This section covers five of the languages available in TIA Portal and demonstrates their usage against the control scenario.

Ladder logic

LAD is the most popular language in the PLC control space. It's likely that programmers of PLCs have never seen or used LAD. Although it's still extremely popular, other languages such as SCL are also becoming very popular.

Overview

Ladder gets its name from the way that the logic *flows* from the left of a network to the right of a network. When multiple networks are placed in series, a *ladder* is formed.

Ladder, like other languages that are programmed in networks, is processed as follows:

- Networks
  - Top to bottom
- Logic inside networks
  - Left to right.
  - Top to bottom.
  - Conditions that come to a common point (the closing of a **branch**) will evaluate all conditions left of the common point before proceeding.
  - Outputs/instructions that are on open branches will be processed top to bottom from the point where the branch opens.

It's important to understand the logic flow, especially when using branches, otherwise logic may not execute in the expected way.

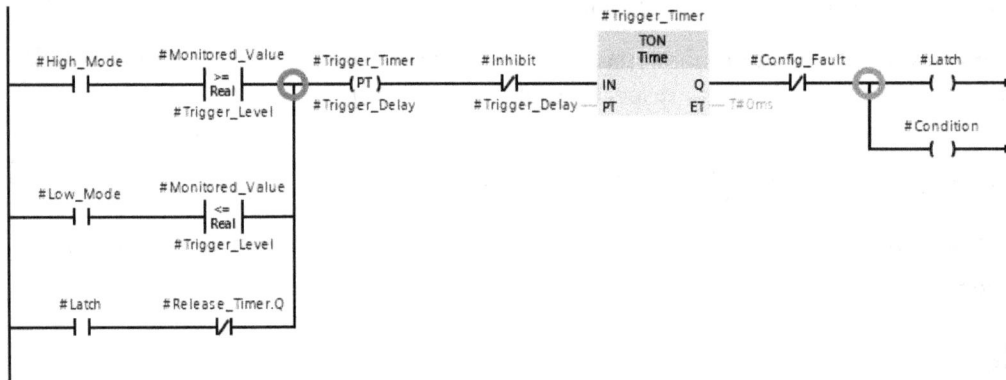

Figure 5.6 – Example of closing and opening branches

Figure 5.6 is an example that demonstrates the flow of logic when branches are involved. Before `Trigger_Timer`, there are three branches of code that close at a common point (indicated by a red circle). At the point of all three branches closing at a common point, all logic to the left will have been evaluated, with the three branches executed from left to right, top to bottom.

After `Trigger_Timer`, the logic opens a branch (indicated by a blue circle) so that there are two branches that contain coils. At this point, the logic is again executed in the order top to bottom.

Instructions in LAD

Every language has roughly the same **basic instruction set**, but they are implemented in different formats. They consist of groups of instructions that are categorized by their function and the data types that the instructions act upon.

LAD's basic instruction set can be found in the **Instructions** panel on the right-hand side of TIA Portal and in the **Basic instructions subsection**.

The basic instructions make up the base instruction set for the language; most of the other options for instructions come in the form of a **function** or **function block**, which can usually be called in any language.

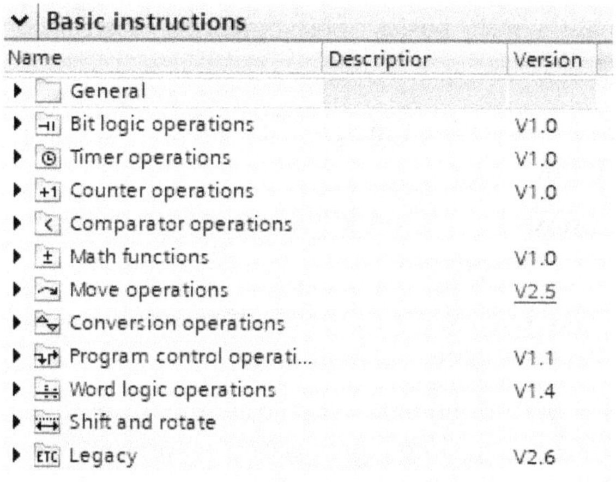

Figure 5.7 – Basic instructions palette for the LAD language

The **Basic instructions** palette may look different in different languages or may not even be available. As TIA Portal versions change, the available instructions inside the folders may also change.

Ladder bit logic operations

Ladder is a perfect choice of language for sections of code that require **bit logic** to be managed as it is configured for the easy reading of bit logic. Ladder was developed to operate in the same way that relay panels operate, so the terminology for the naming of instructions is similar, such as **Normally Open Contact**.

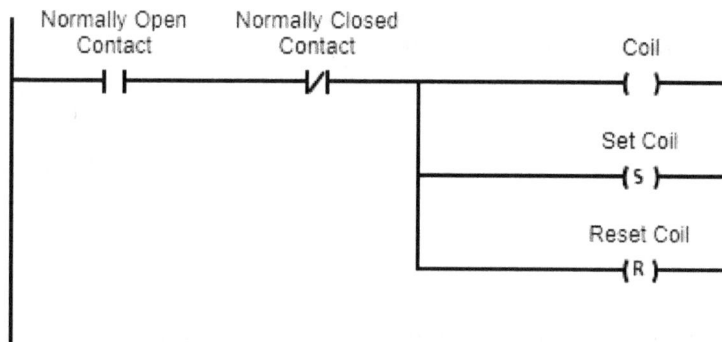

Figure 5.8 – Example of LAD and bit logic operations

Most LAD that programmers will read, write, and work with will look similar to the preceding, with normally open contacts, normally closed contacts, and different variations of coils.

It is these instructions that form the basics of bit logic. Each instruction works with `True` or `False` logic:

- **Normally Open Contact** – When the preceding logic = 1 (`True`) and the assigned variable = 1, the output = 1.

- **Normally Closed Contact** – When the preceding logic = 1 (`True`) and the assigned variable = 0 (`False`), the output = 1.

- **Coil** – When the preceding logic = 1, the assigned variable = 1. When the preceding logic = 0, the assigned variable = 0.

- **Set Coil** – When the preceding logic = 1, the assigned variable = 1. The variable is retained at 1 even if the preceding logic becomes 0 on subsequent scans.

- **Reset coil** – When the preceding logic = 1, the assigned variable = 0. The variable is set to 0 even if it was not set to 1 by a set coil.

These are the most common instructions used in Ladder and are most likely to be used at the lowest levels in the PLC program.

> **Note**
> Further help on instructions can be found by highlighting an instruction and pressing *F1*. This will open TIA Portal's information system where more detailed help is available.

Box instructions

A **box instruction** is an instruction that literally appears inside a box.

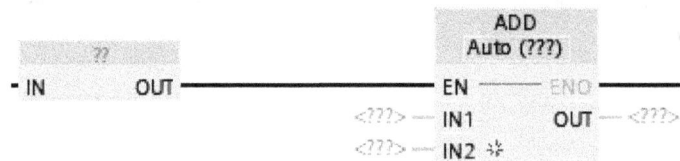

Figure 5.9 – An empty box instruction (left) and an un-parameterized ADD box instruction (right)

These are mainly used when an instruction needs to accept *parameters* (or arguments). There are hundreds of different box instructions that can be called in LAD (and other languages).

> **Note**
> Functions and function blocks can also be called from an empty instruction box. Simply type in the name of the **program block** where the ?? appear.

Comparators

Unlike most other PLC environments, TIA Portal does not display a comparator as a box instruction in Ladder, nor does it use the notation of *LE* for **less than or equal to**, for example. Instead, TIA Portal chooses to display it as a contact with the type of comparison being performed in the center.

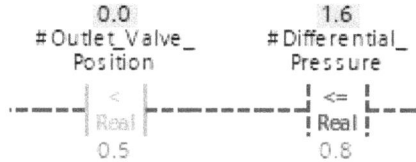

```
        0.0                         1.6
   #Outlet_Valve_              #Differential_
      Position                   Pressure
  ---| < |----------------------| <= |---
     | Real |                    | Real |
        0.5                         0.8
```

Figure 5.10 – Example of a "less than" comparator and a "less than or equal to" comparator

This does not change how the instruction operates in any way; however, it may be unexpected for programmers who use more than one development environment.

Control scenario walk-through

The *control scenario* is relatively simple and LAD would be a good choice to implement for this type of control requirement. Because the filter system is simply waiting for events to occur, with only one element of active control (the **Outlet Valve** calculation), the control is predominantly bit logic based, which suits Ladder perfectly.

Starting the system

As per the *control scenario*, the system can only start when the starting conditions are True and the operator has requested a start via the **Start** button.

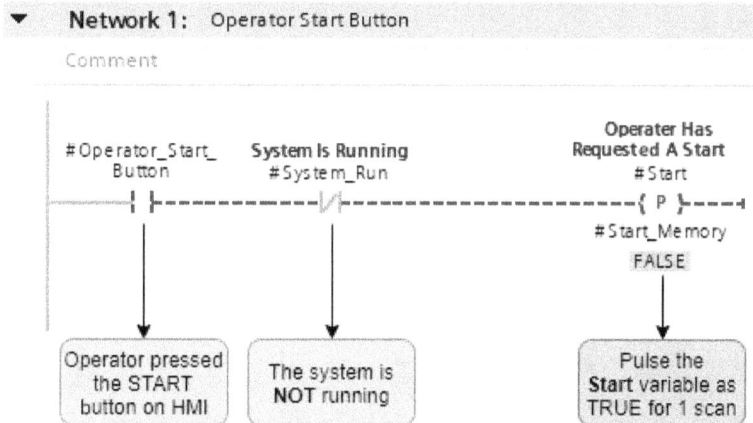

Network 1: Operator Start Button

Comment

```
                                                     Operater Has
                                                  Requested A Start
 #Operator_Start_    System Is Running                 #Start
     Button            #System_Run
  ---| |----------------|/|--------------------------( P )----|
                                                    #Start_Memory
                                                       FALSE
```

| Operator pressed the START button on HMI | The system is NOT running | Pulse the Start variable as TRUE for 1 scan |

Figure 5.11 – Operator Start button logic

Network 1 of the *Ladder solution* manages the operator's **Start** button presses by ensuring the system has not already started and then pulsing a local variable called `Start`.

Local Variables

Local variables are variables that exist in the current scope only. The `Start` variable in *Figure 5.11* is a local variable. This is denoted by the # that appears before the variable.

Notice that all of the variables in *Figure 5.11* are local to the function block as they all appear in the interface or are `temp` variables used by the function block.

When the operator presses the **Start** button on the HMI, `Operator_Start_Button` will be set to `True`. As long as the `System_Run` variable is `False`, the *"P" coil* will pulse the `Start` variable with a `True` value for one scan of the PLC.

Note

The `Start_Memory` variable is a **static** declared variable that holds the previously read value of `Start`. This is how the PLC recognizes that the preconditions have changed from `False` to `True`, and on the next scan remain as `True` (which stops the coil from outputting that scan again).

Once the `Start` variable contains a `True` value, the next network checks the requirements laid out in the control scope.

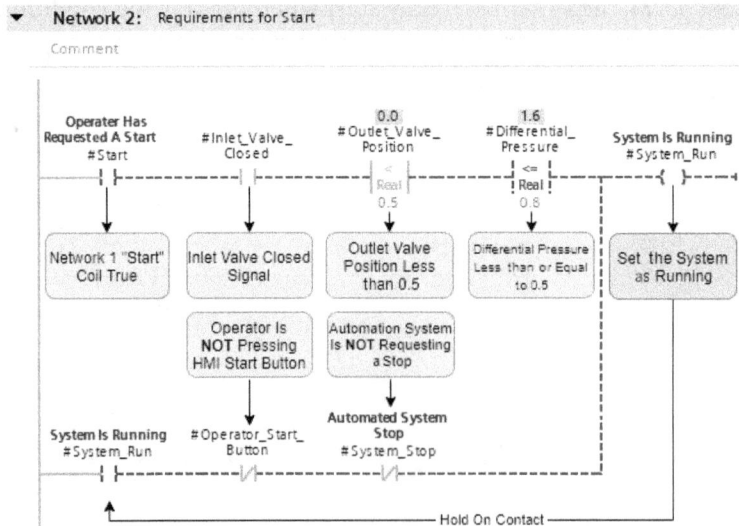

Figure 5.12 – Requirements for starting the system

At this point, the logic has received the input from the operator to start the system (via the `Start` variable) and is now checking the other requirements in the *control scenario*.

The interesting part of this network is the use of a **hold on contact**. This is exactly the same as an electrical panel whereby a relay will close a contact and power its own coil until another condition breaks that hold. This type of logic may also be referred to as a **latching circuit**.

Figure 5.12 shows that once the top branch of the network is `True`, `System_Run` becomes `True`. On the next scan, everything to the left of the coil is evaluated *before* the coil is written. This means that `System_Run` is still `True` on the bottom branch at the point of the second scan. As long as the `Operator_Start_Button` is no longer pressed and the automated system is not requesting `System_Stop`, then `System_Run` feeds its own coil, keeping the coil in a high `True` state until one of the two conditions breaks the hold on contact.

In this particular case, it is essential that the *HMI Start button* can only be active for one scan. The following network ensures that the button is written back to `False` at the end of the function block:

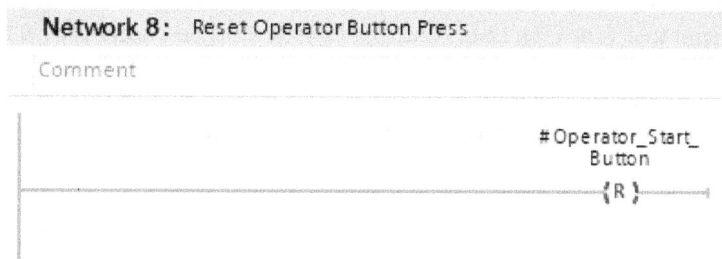

Network 8:   Reset Operator Button Press

Comment

```
                                              #Operator_Start_
                                                  Button
                                                   ( R )
```

Figure 5.13 – Resetting the operator Start push button

Because the HMI is *event-driven*, the condition of a button press can only happen once per press. If an operator keeps their finger on the button, the PLC *wins* and the value is written back to `False` in the same scan cycle.

Opening the inlet valve

Now that the system is started and holding itself in an active state via the `System_Run` variable, assets can be controlled in accordance with the *control scenario*.

The next step is to open the **Inlet Valve**. This valve stays open the entire time the system is running.

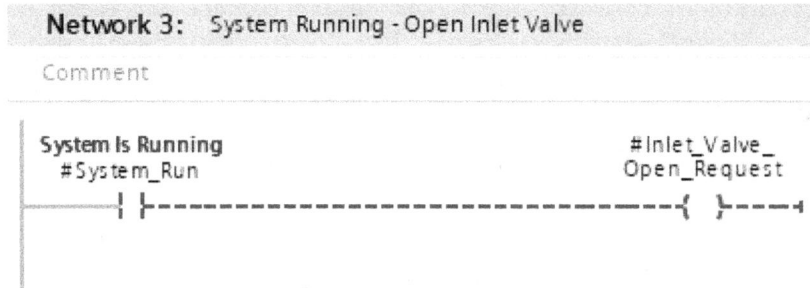

Network 3:   System Running - Open Inlet Valve

Comment

```
System Is Running                                          #Inlet_Valve_
  #System_Run                                              Open_Request
     ─┤ ├──────────────────────────────────────────( )────┤
```

Figure 5.14 – Simple LAD to open the inlet valve

Whenever the `System_Run` variable is `True`, `Inlet_Valve_Open_Request` is also `True`.

Asset Control Requests

Requesting an asset to do something, as opposed to controlling the IO directly, is a good habit to get into. `Inlet_Valve_Open_Request` in *Figure 5.14* is an output from the function block. Anything can be assigned to it outside of the block. This leaves programmers free to focus on the logic and *how* it is supposed to be controlled instead of *what* it is controlling.

It also allows for easy expansion should another block also need to control the same asset. In this case, neither block needs to be changed. The decision of which block is in control of the asset can be made elsewhere after they have both been processed.

The filling operation

Once the **Inlet Valve** is open, the system will begin to fill. The *control scenario* states that the filling operation should wait until the **Tank Level** reaches a particular level before releasing control of the **Outlet Valve**.

Network 4: System Running - Manage Fill

Comment

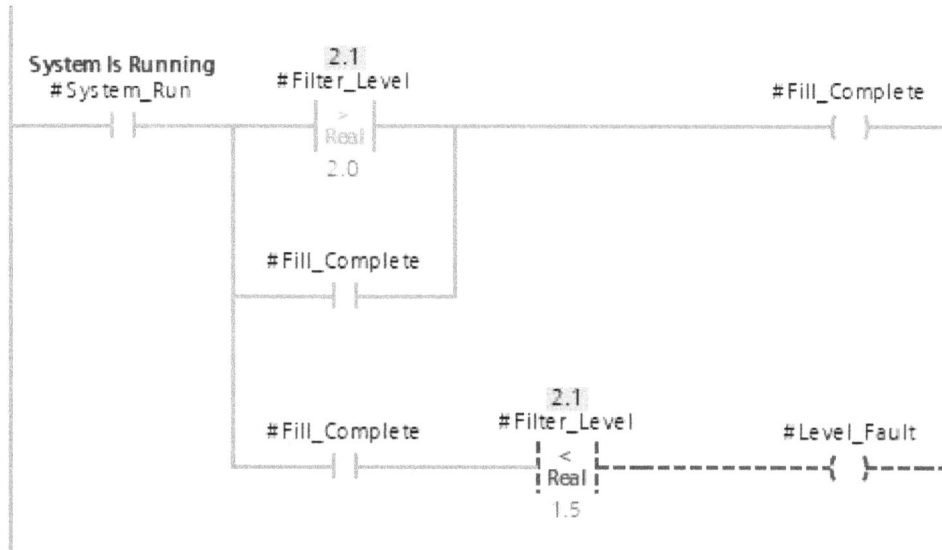

Figure 5.15 – Fill management

The fill management also contains a hold on contact configuration for the Fill_ Complete signal. However, this time, there are no release variables. Once Fill_ Complete is set at the coil, it will hold itself on until System_Run becomes False.

An additional function is also managed in this network; the Level_Fault variable is set if Filter_Level falls below 1.5, but only after Fill_Complete has already been set to True.

Calculating the outlet valve position

The *control scenario* states that the outlet valve's position should be calculated based on the level in the tank after the tank has filled to a specified level.

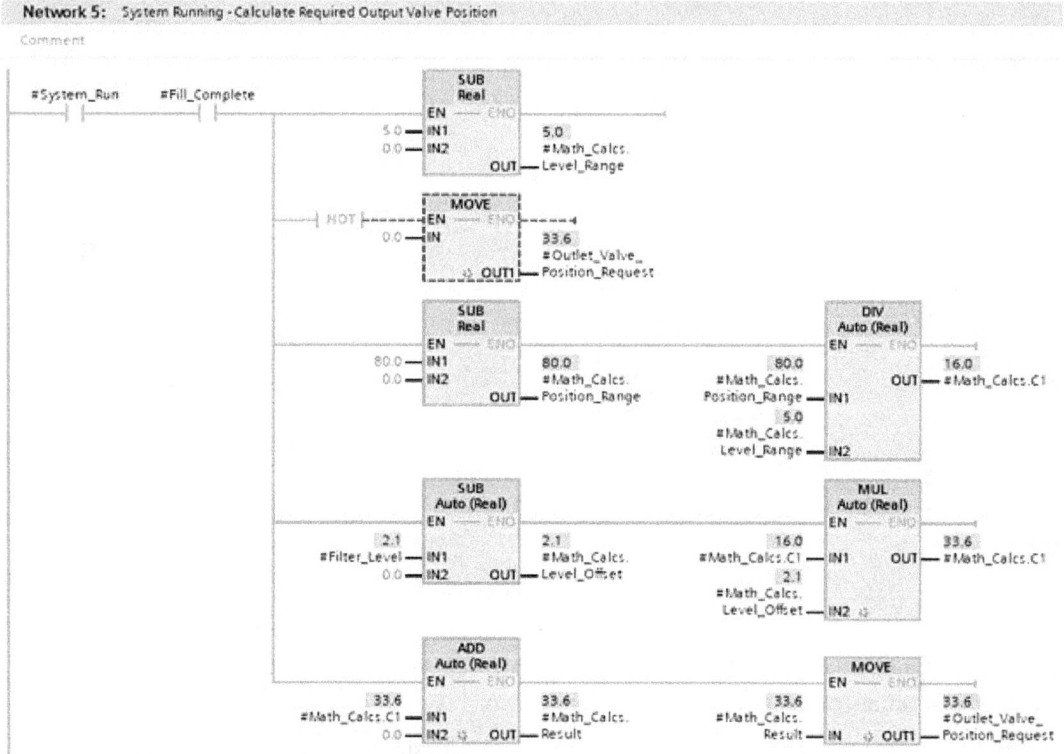

Figure 5.16 – Calculating the outlet valve position based on tank level

This is where LAD shows a weakness against some other text-based languages. The formula for the calculation shown in *Figure 5.16* is as follows:

```
Outlet_Valve_Position_Request =
(((Max Tank Level - Min Tank Level)
/
(Max Valve Position - Min Valve Position))
*
(Current Tank Level - Min Tank Level))
+
Min Valve Position
```

While this isn't a particularly complex formula, LAD has a hard time laying it out nicely and comprehensively.

Programmers will need to remember when doing calculations in Ladder this way how logic flows to ensure calculated values are only used when ready. It's also a good idea to create temporary variables to hold calculation results that are required later to continue the calculation.

Note the NOT instruction, which sets the Outlet_Valve position to 0.0 when the System_Run variable is False or the Fill_Complete variable is False.

> **Tip**
>
> There are better ways to do calculations when using Ladder. Check out the Calculate instruction, or even better, insert an **SCL network** to perform the calculation!

Managing stop conditions

At this point, the system is running completely, the **Inlet Valve** is open, and the **Outlet Valve** is being controlled by the tank level. The next step is to recognize a requirement to stop the system.

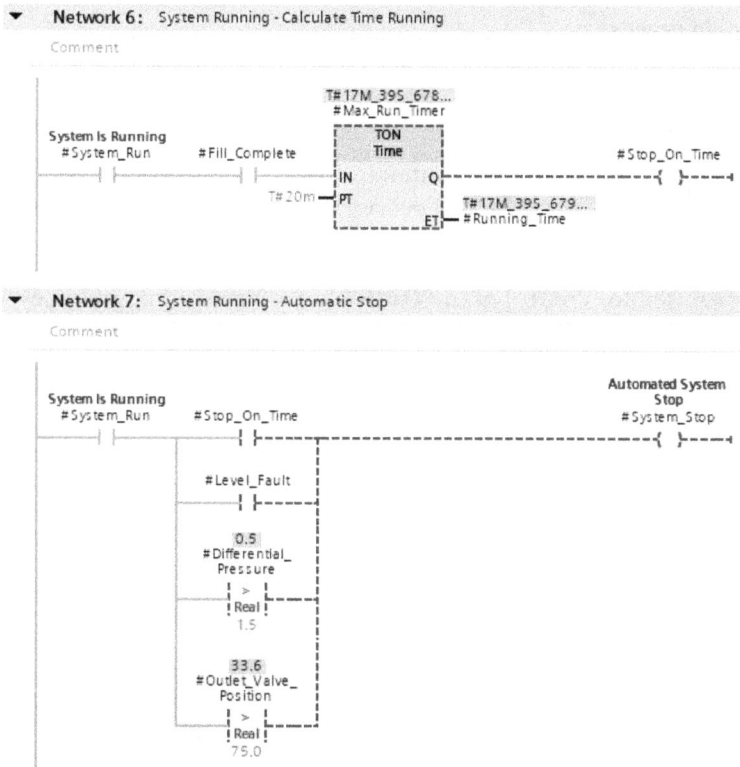

Figure 5.17 – System stop requirements

Figure 5.17 shows how **Network 6** and **Network 7** are used to consolidate stop requirements to a single `System_Stop` variable (which was also used in **Network 2,** in *Figure 5.12*, to break the hold on contact for the system running).

The `TON` instruction stands for **Timer "On" Delay**, meaning that when the `IN` condition is `True`, the timer will begin counting a delay time until **Elapsed Time (ET)** equals **Preset Time (PT)**. Once `ET >= PT`, the `Done (Q)` bit is set to `True` until the `IN` condition is set back to `False`.

In this case, the timer only starts to count once the tank has filled and therefore the system has released the control for the **Outlet Valve**.

If any of the stop conditions on **Network 7** become `True`, `System_Stop` is set to `True` and on the next scan, the system is stopped by breaking the hold on contact on **Network 2**.

> **Remember**
> Static variables must be used if the variable value is required to be retained between scans of the PLC. If `System_Stop` were a `Temp` variable, the system would never stop!

Control summary

In this example, all networks have been written in LAD. For most of the *control scenario*, the language has provided everything that is required for easy and simple control.

The calculations required for control of the **Outlet Valve** have highlighted how Ladder could be weak with very complex calculations.

Some of the key aspects of using Ladder have been demonstrated as the following:

- Easy to read and follow, especially when monitoring online
- Network-managed sections of code, further enhancing the usability of the language
- Best used for Boolean (`True`/`False`) logic

Function Block Diagram

FBD is a graphical language that takes the form of grouped *blocks* that perform specific functions.

These blocks make up the logic, much in the way LAD works, by connecting to each other from left to right toward an endpoint such as an assignment or instruction. FBD is programmed in networks, the same way as LAD is.

Overview

FBD is very graphical and is designed to help guide the user through the logic while making it easy to perform more complicated functions hidden away in blocks. Essentially, FBD calls instructions in the same way that any other language would call a function block. Even simple instructions such as **greater than** are still called in the same format and interface style as that of a user-created function block.

FBD is processed as follows:

- Networks

  - Top to bottom

- Logic inside networks

  - Left to right.

  - Top to bottom.

  - All inputs to a block must be evaluated before the block is executed.

  - Branches are evaluated top to bottom from the point at which the branch opens.

The logic flow for FBD is the simplest of the graphical languages as branches cannot be closed. When a branch is open, it must complete with its own end instruction or assignment, it may not rejoin the branch above (this would create an OR logic path; FBD provides an OR block instead).

Instructions in FBD

FBD has almost exactly the same instruction set as Ladder, however, it is implemented slightly differently due to the nature of the language.

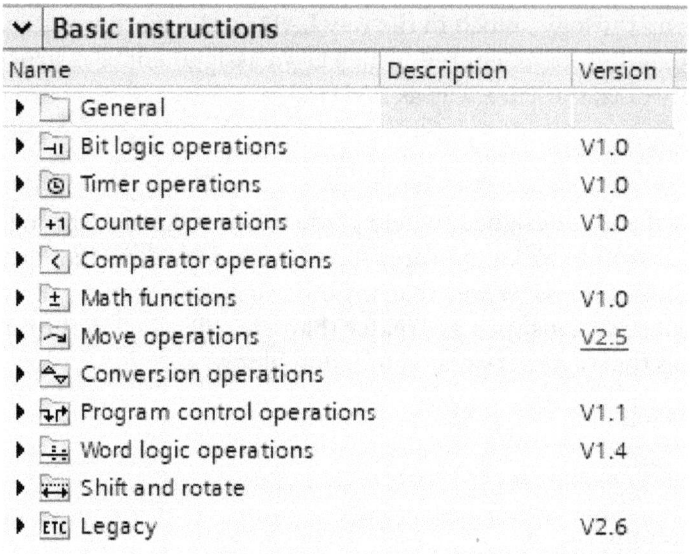

✔ Basic instructions		
Name	Description	Version
▶ 🗆 General		
▶ ⊣⊢ Bit logic operations		V1.0
▶ ⊙ Timer operations		V1.0
▶ +1 Counter operations		V1.0
▶ < Comparator operations		
▶ ± Math functions		V1.0
▶ ⤳ Move operations		V2.5
▶ ⤵ Conversion operations		
▶ ⤶ Program control operations		V1.1
▶ ⫦ Word logic operations		V1.4
▶ ⫣ Shift and rotate		
▶ ETC Legacy		V2.6

Figure 5.18 – Basic instructions palette for the FBD language

While all of the top folders in the **Basic instructions** palette are the same, the content of the folders is different to suit the FBD language.

FBD is also designed as a language for bit logic, focusing on AND and OR blocks in order to build the logic.

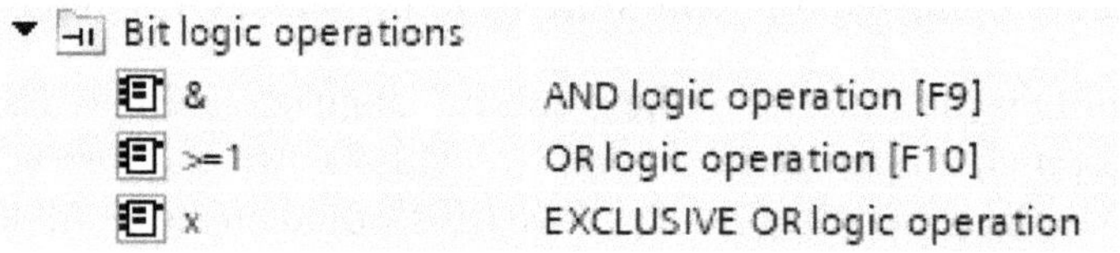

▼ ⊣⊢ Bit logic operations
    🔲   AND logic operation [F9]
    🔲 >=1   OR logic operation [F10]
    🔲 x   EXCLUSIVE OR logic operation

Figure 5.19 – Logic gate blocks

As with all instructions, these can be found from the **Instructions** tab on the right-hand side of TIA Portal. The AND and OR blocks are the most basic blocks provided in FBD.

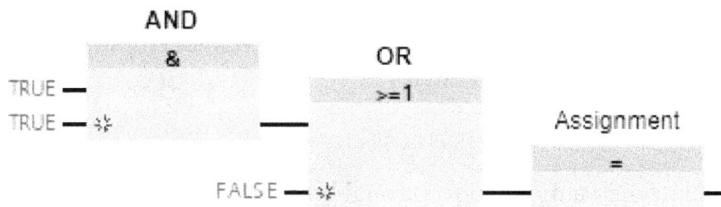

Figure 5.20 – AND and OR blocks, providing an Assignment with a true value

In *Figure 5.20*, the AND and OR instructions are connected to an **Assignment** instruction, which behaves the same way that a **coil** behaves in LAD. When a logical 1 (True) is passed to it, the variable associated with the **Assignment** is set to 1, otherwise, it is set to 0 (False).

Set and reset assignments are also possible in FBD. After placing an **Assignment** block, a dropdown on the block itself can be used to select a variety of options. Choosing **S** will create a **Set Assignment** and choosing **R** will create a **Reset Assignment**.

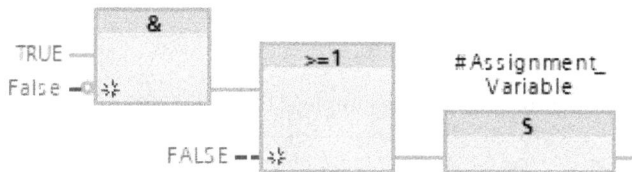

Figure 5.21 – Example of a Set Assignment

Box instructions

In FBD, all instructions are box instructions, however, calling instructions that have an interface work in the same way as all instructions. The difference is that an interface is also offered.

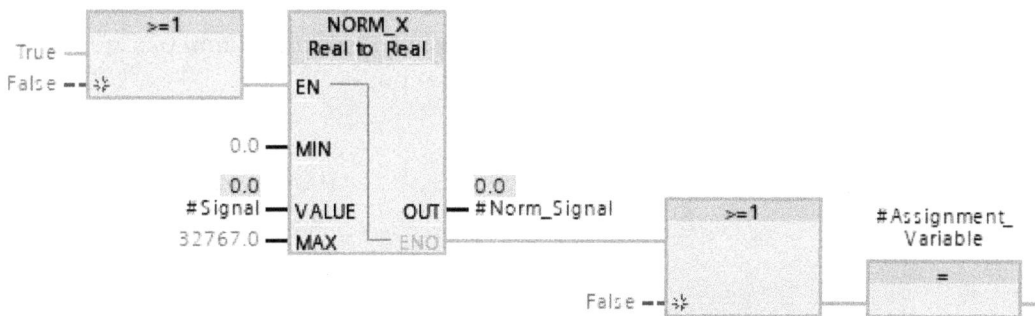

Figure 5.22 – Box instructions in FBD

Figure 5.22 shows how box instructions look and behave in FBD. The most noticeable difference from a box instruction in Ladder is that the EN (enable input) and ENO (enable output) output are not on the same level.

Comparators

FBD displays comparators in a more conventional box instruction but still uses symbolic notations for the instruction type.

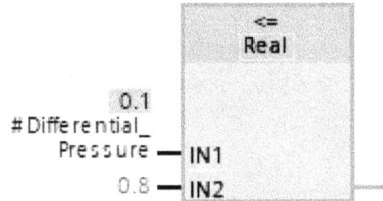

Figure 5.23 – Comparator for "less than or equal to" in FBD

This notation of <= keeps consistency between similar languages such as Ladder.

> **Note**
> IN1 is compared against IN2. For example, IN1 <= IN2.

Control scenario walk-through

FBD fits well for most of this *control scenario* as the filter system is mostly basic Boolean logic.

The logic for the FBD solution results in the same operation of the filter system as all other languages.

Starting the system

As per the *control scenario*, the system can only start when the starting conditions are True, and the operator has requested a start via the **Start** button.

Network 1: Operator Start Button

Comment

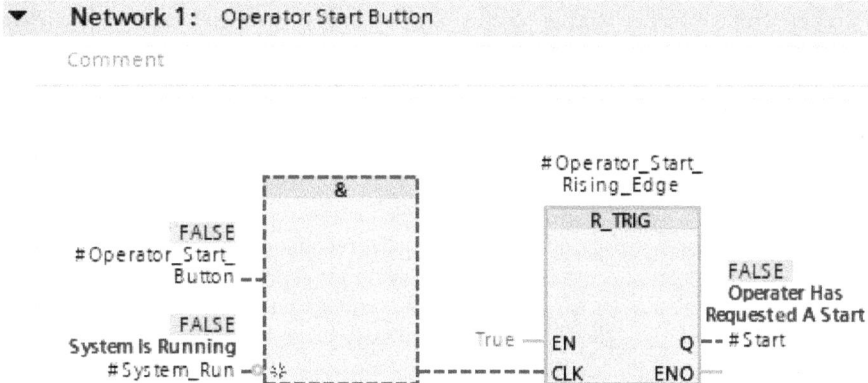

Figure 5.24 – Operator Start button logic

Network 1 of the *FBD solution* uses an AND instruction to check that Operator_
Start_Button is True, and the system is *not* already running (an inverted input into
the AND instruction is used, indicated by the small circle between the AND instruction and
the System_Run variable). If the conditions are True, a logical 1 is output from the AND
instruction to CLK of the R_Trig instruction.

R_Trig is a **Rising Edge Trigger that** outputs a True value on the Q output for one scan
and will not do so again until CLK has been scanned as a False value.

> **Note**
>
> A negated input is displayed with a small circle at the end of the input line.
> Negating an input can be achieved by selecting **Invert RLO** from the **General**
> folder in the **Instructions** panel.

Once R_Trig sets the Start variable to True, the system is required to check more conditions prior to opening the **Inlet Valve**.

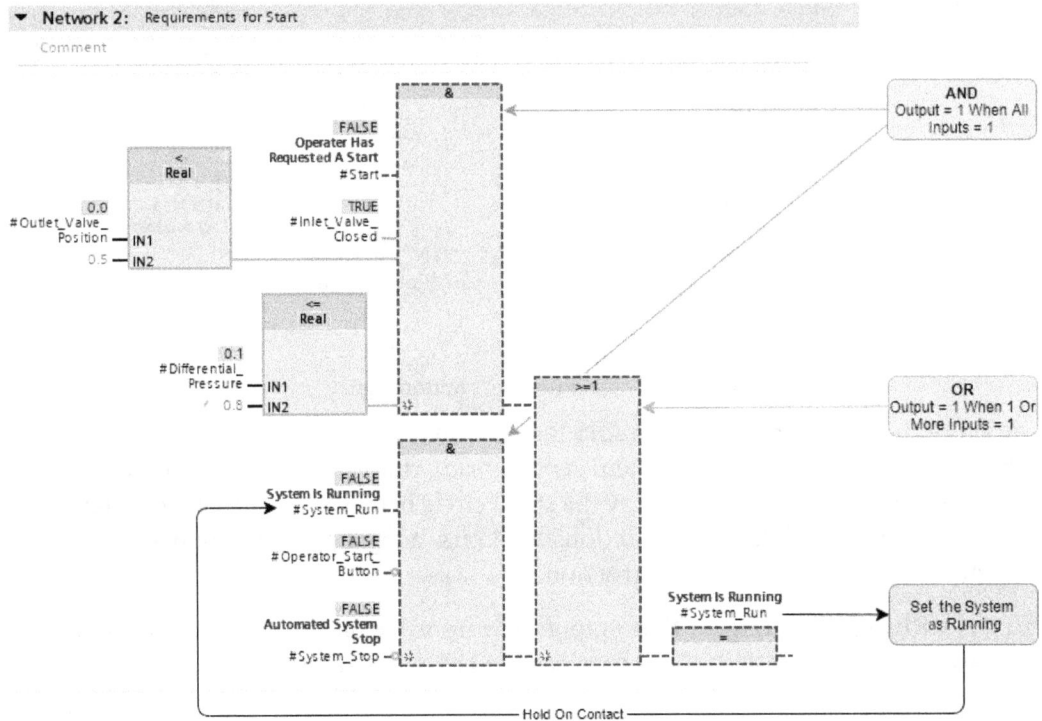

Figure 5.25 – Requirements for starting the system

At this point, the logic has received the input from the operator to start the system (via the Start variable) and is now checking the other requirements in the *control scenario*.

Figure 5.25 demonstrates how this logic is split into two areas. Before System_Run is written to by the Assignment instruction, an OR instruction is used (>=1). This instruction segregates the requirements for a start and the **hold on contact** requirements that keep the System_Run variable held as True.

When all conditions are True on the first AND instruction, the OR instruction allows the logic to set System_Run to True. On the next scan, the second AND instruction (providing the conditions are correct) will hold the System_Run instruction high until either the operator stops the system with the button or an automated system stop occurs.

Because `Operator_Start_Button` is used again to unlatch `System_Run`, it's important to make sure that the value is reset to `False` before the next scan.

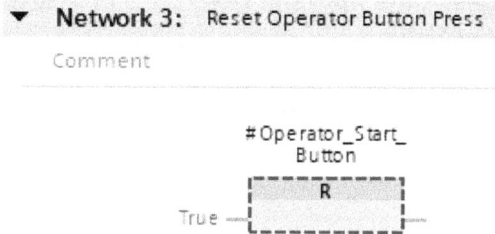

Figure 5.26 – Reset output

Using a `Reset Output` instruction ensures that `Operator_Start_Button` is always `False` before **Network 2** is next scanned.

> **Remember**
>
> The HMI is event-driven. Holding the button in does not send another `True` value to the PLC.

Opening the inlet valve

Opening the `Inlet_Valve` is as simple as assigning a `TRUE` value to it when the `System_Run` variable is `True`.

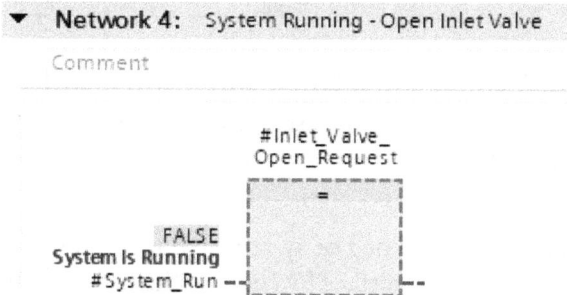

Figure 5.27 – Opening the inlet valve

This can be done with a single `Assignment` instruction.

The filling operation

Once the **Inlet Valve** is open, the system will begin to fill. The *control scenario* states that the filling operation should wait until the **Tank Level** reaches a particular level before releasing control of the **Outlet Valve**.

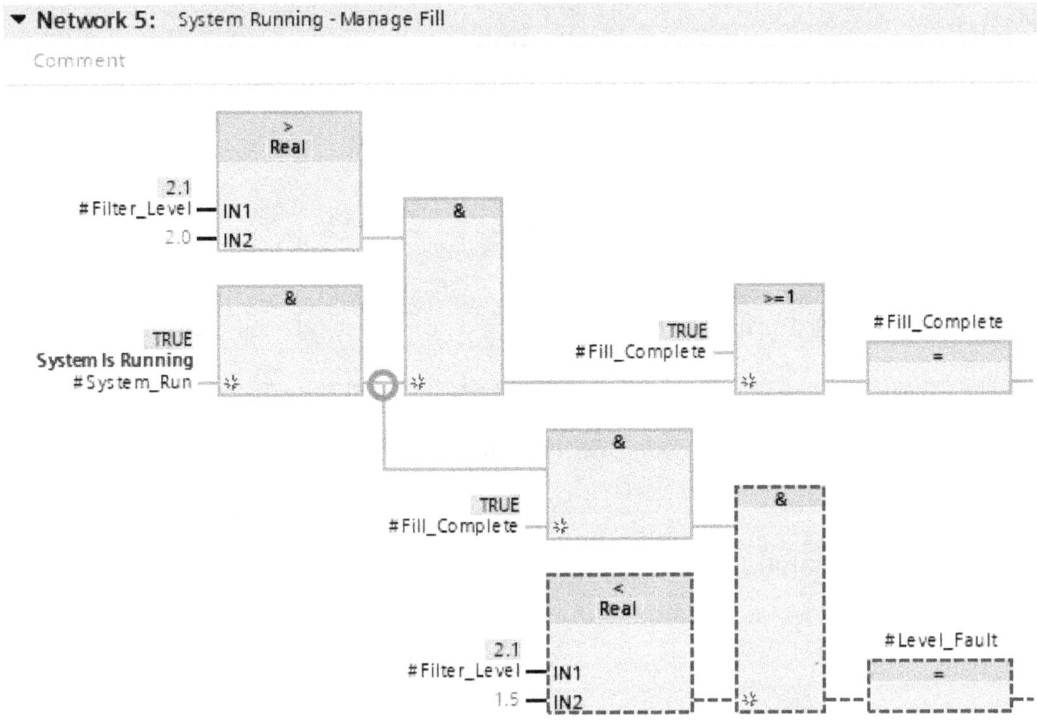

Figure 5.28 – Filling management and level fault

Figure 5.28 shows an example of a branch in use in FBD (shown with a blue circle). In order to branch the System_Run variable, an AND instruction has been used that only has a single input.

Most instructions (that aren't user-defined or system-defined function blocks) can have additional inputs added to the interface by clicking the little yellow star icon. Interface elements can also be deleted by selecting the input line and pressing *Delete* on the keyboard.

When an AND instruction only has a single input, it is always the same state as the variable assigned to the input. In the case of *Figure 5.28*, this has been done to allow a branch to be added immediately after the output. It is not possible to add a branch to an input unless it is already connected to the output of another block.

Note that this network also contains a **hold on contact** at the OR block before
Fill_Complete. This will keep Fill_Complete set to True until logic before
the OR block becomes False.

Calculating the outlet valve position

The *control scenario* states that the outlet valve's position should be calculated based on
the level in the tank after the tank has filled to a certain level. Just like the LAD solution,
FBD does a poor job at laying out this calculation.

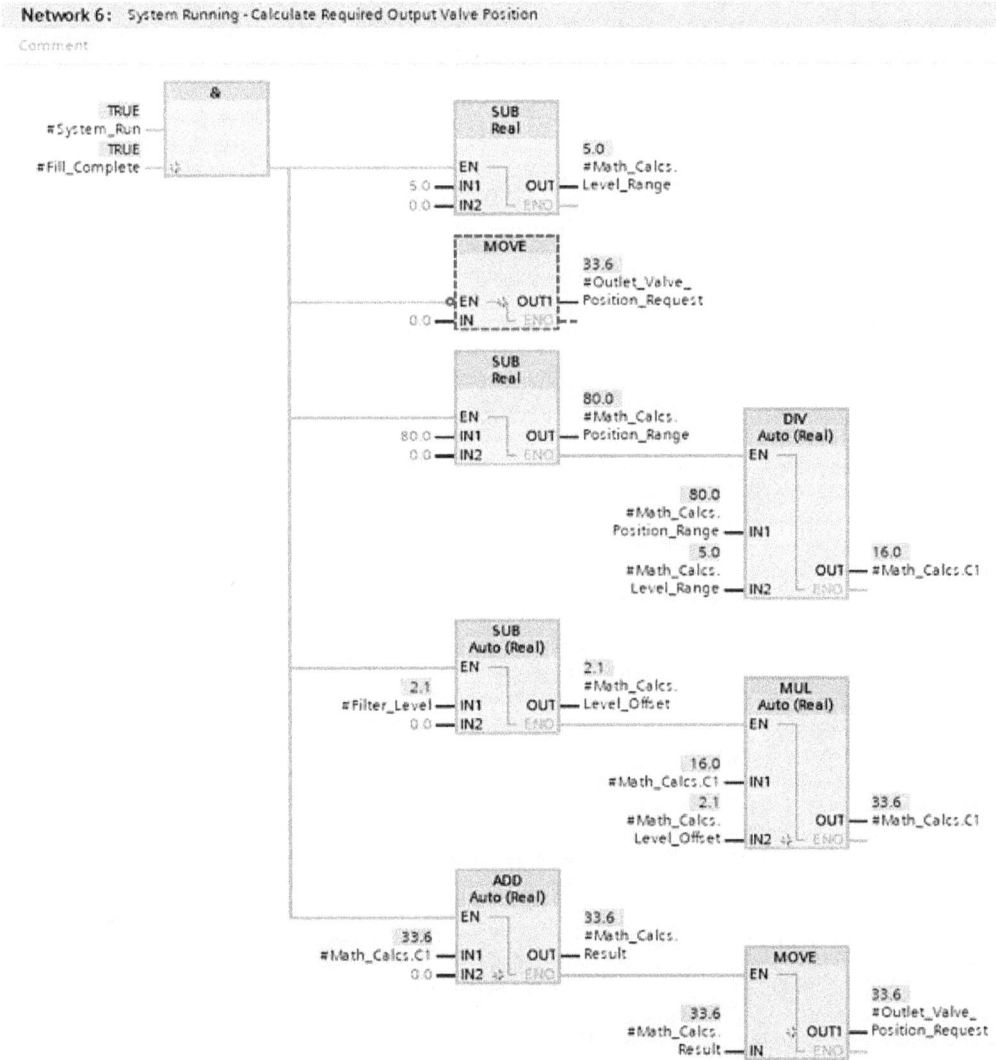

Figure 5.29 – Calculation of the Outlet Valve position

The way FBD lays out the math instructions results in large areas of unused space, which can make things difficult to read, especially on smaller screens. Note that the MOVE instruction's position is right at the bottom, despite there being enough room to have it much higher and visible. This is something to watch out for.

As with Ladder, programmers will be required to create holding variables to move results from one instruction to the next.

> **Note**
> Remember to observe the flow of logic with branches. Each branch will execute top to bottom, left to right. Also note that some inputs are inverted.

Managing stop conditions

When the system is filled, the outlet valve is open, and the system is running, the stop conditions need to be evaluated.

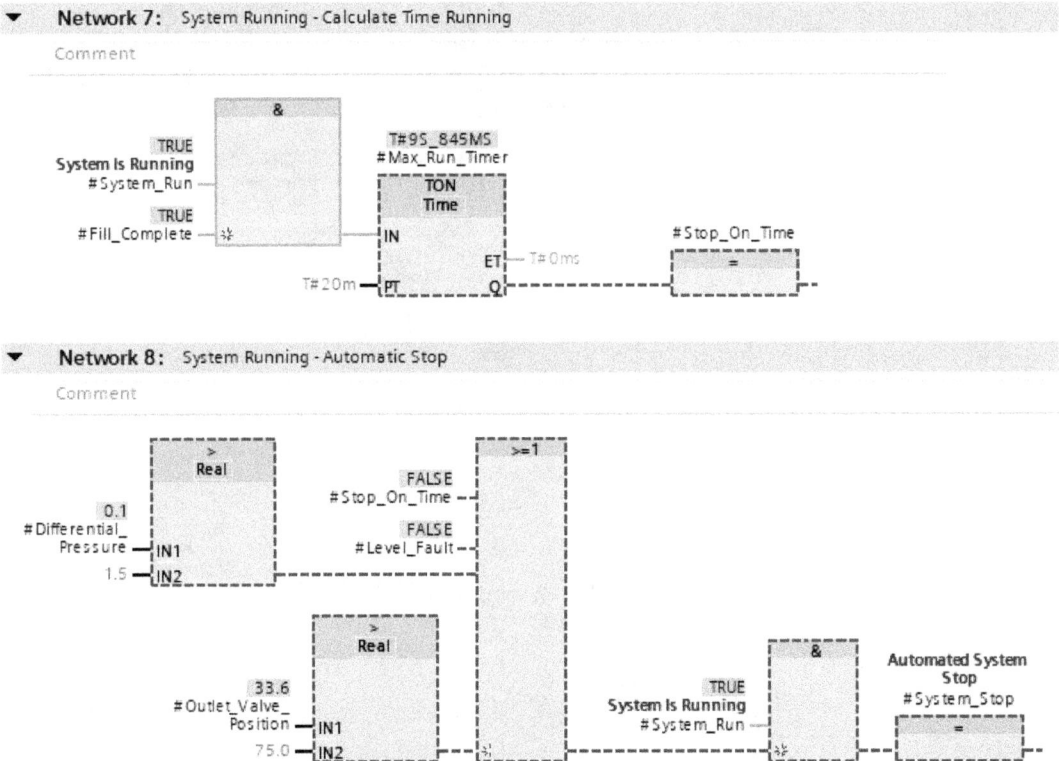

Figure 5.30 – Stop conditions

Just like the Ladder solution, **Network 7** and **Network 8** are being used to consolidate the required variables into a single System_Stop variable that stops the system.

The FBD layout does a better job at keeping things neat with the OR logic required for the conditions than that of the Ladder solution (*Figure 5.30*).

Control summary

In this example, all networks have been written in FBD. For most of the *control scenario*, the language has provided everything that is required for easy and simple control.

The calculations required for control of the outlet valve have highlighted how FBD could be weak with very complex calculations.

Some of the key aspects of using FBD have been demonstrated as the following:

- Easy to read and follow, especially when monitoring online
- Network-managed sections of code, further enhancing the usability of the language
- Best used for Boolean (True/False) logic

The similarities between FBD and Ladder are very strong, despite them looking different on the surface.

Structured Control Language

SCL is a textual language, meaning it is written in a text-based language. Because of this, it can actually be written outside of TIA Portal and then copied in, however, access to variables and tags would not be possible if written elsewhere.

Text-based languages operate differently from graphical languages, and if programmers are new to a language, it can seem busy and hard to read at first. SCL is a very popular language due to its extended instruction set and flexibility to perform well in almost all areas.

Overview

SCL is more commonly referred to as **structured text** as nearly every other PLC environment refers to it as ST and not SCL as Siemens does. The *structured* part of the name simply means that the language is based on instruction sets. SCL is still governed by the same basic instructions that other languages use, however, they are used differently.

Unlike Ladder and FBD, SCL does not have networks to group code together. Instead, SCL uses line numbers, and instructions may be spread over many lines to complete block interfaces and other statements.

The scan of the language is far more simple:

- SCL code

  - Line by line

- Statements

  - Statements such as the `IF` statement may cause lines to be skipped over if conditions are not `True`. This is handled automatically.

There are no *branches* in SCL either. Pathways are created by using *statements* that change which area of code is executed at any one time.

```
IF #Condition = TRUE THEN
    //Condition = True Code
    #ConditionResult := True;
ELSE
    //Condition = False Code
    #ConditionResult := False;
END_IF;
```

Figure 5.31 – Example of flow-changing IF statement

Figure 5.31 shows an `IF` statement that executes different areas of code depending on the state of the `Condition` variable. When the `Condition` variable is `True`, the section above the `ELSE` condition executes. When the `Condition` variable is `False`, the section of code below the `ELSE` condition executes.

> **Note**
>
> If `ELSE` is not included, no code would be executed if the condition was not `True`. SCL would continue from the `END_IF` part of the statement.

Instructions in SCL

The SCL instruction set is handled differently from other languages and is more closely related to that of **Statement List (STL)**. The **Basic instructions** panel has a greatly reduced offering of **Bit logic operations** and the **General** palette doesn't exist at all.

Basic instructions

Name	Description	Version
Bit logic operations		V1.0
R_TRIG	Detect positive signal edge	V1.0
F_TRIG	Detect negative signal edge	V1.0

Figure 5.32 – Basic instructions panel of SCL

This is because all of the general instructions and most of the bit logic operation instructions can simply be written out as text or are completed through different statements.

SCL bit logic operations

Bit logic can still be achieved easily in SCL; however, it can become difficult to read when the logic starts to become complex and brackets are introduced. Because branches do not exist in SCL either, writing to multiple outputs can be difficult to read.

```
#Condition_1 := TRUE;
#Condition_2 := True;

#Result_1 := #Result_2:=  #Condition_1 AND #Condition_2;
```

#Condition_1	TRUE
#Condition_2	TRUE
#Result_1	TRUE
#Result_2	TRUE
#Condition_1	TRUE
#Condition_2	TRUE

Figure 5.33 – Example of writing to two variables with online monitoring displayed to the right

Figure 5.33 shows that both `Condition_1` and `Condition_2` are set to `True` values (notice that capitalization of the `True` value does not matter).

`Result_1` is set to the value of `Result_2`, which is set to `True` when both `Condition_1` AND `Condition_2` are `True`.

SCL is flexible as a textual language and can be written in many different styles. The same logic in *Figure 5.33* can be written in a different way, with the same outcome.

```
#Condition_1 := TRUE;
#Condition_2 := True;

#Result_2 := #Condition_1 AND #Condition_2;
#Result_1 := #Result_2;
```

#Condition_1	TRUE
#Condition_2	TRUE
#Result_2	TRUE
#Result_1	TRUE

Figure 5.34 – Another example of writing to two variables with online monitoring displayed to the right

This approach is probably the preferred approach as it's easier to read.

Unlike Ladder and FBD, all variable assignments are done with the same assignment instruction, `:=`. There are no `Set` or `Reset` assignments in SCL; all assignments are permanent. For example, if a variable is set to `True`, it will remain `True` until set to another value (as long as the variable is not a `Temp` variable).

Box instructions

As expected, there are no box instructions in SCL as box instructions are a graphical concept. When calling an instruction that requires inputs or has outputs, a particular style of layout is used to allow interfacing to the instruction.

```
#Result := NORM_X(MIN:= int_in , VALUE:= int_in , MAX:= int_in )
```

Figure 5.35 – Example of placing an instruction with an interface

When typing `Norm_X`, the SCL editor will offer solutions as to what you may wish to insert. Selecting `Norm_X` from the list will cause TIA Portal to populate the instruction with the interface ready to be entered with variables.

```
#Result := NORM_X(MIN := 0, VALUE := #Variable_1, MAX := 100);
```

Figure 5.36 – Instruction with interface populated

After filling in the highlighted areas (and adding the termination character, `;`, to the end of the instruction), the instruction's interface is populated.

> **Note**
>
> When calling a function block or function that also contains output variables, the syntax used is `=>`, as opposed to `:=` by the Input and InOut variables.

Comparators

In SCL, comparators are simply written in line with variables. This approach is what makes SCL so quick and easy to work with and it also keeps the SCL language in line with most other languages in TIA Portal.

```
#Result := #Variable_1 >= 20;
```

Figure 5.37 – Comparator example for "greater than or equal to"

By using symbolic syntaxes, SCL ensures that the code is easy to write and read when using comparators.

Control scenario walk-through

The *control scenario* in SCL is constructed differently from that of Ladder and FBD. This is because networks do not exist, so logic has to be structured in a different manner.

It is a good practice to comment in SCL. Having well-laid-out and -commented logic is the difference between being able to revisit SCL and understand it easily and having to spend time understanding how it has been written.

```
1    // //===================================\\
2    // || SCL - Structured Control Language ||
3    // ||===================================[|
4    // || (Structured Text)                 ||
5    // \\===================================//
```

Figure 5.38 – Example of an important comment

Comments are free to be structured in whatever manner desired. Writing comments that head up sections or areas of code is a good habit to get into.

Starting the system

The control system is started when the **Start** button is pressed and some additional conditions are in the correct state.

```
8    // //========================================================================\\
9    // ||                         Detect Start                                   ||
10   // ||========================================================================[|
11   // || As long as conditions are within limits, and the operator requests     ||
12   // || a start, start the system                                              ||
13   // \\========================================================================//
14
15
16 ⊟REGION Detect Start
17       //Get Start Conditions From Assets
18       #Start_Conditions_OK :=
19       #Inlet_Valve_Closed AND
20       #Outlet_Valve_Position < 0.5 AND
21       #Differential_Pressure <= 0.8;
22
23       //Set Start Condition On Operator Press
24       #System_Run :=
25       (#Start_Conditions_OK AND #Operator_Start_Button AND NOT #System_Run) OR
26       (#System_Run AND NOT #Operator_Start_Button AND NOT #System_Stop);
27
28       //Reset Operator Button Press
29       #Operator_Start_Button := False;
30 END_REGION
```

Figure 5.39 – Starting the system in SCL

The comments help understand which areas of logic are fulfilling different requirements. The entire section of logic is encapsulated in a **region**. Regions help to structure logic and allow the TIA Portal programmer to collapse areas of logic. Effectively these act as **networks**, however, the programmer must implement it themselves.

In SCL, the variable that is being updated (written to) is declared first, before the logic that will update it. *Figure 5.39* shows that when the start conditions are all in the correct state, the Start_Conditions_OK variable is updated to True. Notice that Start_Conditions_OK is the first variable declared, then the logical components are used to create the True or False value based on the outcome.

The System_Run variable is set to True by one of two situations; either the first set of brackets (parentheses) resolve to a True result, OR the second set of brackets resolve to a True result.

Notice that the second set of brackets also contains a *hold on contact*. Also note the NOT instruction that comes before variables; this inverts the variable's value requiring a False value to allow a logical value of 1 to be output.

The Operator_Start_Button variable is written as False on every scan by the logic on *line 29*.

Opening the inlet valve

Opening the inlet valve is just as simple in SCL as in Ladder or FBD.

```
32  // //================================================\\
33  // ||                  Open Inlet Valve              ||
34  // ||================================================[|
35  // || When the system is running, open the Inlet Valve ||
36  // \\================================================//
37
38  REGION Open Inlet Valve
39       //Open the Inlet Valve
40       #Inlet_Valve_Open_Request := #System_Run;
41  END_REGION
```

Figure 5.40 – SCL Open Inlet Valve

The logic on *line 40* is effectively copying the data in `System_Run` into the
`Inlet_Valve_Open_Request` variable.

The filling operation

Managing the filling sequence requires another *hold on contact*.

```
43   // //=====================================================================\\
44   // ||                         Manage System Fill                          ||
45   // ||=====================================================================[|
46   // || On first system start, filling is required. If the level is below 2.0 ||
47   // || the Outlet Valve control may NOT start                              ||
48   // \\=====================================================================//
49
50  ⊟REGION Manage System Fill
51       //Set Fill Complete
52       #Fill_Complete :=
53       (#System_Run AND #Filter_Level > 2.0) OR
54       #Fill_Complete AND #System_Run;
55
56       //Monitor Level Fault (Low Level When In System Run)
57       #Level_Fault := #Filter_Level < 1.5 AND #Fill_Complete;
58  END_REGION
```

Figure 5.41 – Managing the filling of the system in SCL

Note that only one set of brackets is being used for the `Fill_Complete` variable logic.
It's important to be careful with AND/OR logic in SCL when brackets are being used to
ensure that the logic is grouped and evaluated correctly, otherwise incorrect values could
be returned.

There are two ways that the `Fill_Complete` logic could be interpreted by
a programmer:

- `((System_Run AND Filter_Level > 2.0) OR Fill_Complete) AND System_Run`
- `(System_Run AND Filter_Level > 2.0) OR (Fill_Complete AND System_Run)`

In the case of SCL, the second option is what is processed. It would be better to write the brackets in though, to improve readability.

Calculating the outlet valve position

Calculation of the outlet valve position is much easier in SCL thanks to the text-based language.

```
60  // //========================================================\\
61  // ||                  Manage Outlet Valve Position         ||
62  // ||]=======================================================[|
63  // || The system is now filled and in System Run Mode, Calculate the ||
64  // || Outlet Valve Position                                  ||
65  // \\======================================================//
66
67  REGION Calculate Outlet Valve Position
68      //Calculate Outlet Valve Position
69      //
70      //((Max Outlet Valve Position - Min Oulet Valve Position)                        )
71      //(----------------------------------------------  * (Filter Level - Min Filter Level)) + Min Valve Position
72      //(        (Max Filter Level - Min Filter Level)                    )
73      #Outlet_Valve_Position_Request := (((80.0 - 0.0) / ( 5.0 - 0.0)) * (#Filter_Level - 0.0)) + 0.0;
74
75      IF NOT #Fill_Complete THEN
76          #Outlet_Valve_Position_Request := 0.0;
77      END_IF;
78  END_REGION
```

Figure 5.42 – Calculation of the outlet valve position in SCL

The comment explains the calculation that is taking place. *Line 73* performs the calculation and sets the Outlet_Valve_Position_Request variable to the result.

While this is far more compact than the graphical language solutions, the only information given in the monitoring pane is the end result.

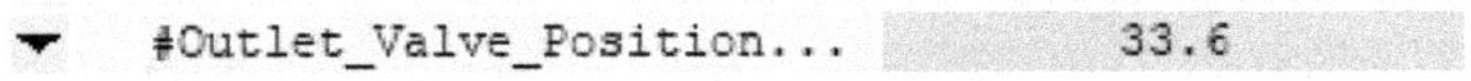

▼   #Outlet_Valve_Position...   33.6

Figure 5.43 – Monitoring pane result

This means that calculations that have many parts to them, where seeing interim results may be beneficial, will still need to be broken down into different elements.

Managing stop conditions

The system now needs to manage the stop conditions while the system is actively running.

```
80   // //======================================\\
81   // ||          Manage Stop Conditions      ||
82   // ||=====================================[|
83   // || Conditions that will stop the system ||
84   // \\======================================//
85
86 ⊟REGION Manage Stop Conditions
87  │       //Call Max Run Timer
88 ⊟│       #Max_Run_Timer(IN:=#Fill_Complete,
89  │                      PT:=T#20m,
90  │                      Q=>#Stop_On_Time);
91  │
92  │       //Manage Stop Conditions
93  │       #System_Stop :=
94  │       (#Stop_On_Time OR
95  │       #Level_Fault OR
96  │       #Differential_Pressure > 1.5 OR
97  │       #Outlet_Valve_Position > 75.0) AND
98  │       #System_Run;
99  │
100  END_REGION
```

Figure 5.44 – SCL stop conditions

In *Figure 5.44* a TON timer is called on *line 88*. One of the drawbacks of SCL is that it is not immediately obvious *what* is actually being called. Function blocks with large interfaces may not be as easily recognizable as a TON timer. To check what is actually being called, click on the instance name (before the interface variables – Max_Run_Timer in this case) and press *Alt + Enter*. This will display the properties for the instance. In the general area, the data type value will contain the type of function being called.

Control summary

The SCL solution demonstrates how textual languages are still capable of basic Boolean logic as well as showcasing the strengths that come with being able to simply write out comparators and math calculations.

Some of the key considerations around using SCL include the following:

- The online monitoring of code is poor.
- Finding the instance type of a called function is more involved than graphical languages.
- Comments are free-text style, which allows for nicer and better-structured comments than in graphical languages.
- Using regions can help segment the logic into manageable areas.
- Bit/Boolean logic using brackets needs to be carefully written to ensure the grouping of the variables is correct.

GRAPH

GRAPH is another graphical language that is designed to handle sequences. It's often linked to **sequential function chart**, however, they are not the same language. Siemens has a similar concept but has introduced interlocks and supervisory elements that offer additional controls around when and how transitions and actions take place.

Overview

GRAPH is a special language that is centered around the management of sequences and transitions between sequence steps. It is simplistic in nature but capable of controlling complex and parallel sequences.

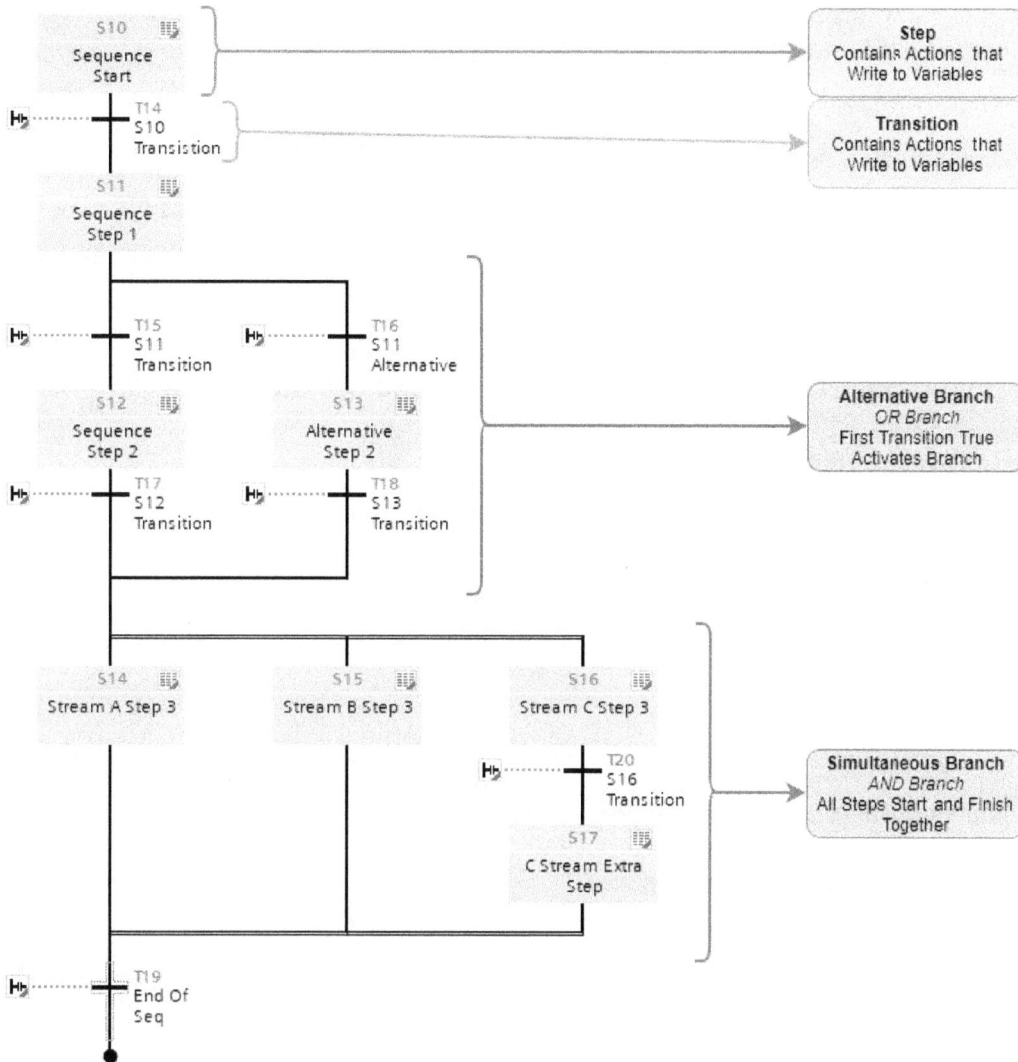

Figure 5.45 – Example of GRAPH sequence view

Figure 5.45 shows an example of a GRAPH sequence diagram. There are four basic components in this example:

- **Steps** – This is the part where the sequence writes outputs.
- **Transitions** – Allows the movement to other steps.
- **Alternate branches** – GRAPH's OR branch; only one of the pathways can be taken.
- **Simultaneous branches** – GRAPH's AND branch; all the pathways are taken and executed together, and the sequence continues when the transition after the branch is True.

Logic flow in GRAPH is top to bottom, in the order that the transition gates allow.

Instructions in GRAPH

There are two different types of instructions in GRAPH:

- Step instructions
  - Instructions are action instructions.
- Transition instructions
  - Instructions are LAD instructions.

Depending on where in the GRAPH program is being edited dictates what instructions are available.

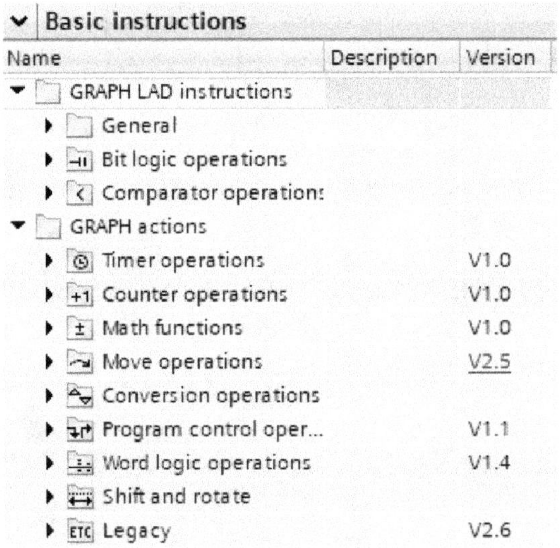

Figure 5.46 – GRAPH's Basic instructions view

When modifying transitions, the **GRAPH LAD instructions** set is used. When modifying step actions, **GRAPH actions** is used.

Step actions

These are actions that appear as part of a step. Steps are capable of writing outputs to a variable that appears in the interface, or globally. Each action has a set of parameters:

- **Interlock** – If the interlock is `True`, the action is enabled for execution.

- **Event** – The event that must occur for the action to be executed.

- **Qualifier** – The conditions in which the action is executed.

- **Action** – The logic that action will execute.

These parameters come together to form an action within a step.

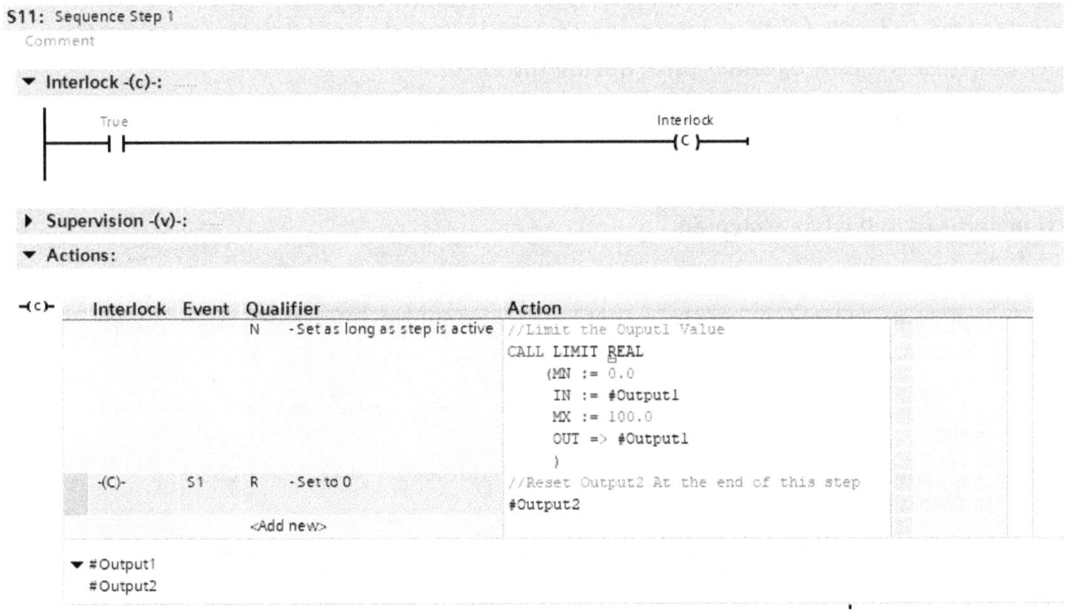

Figure 5.47 – Step actions in GRAPH

By double-clicking on a step in the **Sequence** view, the **Step** view is opened where **Interlock**, **Supervision**, **Actions**, and **Transition** can be viewed and modified.

Figure 5.47 shows an example of two actions within a single step. The **Interlock** value for *Action 1* is not set so, therefore, is `True` by default (enabling) but for *Action 2* it is defined as `-(c)-`. The interlock can be modified in the **Interlock** rung at the top of the **Step** view. In this example, it is simply set to `True`, however, the interlock variable can be written as `True` to enable actions with the required logic for the application being developed.

> **Note**
>
> If no logic is provided to drive **Interlock** to `True`, then **Interlock** is assigned `True` by default due to the permanent connection to the power rail on the left. The **Interlock** coil cannot be deleted.

The **Event** column can be set to a variety of different events that trigger the execution of the action. *Action 1* has no **Event** assigned, so only **Qualifier** is used to trigger the execution of the action. *Action 2*, however, has the **S1** event assigned. **S1** is the ID for *Incoming Step* and calls for the action to be executed when the step is entered for the first time. This means it only executes once during the event.

The **Qualifier** column contains additional conditions for the action. *Action 1* has the ID of **N**, which keeps the action executing every cycle of the step until the step is no longer active. *Action 2* has the ID of **R**, which is **Reset**, and sets the assigned variable in the **Action** column to **0** when executed.

The **Action** column contains the logic to be executed. The language that the action is written in appears to be SCL, however, there are some subtle differences, such as the termination character, `;`, is not required.

> **Note**
>
> Right-clicking the **Action** table allows for two additional options to be checked that are not checked by default.
>
> The **Show event descriptions** option displays the description of the ID for events, making it easier to identify how an event is configured.
>
> **Allow multi-line mode** allows for the `Return` key to drop the action code down a line. It also enables comments in the **Action** column (by using `//` as in SCL).

Transitions

Transitions are simply a single LAD network that results in a transition gate either being True or False.

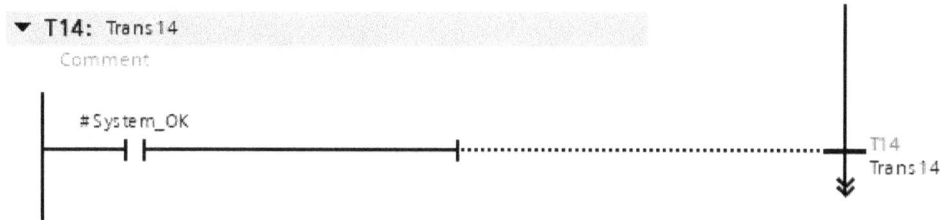

Figure 5.48 – Transition Ladder network

If the network ends with a logical 1 (True) then the transition gate allows the sequence to advance to the next step. If it is a logical 0 (False), then the step does not advance.

> **Note**
> Variables cannot be written to in transition networks. Calls to functions cannot be made in transition networks either.

Supervision coils

These are used to raise alarms and indicate an issue in the sequence. In the interface of the GRAPH block, the supervision coils can be acknowledged and reset to either continue the sequence or reset it to the initial step.

Control scenario walk-through

GRAPH is by far the easiest language to read at a high level and see exactly what point the *control scenario* is at and what steps are executing.

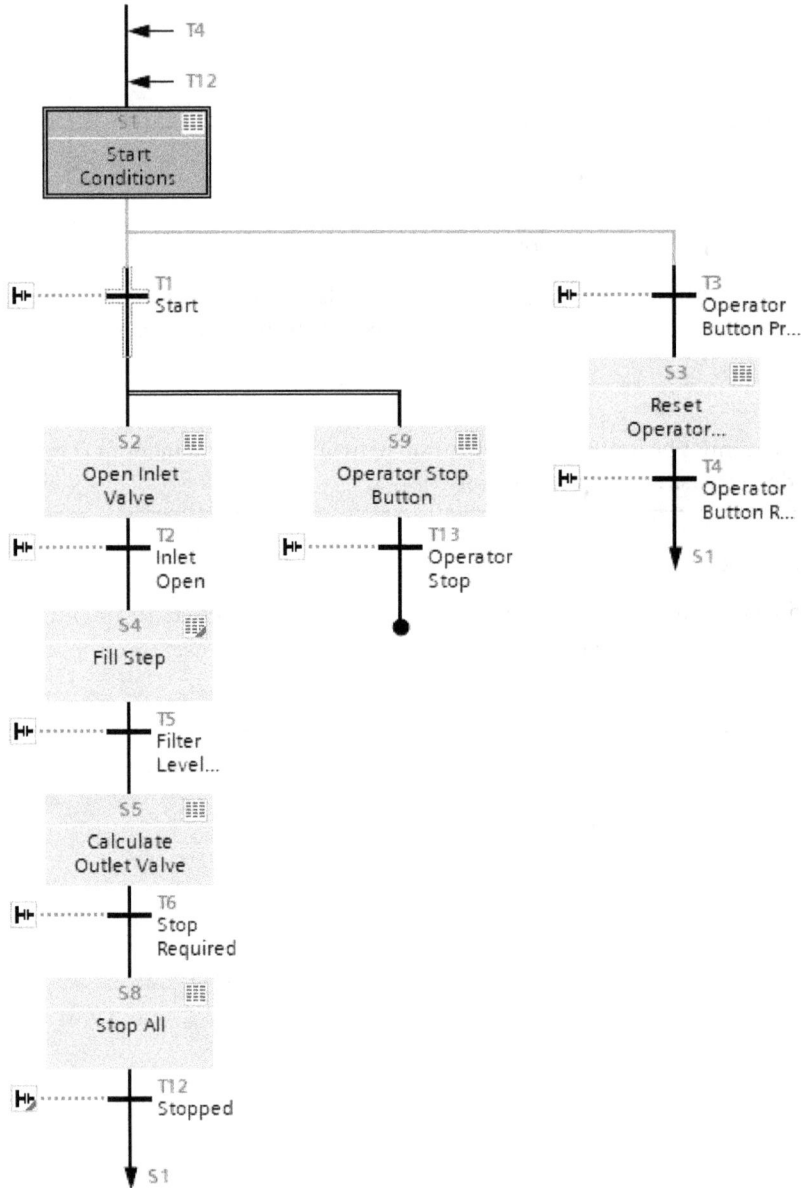

Figure 5.49 – Control scenario overview in GRAPH's Sequence view

This view allows for quick and easy monitoring of the system, without having to read the code directly to figure out what the current step is doing and how the next step is activated.

It's also possible to quickly jump into *on-the-fly* views of **Actions** and **Transitions** to monitor values and check logic. This can be done by clicking on the table and ladder icons within **Steps** and next to **Transitions**.

Starting the system

To start the system, the step **S1** is monitored for the required conditions set out in the *control scenario* and also the **operator Start button**.

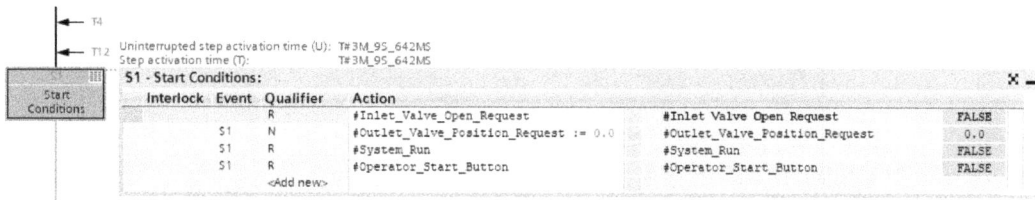

Figure 5.50 – Start conditions step, with the action table expanded

This step has four actions, three of which only occur on entry into the step (due to the **S1** event). The `Inlet_Valve_Open_Request` variable is set to `False` while in this step because, in later steps, it is set to a permanent `True`.

This step and associated actions are simply to condition the system ready for starting. It's the transition **T1** that starts the system.

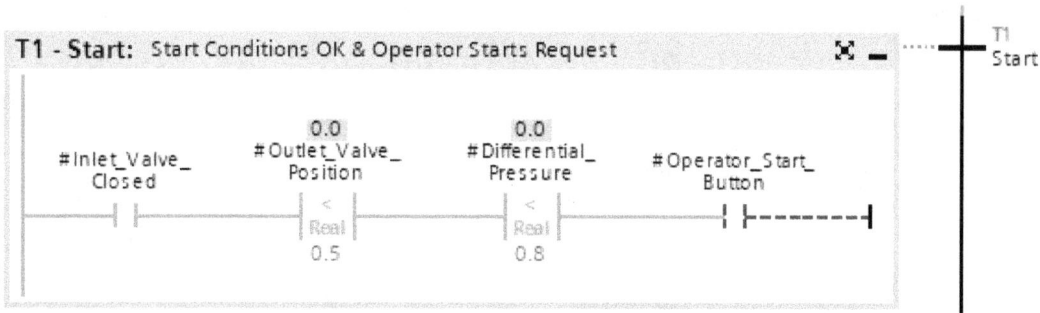

Figure 5.51 – Transition logic for the start signal

In *Figure 5.51*, the LAD demonstrates that the system is currently waiting for `Operator_Start_Button`, and when this becomes `True`, the **T1** transition gate will allow the sequence to continue.

There is also an *alternative branch* that follows step **S1**. Transition **T3** is `True` when `Operator_Start_Button` is pressed and calls step **S3**, which resets the button so it can be pressed again from the HMI system. **T4** is a simple *fall-through* transition gate.

Figure 5.52 – Alternative branch that resets the Operator_Start_Button variable

Note that this alternative branch can only execute in the **T1** transition gate that did not allow the sequence to advance. The arrow at the end of the sequence is a `Jump` command back to step **S1**.

Opening the inlet valve

Opening the inlet valve is a straightforward `Set` command on an **S1** event (on entry to the step).

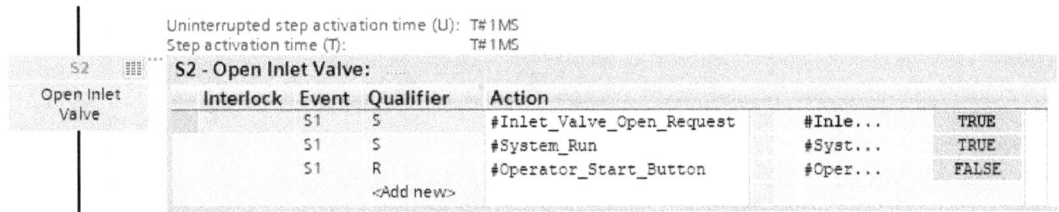

Figure 5.53 – Step S2 – opening the inlet valve

Note that some additional actions are also required, resetting the `Operator_Start_Button` variable back to `False` and setting the `System_Run` variable to `True`.

Note that the complete execution time of step **S2** was 1 ms. This is because of the configuration of the transition below the step.

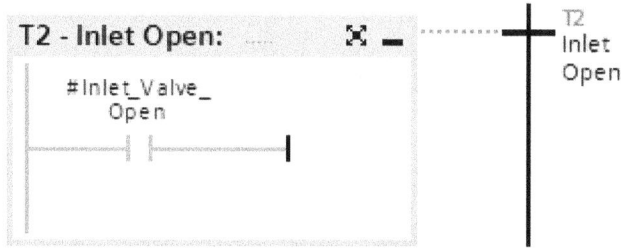

Figure 5.54 – Inlet open transition T2

As soon as the Inlet_Valve_Open signal is received, the sequence continues.

The filling operation

The filling of the system is much simpler in GRAPH. There is no *hold on contact* required as in other languages.

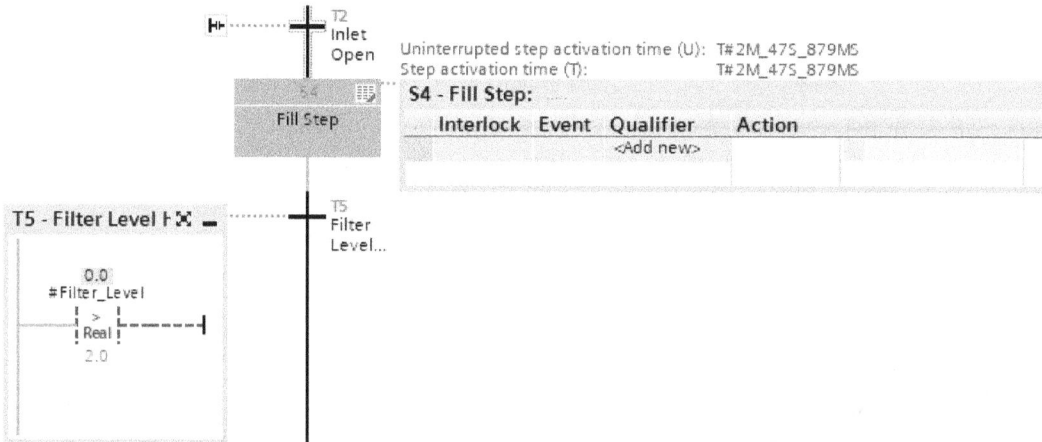

Figure 5.55 – Fill step S4

The **fill step (S4)** does not contain any actions, which is identified by the small gray band on the table icon on the step. There are no actions because the system does not actually need to do anything other than wait for the Filter_Level variable to be above 2.0.

The transition **T5** waits for Filter_Level to be greater than 2.0 and then allows the sequence to continue.

> **Note**
>
> In GRAPH, it would be very easy to add a timeout here by using the CMP >
> T instruction from the **Comparator operations** folder in the **Instructions**
> panel. This would compare the execution time of the current step
> against a value. This could be used to continue the sequence or raise
> a supervision issue.

Calculating the outlet valve position

GRAPH shows a weakness when calculations are required. Like LAD and FBD, attempting to do calculations in GRAPH is cumbersome.

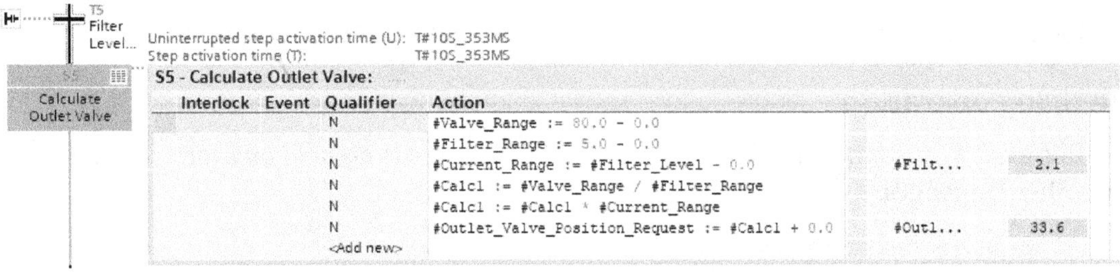

Figure 5.56 – Calculating the outlet valve position in GRAPH

Figure 5.56 demonstrates how the outlet valve position is calculated in GRAPH. A series of actions are called, with the **N** qualifier (when **Step** is active), to calculate the position in stages.

> **Note**
>
> Function blocks and functions can be dragged into the **Action** column. With
> calculation requirements like this, it's far easier to create a function in SCL and
> perform the calculation in the function. The function can then be called on
> a single action to complete the calculation easily.

Managing stop conditions

The system will continue to run step **S5** until one of the conditions set out in the *control scenario* is True.

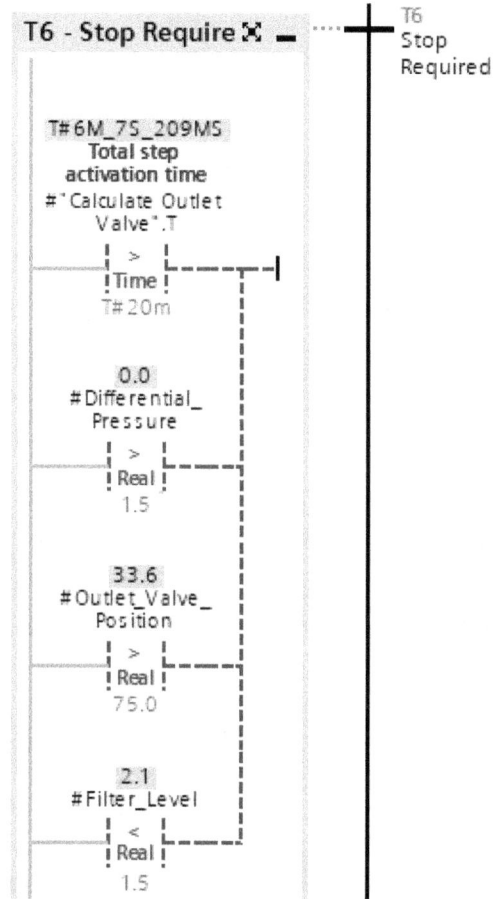

T6 - Stop Require ✕ ▬   T6
                                  Stop
                                  Required

T#6M_7S_209MS
Total step
activation time
#"Calculate Outlet
Valve".T
      | > |
      !Time!
      T#20m

     0.0
#Differential_
Pressure
      | > |
      ! Real !
       1.5

     33.6
#Outlet_Valve_
Position
      | > |
      ! Real !
       75.0

     2.1
#Filter_Level
      | < |
      ! Real !
       1.5

Figure 5.57 – Stop conditions

Note that step **S5** – Calculate Outlet Valve – is directly referenced as instance data and the T variable is compared to T#20m. GRAPH does an excellent job of allowing all steps to report their **activation time** without the programmer having to set anything up, making these sorts of conditions very easy to configure or add in at a later date.

Once transition **T6** is `True`, the final step, **S8**, is called.

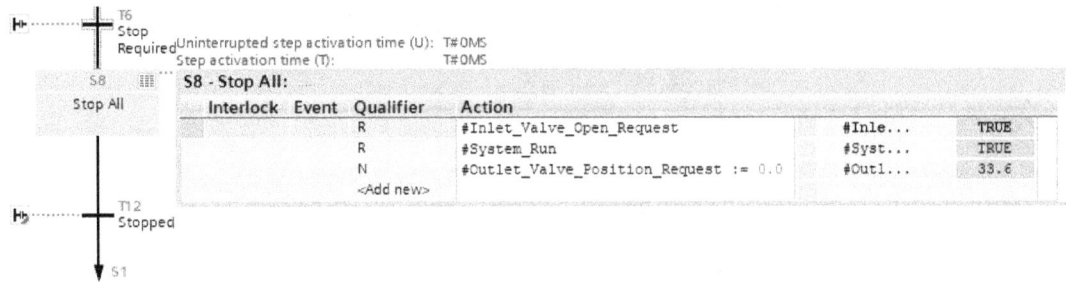

Figure 5.58 – Stop step

This step conditions the variables so that the system stops correctly and is ready for the next start event.

The final transition gate, **T12**, contains no logic, noted by the small gray band on the icon to the left of the transition gate. This means that step **S8** will simply execute once and fall through the gate.

After **T12**, a *jump* is executed back to step **S1**, completing the sequence.

Parallel stop step

Between steps **S1** and **S2**, there is a *simultaneous branch* that calls for step *S9* to become as active as the rest of the sequence.

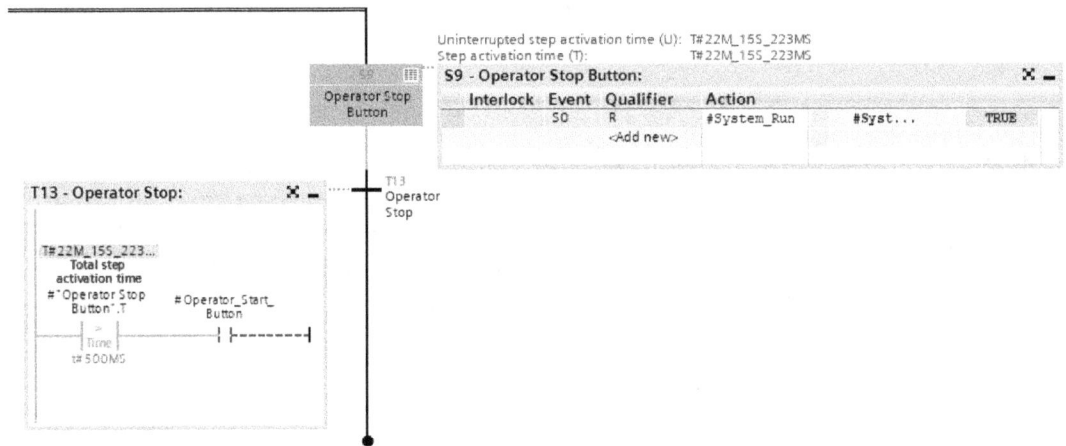

Figure 5.59 – Additional stop step

This step is configured so that if `Operator_Start_Button` is pressed again, `500 ms` after the step becomes active, then a `Sequence End` command is given and the whole sequence immediately stops (represented by the circle after **T13**).

When a sequence is stopped in this way, inputs to the call of the GRAPH function block must be used to re-initialize the sequence.

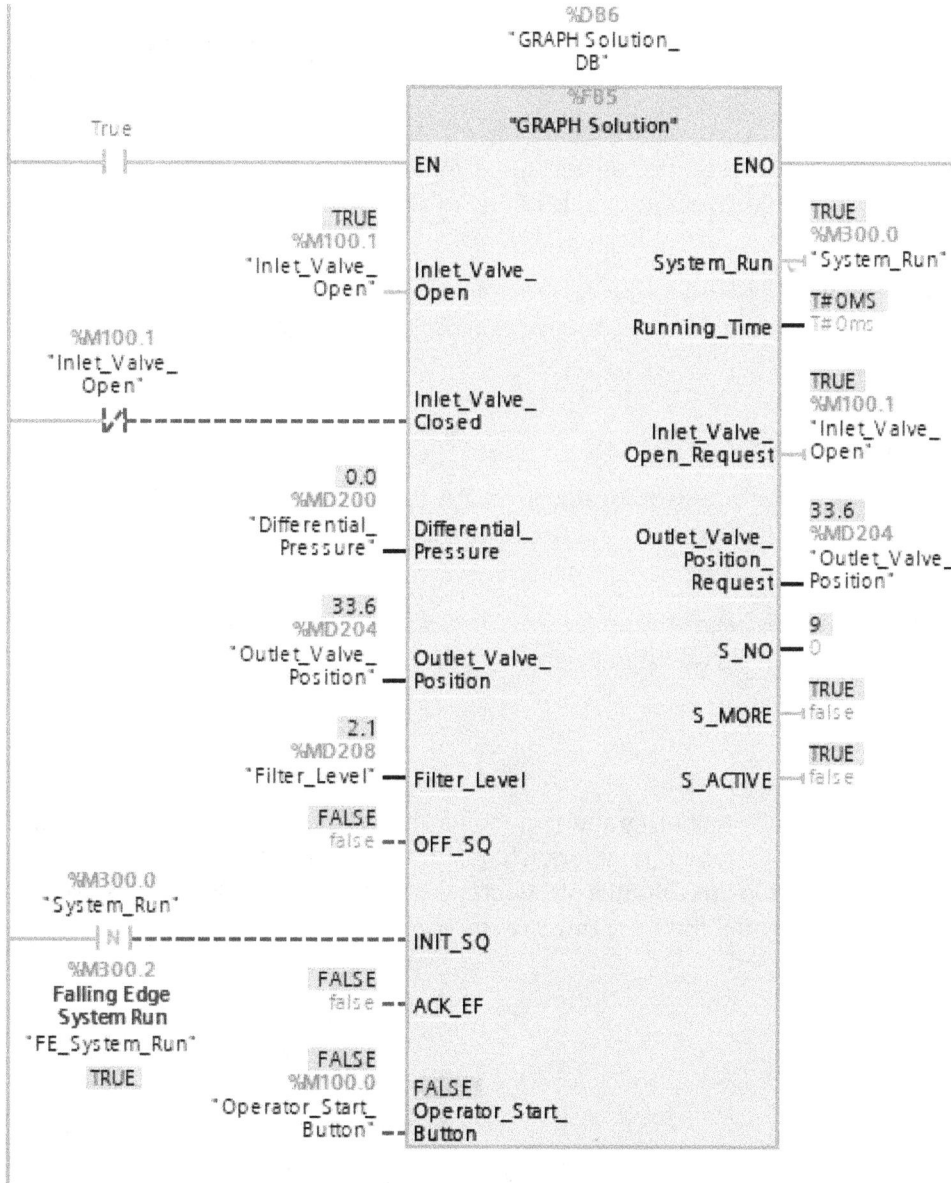

Figure 5.60 – GRAPH solution function block call in main (OB1)

Figure 5.60 shows that the INIT_SQ (initialize sequence) input is pulsed to True when the System_Run variable becomes False. Because the System_Run variable becomes False when the system is stopped, this re-initializes the sequence back to step **S1** whenever transition gate **T13** stops the system.

Control summary

GRAPH is a good choice for the types of controls where sequences help manage the control, and the graphical overview allows the maintenance and observation of the logic to be simple.

The downside to GRAPH is that it can become complicated with supervisions and interlocks, but TIA Portal gives enough flexibility to allow the programmer to make these as simple or as complex as the application requires.

Overall, GRAPH should be used where a sequence is required and avoided for event-driven or reactive logic where *steps* are not executed in a linear order most of the time.

Cause and effect matrix

CEM is the newest of the languages available in TIA Portal. Its design is specifically targeted at making the management of interlocks and system signals easier to manage via a matrix-style grid.

CEM is not new and has been around for many years. It is a great way to quickly and easily identify how a signal condition goes on to affect different equipment in a process or factory.

Overview

CEM may not make sense to a programmer who has never seen a cause-and-effect matrix before, however, it's an extremely simple concept. On the left-hand side are rows of **Causes** and across the top are columns of **Effects**. Where **Causes** and **Effects** intersect, an **action** can be placed that links the cause to the effect.

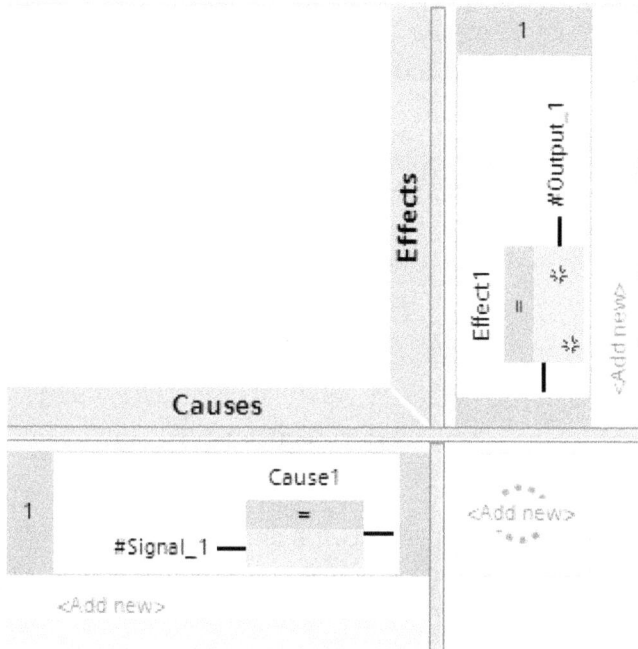

Figure 5.61 – CEM example

Figure 5.61 shows an example of a cause and effect and the intersection point. At this point, actions can be created, similar to **Qualifiers** in GRAPH.

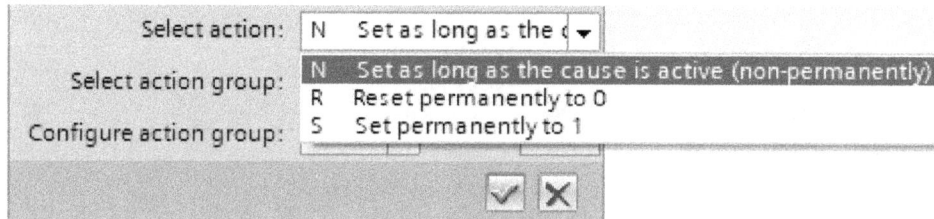

Figure 5.62 – Action selection menu

Only three options are available in CEM as actions. The action selected changes the logical value passed to the effect.

Instructions in CEM

The CEM instruction set is simplified and grouped into areas in which the instructions can be used.

˅	Basic instructions		
Name		Description	Version
▼ ☐ General			
	🔳 Empty box	Empty box [F8]	
	▬ ⊣	Add input/output [Ctrl+Shift+3]	
	▬ ⊸	Invert pin [Ctrl+Shift+4]	
▼ ☐ Cause instructions			
	Bit logic operations		
	🔳 -[=]	Assignment [Shift+F7]	
	🔳 &	AND logic operation [F9]	
	🔳 >=1	OR logic operation [F10]	
	🔳 x	EXCLUSIVE OR logic operation	
	Comparator operations		
	🔳 CMP ==	Equal	
	🔳 CMP <>	Not equal	
	🔳 CMP >=	Greater or equal	
	🔳 CMP <=	Less or equal	
	🔳 CMP >	Greater than	
	🔳 CMP <	Less than	
	Timer operations		
	🔳 OnDelay	Delay activation	
	🔳 OffDelay	Delay deactivation	
	🔳 Pulse	Activate for a limited time	
▼ ☐ Effect instructions			
	🔳 -[=]	Assignment [Shift+F7]	
	🔳 -[S]	Set output	
	🔳 -[R]	Reset output	
▼ ☐ Intersection actions			
	☐ N	Set as long as the cause is active	
	☐ S	Set permanently to 1	
	☐ R	Set permanently to 0	

Figure 5.63 – CEM instruction set

Causes and effects have different instructions available. Causes can compare variables or call timers, as well as performing basic Boolean logic. Effects simply write outputs to variables and are capable of using Set or Reset assignments.

Actions

An **action** is indicated by a circle at the intersection of a cause and effect. Inside the circle, the **ID** of the type of action is displayed.

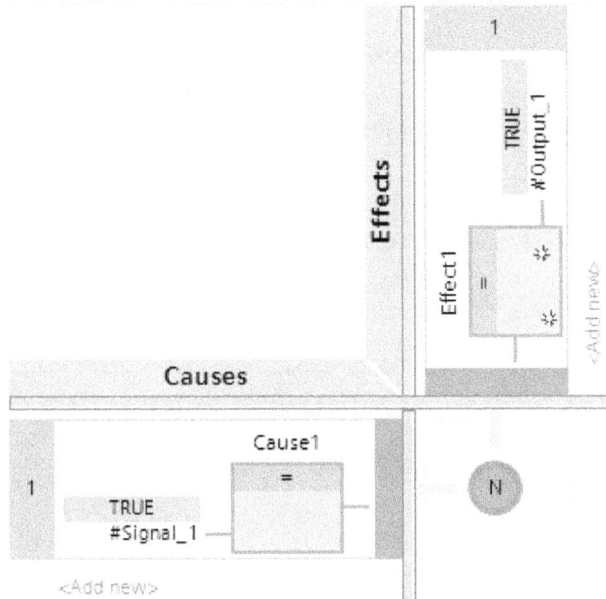

Figure 5.64 – CEM with active action

When the Cause is True, the action **N** passes a logical 1 (True) to the **Effect**. If the **Cause** is False, the **Effect** will receive a logical 0 (False).

The Set and Reset actions perform in the same manner as other languages, permanently passing a logical 1 or 0 to the **Effect**. A Set and Reset action should always be used together on the same intersection column.

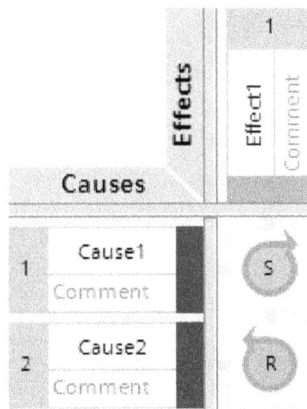

Figure 5.65 – Set and reset actions in use (Causes and Effects minimized view)

When a **Set** action has previously been active and its associated cause is now `False`, a light blue background is displayed inside the **Set** action. This indicates that **Set** is still outputting a `True` value and the **Effect** is indicated with a green indication in monitoring mode.

When a **Set** and **Reset** are both available as actions to an **Effect**, the **Set** is dominant. This means if both causes are `True` for the **Set** and **Reset**, the **Effect** will receive a `True` signal, despite the **Reset** action being active.

> **Note**
>
> In the top left of the CEM window, two buttons are provided for quickly minimizing the cause and effect areas. This is useful if the CEM matrix is large, however, understanding what the causes and effects represent can be more difficult.

Groups

Actions can be grouped so that all actions in the group need to be in a `True` state before the effect receives a `True` value.

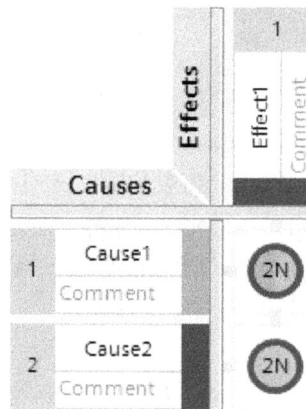

Figure 5.66 – Example of a CEM action group

Figure 5.67 shows an example of a two-group configuration. Both **Cause1** and **Cause2** belong to the same group. Because only **Cause1** is in a `True` state, the effect does not receive a `True` value.

> **Note**
>
> **R** actions (**Resets**) cannot be grouped with **N** or **S** actions as this would be contradictory. **R** actions can, however, be grouped together, meaning all would need to be active to result in a `False` value being passed to the effect.

Intersection columns

Effects can have additional intersection columns added by right-clicking on the effect and choosing **Add intersection column**.

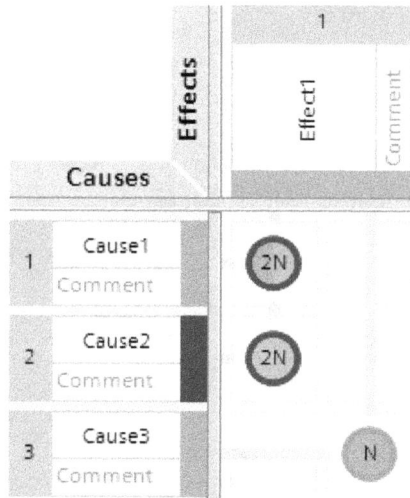

Figure 5.67 – Effect with multiple intersection columns

When another intersection column is added, this enhances the effect's action set, allowing more than one action result to interact with the effect. *Figure 5.68* demonstrates this as the group **2N** in intersection column 1 results in a `False` value, however, intersection column 2 results in `True`. The end result for the effect is that `True` is passed.

> **Note**
>
> If an **R** action is passed in a separate intersection column and another intersection column is passing `True` to the effect, it does *not* result in the effect being `False`. The **R** action only affects the **Set** action in the same column.

Causes

Causes can also contain additional comparisons and instructions. These include basic Boolean logic and numerical comparisons. Time delays can also be added to causes, delaying the output of the cause for a defined period.

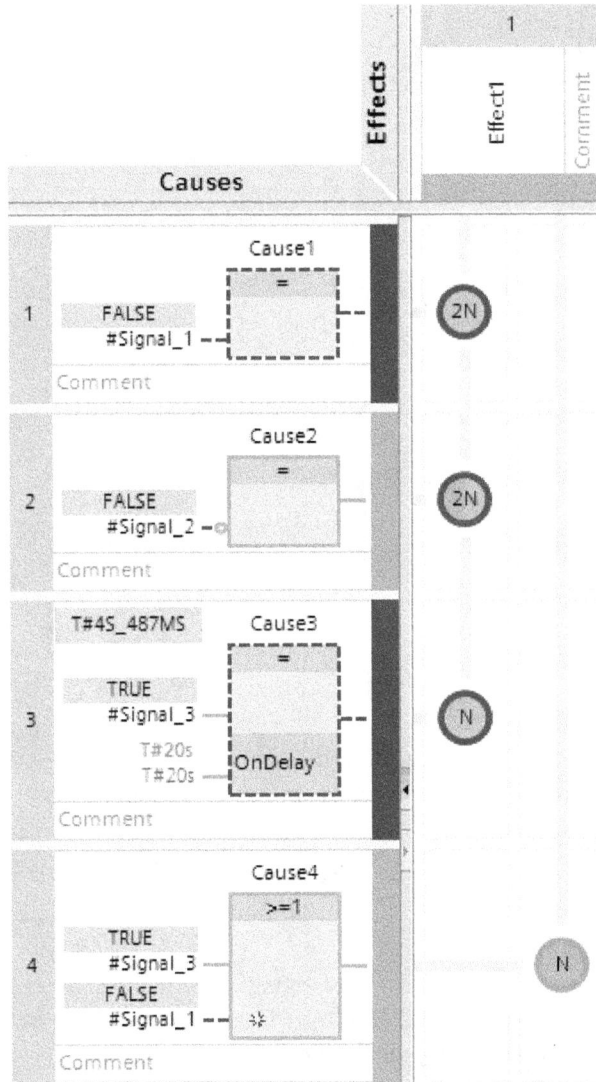

Figure 5.68 – Example of causes with delays and OR instructions

Figure 5.68 shows an example of an OnDelay timer being used on **Cause3** and an OR block being used on **Cause4**. It also shows that intersection column 1 can cause True to be passed to the effect if either group **2N** becomes True or **Cause3** becomes True.

Control scenario walk-through

CEM is not a language that is designed for this type of control scenario. This does not mean it is not possible, but it is not the easiest to read, write, or maintain. The main difference for this *control scenario* is that CEM is not capable of performing mathematical calculations. This has to be done outside of the function block and passed back in on the next scan.

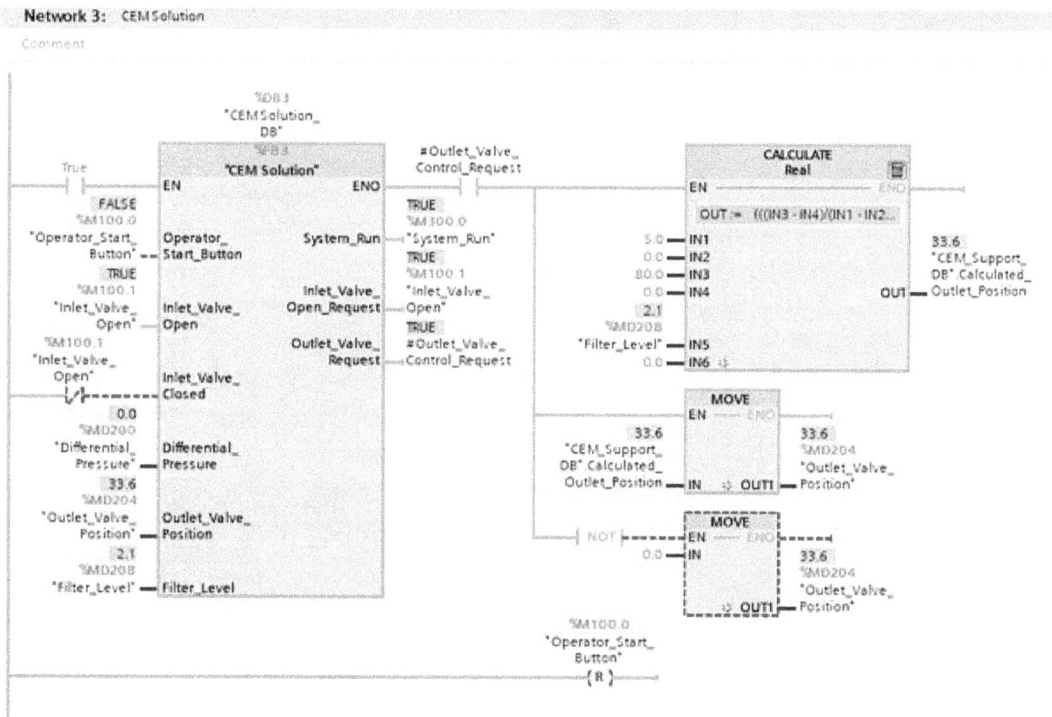

Figure 5.69 – Calculation of outlet valve position outside of the CEM block

Figure 5.69 is an example of how mixing two languages (CEM and LAD) can meet requirements where one language is not capable of doing so.

Starting the system

The starting of the system is managed by **Effect 1**, which consists of two intersection columns. The first column manages the starting of the system from a stopped position and the second column maintains a *hold on contact*.

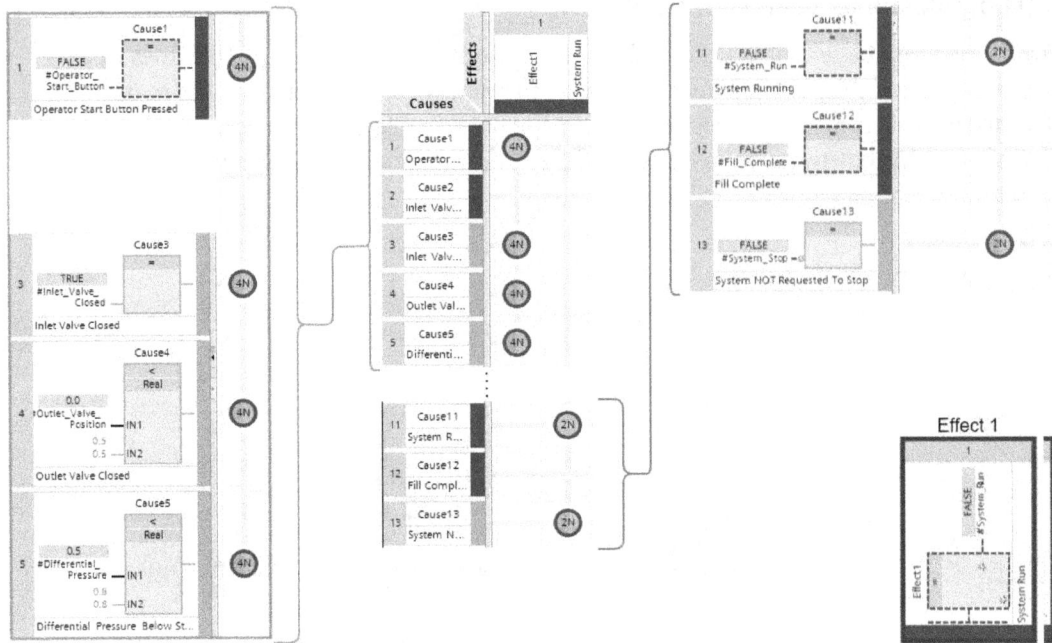

Figure 5.70 – Starting the system in CEM

Groups are used to start and maintain the *hold on contact*. The first column checks the required start conditions and `Operator_Start_Button`. Once the `System_Run` variable has been set to `True`, **Cause11** and **Cause13** maintain `System_Run` as `True` until a stop condition is raised.

Opening the inlet valve

Opening the inlet valve is a simple **N** action (True when the cause is True).

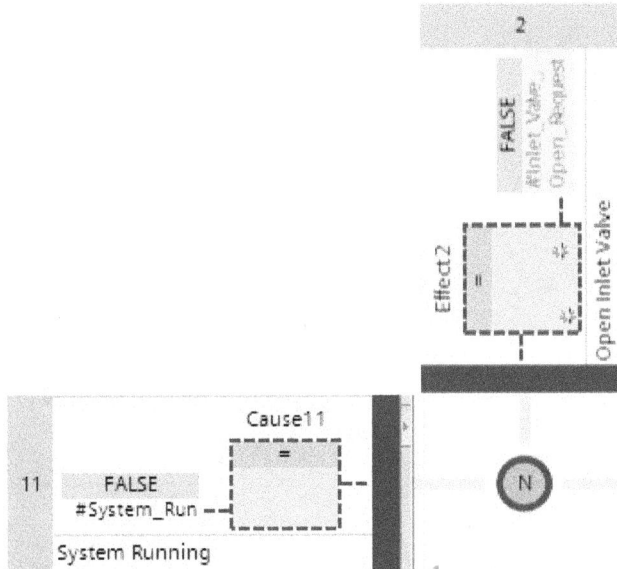

Figure 5.71 – Opening the inlet valve when the system is running

Effect 2 is only called when the system is running. This is easily achieved with a simple **N** action (True when **Cause** is True).

The filling operation

The filling operation is split across the different causes that are all part of the same group.

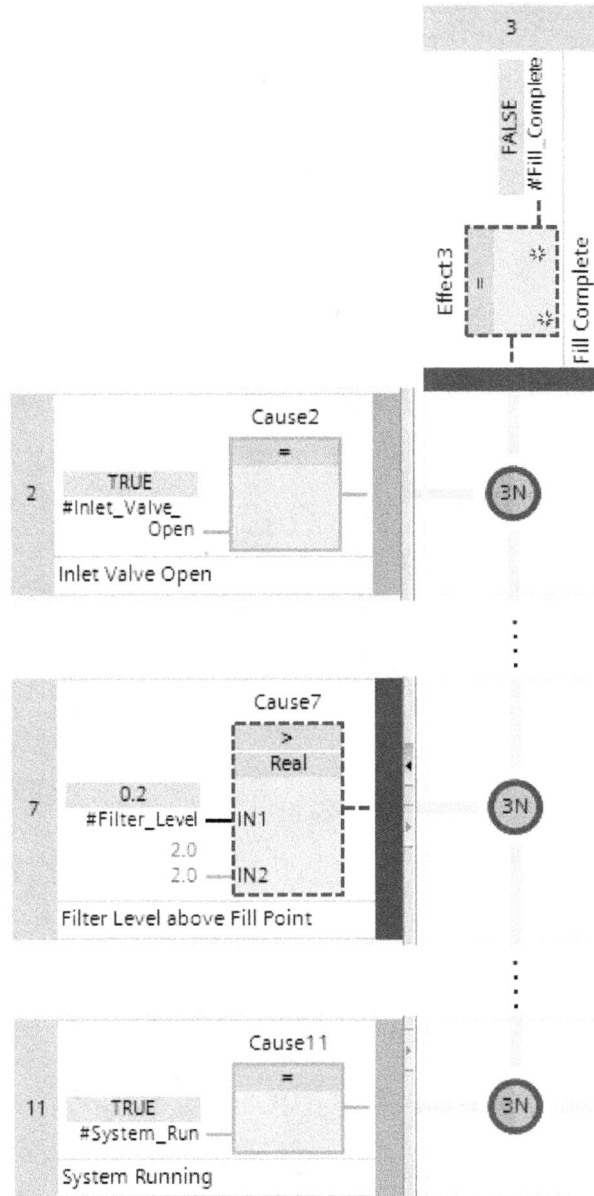

Figure 5.72 – Filling the system

All three causes are part of the *3N group*, meaning all three must be active in order to set the Fill_Complete variable to True. *Figure 5.73* shows the system in a filling state. When *Cause7* becomes True, all three actions in the group become True and Fill_Complete is set to True.

Calculating the outlet valve position

Calculations cannot be made in CEM, so a different approach is required. In the CEM logic, the system sets an output that calls for the calculation to take place.

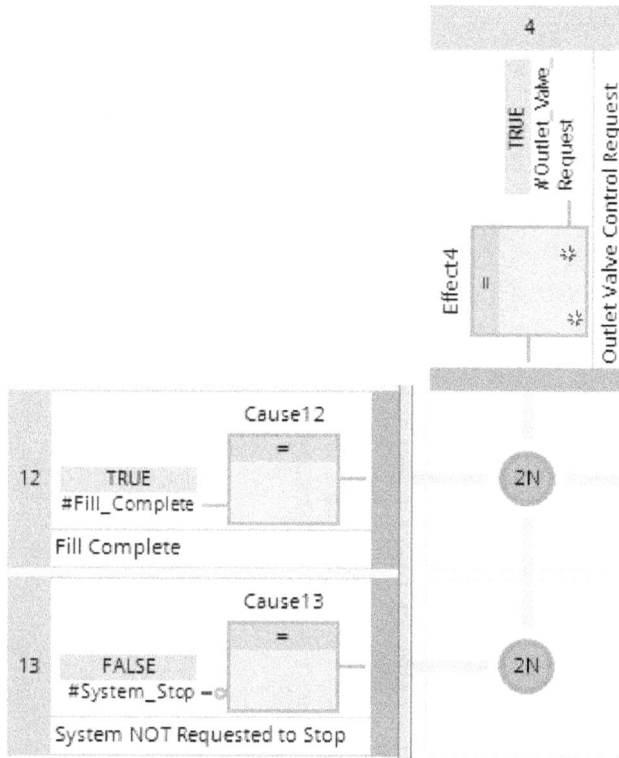

Figure 5.73 – Call for outlet valve control request

This calculation then takes place outside of the CEM function block. In the example provided, this takes place in the *Main (OB1)* **organization block**.

Managing stop conditions

Managing stop conditions with CEM is simple. Each of the stop requirements is listed as a cause, and some of them are grouped where required. For the operator stop button, the signal is grouped with the System_Run signal in a **secondary interception column**.

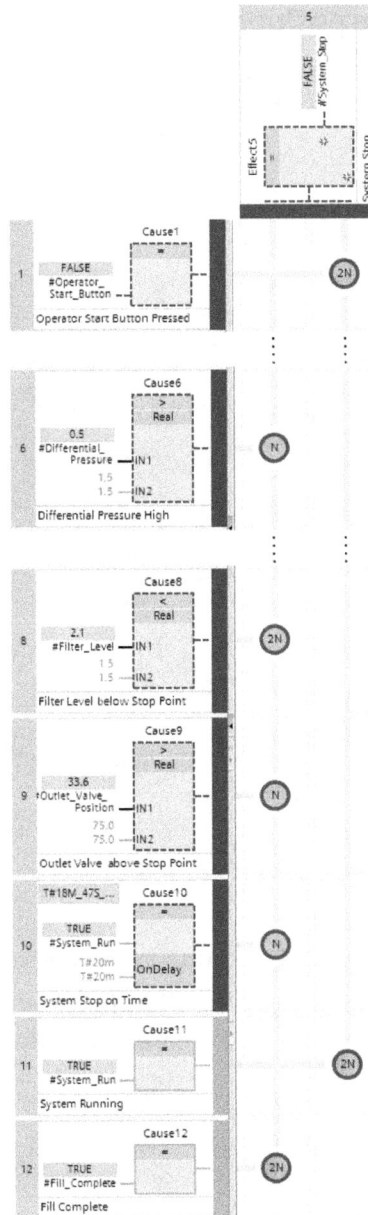

Figure 5.74 – System stop conditions

Because the `System_Stop` signal is used in **Cause13** as a *hold on contact*, the system no longer runs due to the `System_Run` signal being set to `False`.

Control summary

While CEM is the least function-rich language, it is still capable of being used for this type of control with support from other languages. Control with CEM is not recommended as CEM is weighted toward creating interlock matrixes quickly and easily.

With CEM's highly visual interface, finding the reason why an effect is active or inactive is extremely simple and easy to do. With all inputs on the right and all outputs at the top, fault finding is simplified to finding the offending action, which can easily be done by simply tracing the action lines to the appropriate action.

Summary

This chapter has explored a wide variety of different programming languages and techniques. Programmers who are capable of leveraging all of the available languages will find that they are more flexible with the approaches that they have at their disposal.

It is important to understand the differences between the languages and what the strengths and weaknesses of each language are. Writing a project in a language that doesn't best suit the control requirements can significantly increase the time required to develop, test, and maintain the project.

While different languages exist, in most cases, it is still possible to achieve what is required in any language (CEM excluded).

This chapter, along with the provided project, should have helped you gain insight into each of the available languages, how to write basic logic, what each language looks like when monitored, and what the strengths of each language are.

The next chapter introduces standard control objects into projects, tying structured data, standard function blocks, and library management together. Standard control objects help shift the focus from individual data and logic into object-based data and management, which can significantly reduce the amount of time required to develop when utilized to its fullest.

6
Creating Standard Control Objects

This chapter explores using function blocks and functions to create standardized control objects.

This approach takes the structured aspect of **programmable logic controller** (**PLC**) programming to an additional level, where data and control are brought together to act as a standard method. By utilizing this approach, confidence in the reusability of code is greatly increased, while management and maintenance of code are greatly reduced.

In this chapter, the following topics will be covered:

- Planning standard interfaces
- Creating control data
- Creating HMI data
- Structuring logic
- Considerations that have an impact on usability

> **Note**
> This chapter builds on information contained in *Chapter 3, Structures and User-Defined Types*.

Planning standard interfaces

An interface to a function block or function is the first point at which a programmer can start to control how an object is interacted with. The interface serves as a ruleset of what is allowed to be passed in and out of this object.

You can see an example of an interface in the following screenshot:

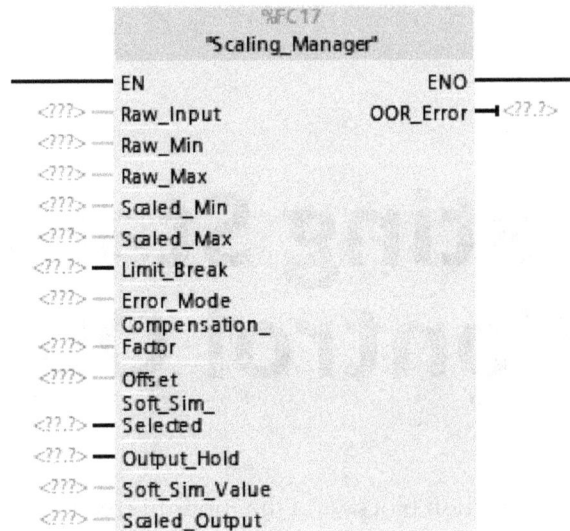

Figure 6.1 – Basic interface example

By offering an interface, programmers (and other people who may have to maintain or modify the project) have a better chance at reducing mistakes when using the object.

Defining variables in an interface

An interface is made up of three different key areas, as outlined here:

- Inputs
- Outputs
- InOut variables

Each behaves differently and performs different functions.

Inputs

Inputs are variables that are *read-only*. Once an input interface variable is passed into an object, it is not recommended to modify it. TIA Portal does allow input variables to be written to; however, changes made to the input will only exist inside the object, and the outside variable data will not change.

You can see an example of an input interface in the following screenshot:

	Input		
	Raw_Input	Int	
	Raw_Min	Int	
	Raw_Max	Int	
	Scaled_Min	Real	
	Scaled_Max	Real	
	Limit_Break	Bool	Allows the scaled value to breach min / max limits
	Error_Mode	Int	0 = Last Known, 1 = Force High, 2 = Force Low
	Compensation_Factor	Real	
	Offset	Real	
	Soft_Sim_Selected	Bool	
	Output_Hold	Bool	

Figure 6.2 – Example of an input interface

Inputs should consist of variables that fit one of the following criteria:

- Variables that bring singular required data through the interface and into the block object

- Variables that allow configuration of the block object including hardcoded options

All inputs in *Figure 6.2* fit these two criteria; the variables defined in the input interface are either singular inputs or inputs that require hardcoded elements for configuration (`Error_Mode`, for example).

Outputs

Outputs are opposite to inputs; they are variables that have permission to be written to the outside variable assigned on the interface of the function or function block.

You can see an example of an output interface in the following screenshot:

	Output	
	OOR_Error	Bool

Figure 6.3 – Example of an output interface

Outputs should consist of variables that have been updated by the block object. However, this can depend on the size of the variable. As with inputs, outputs should ideally be kept to singular variables. This means that structs and **user-defined types** (**UDTs**) should be kept to a minimum when being passed as **inputs** or **outputs**.

> **Note**
>
> Both input and output variables *copy* the information from outside the object to inside the instance data (or temporary data for a function). As projects grow larger, this can have an impact on memory usage.

InOut variables

InOut variables are slightly different from inputs and outputs, both in functionality and how data is managed. An InOut variable creates a **pointer** to a location of memory and passes the pointer into the object. This pointer is then **dereferenced** by TIA Portal inside of the function block/function to obtain the actual data represented by the variable. This has a significant memory improvement for large data types as pointers are 6 bytes in length, even if the actual data being pointed to is 200 bytes in length.

> **Note**
>
> If the data being passed via InOut is less than 6 bytes, the actual data length required is the same length as the original data.

InOut interface variables also act upon the data immediately, meaning that when interface data that is InOut is written to, the data that the pointer is pointing to is updated immediately. The data is updated before the function block/function finishes execution. Data is not copied at the interface.

You can see an example of an InOut interface in the following screenshot:

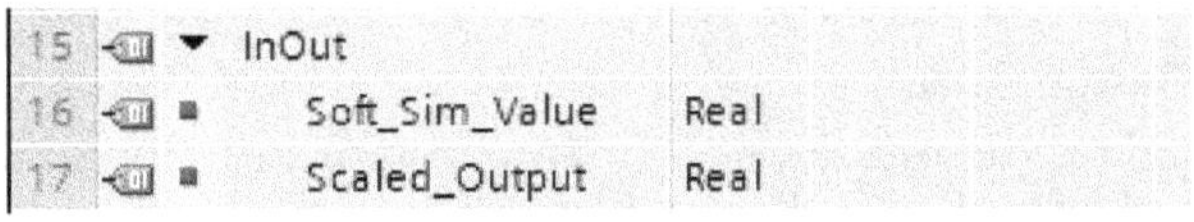

15	▼	InOut	
16	▪	Soft_Sim_Value	Real
17	▪	Scaled_Output	Real

Figure 6.4 – Example of an InOut interface

Figure 6.4 shows two variables being passed via InOut. These variables are both read and written in the block object. Because of this, it makes no sense to copy them in and then write them out. It's easier (and better for memory) to have a single interface with both read and write access.

Large variables in the interface

If a variable is large, passing it into the interface as input or writing it out as output can consume additional memory and other resources. If many instances of function blocks or functions are in use that behave this way, this can cause performance issues as the project grows.

An example can be observed by creating a large interface with a function block, as follows:

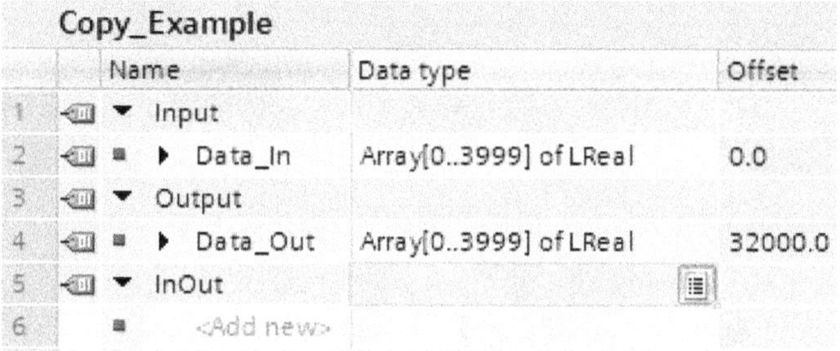

Figure 6.5 – Function block with large input and output variables

The example given in *Figure 6.5* shows that a function block called Copy_Example has a very large input and output interface. An Array[0..3999] of LReal data type would be 32000 bytes. This means that this particular function is copying 64000 bytes of information on every scan.

The logic that this function block is performing is simply to iterate in a loop over every element in the Data_In array, add 5 to the value, and output on the Data_Out array, as illustrated in the following screenshot:

```
1 ⊟FOR #i := 0 TO 3999 BY 1 DO
2        #Data[#i] := #Data[#i] + 5;
3  END_FOR;
```

Figure 6.6 – Logic being performed

This isn't particularly taxing on the **central processing unit** (**CPU**); however, the memory management of this function block is poor.

Comparing against a function block that has a better-constructed interface highlights the performance issue, as illustrated in the following screenshot:

		Name	Data type	Offset
Pointer_Example				
1	▼	Input		
2	■	<Add new>		
3	▼	Output		
4	■	<Add new>		
5	▼	InOut		
6	■ ▶	Data	Array[0..3999] of L...	0.0
7	■	Example	Bool	6.0

Figure 6.7 – Function block with large InOut variable

The immediate difference is the amount of memory the interface requires. `Copy_Example` is using `64000` bytes and `Pointer_Example` is using only 6 bytes (excluding the `Example` Boolean that is used to display the offset for the `Data` variable).

This is further reflected in the **Program Info** object found in **Project tree**. This object allows programmers to view the resource requirements of the PLC and which program blocks/objects are utilizing the resources, as illustrated in the following screenshot:

	Objects	Load memory	Code work-memory	Data work-memory
	Resources of PLC_1			
1		6 %	0 %	3 %
2				
3	Total:	2 MB	512000 bytes	3145728 bytes
4	Used:	118350 bytes	902 bytes	96212 bytes
5	Details			
6	▼ OB		6967 bytes	509 bytes	
7	Main [OB1]	6967 bytes	509 bytes	
8	FC	-	-	
9	▼ FB	10322 bytes	393 bytes	
10	Copy_Example [FB1]	5306 bytes	205 bytes	
11	Pointer_Example [FB2]	5016 bytes	188 bytes	
12	▼ DB	99816 bytes		96212 bytes
13	Copy_Example_DB [DB1]	65359 bytes		64068 bytes
14	Datablock [DB2]	33175 bytes		32070 bytes
15	Pointer_Example_DB [DB...	1282 bytes		74 bytes

Figure 6.8 – Program Info displaying resource usage

This view is split into **program blocks**. *Figure 6.8* shows that the **instance data** for
`Copy_Example` (`Copy_Example_DB`) uses `65359` bytes of data on the memory
card, and the instance requires `64068` bytes of work-memory to process the instance
when running.

Compared to the **instance data** for `Pointer_Example` (`Pointer_Example_DB`),
there is a significant difference. `Pointer_Example_DB` occupies `1282` bytes of data on
the memory card, and a tiny `74` bytes of data work-memory (memory required to store
information) to process.

This means that the amount of executable work-memory required when using an InOut
interface for this example is reduced by 99.88% of that required by using input and output
interfaces.

> **Note**
>
> The actual **code work-memory** (memory required to perform logical
> functions) is a much less significant improvement of only 8.29%. This is
> because the actual logical operation is still the same but only one variable is
> involved.

Planning standard control interfaces

Creating standard control objects (function blocks or functions that are standardized and
stored in a library) will always require more thought than a project bespoke object. The
interface is the key to making a standard control object easy to use.

Have a look at the following diagram:

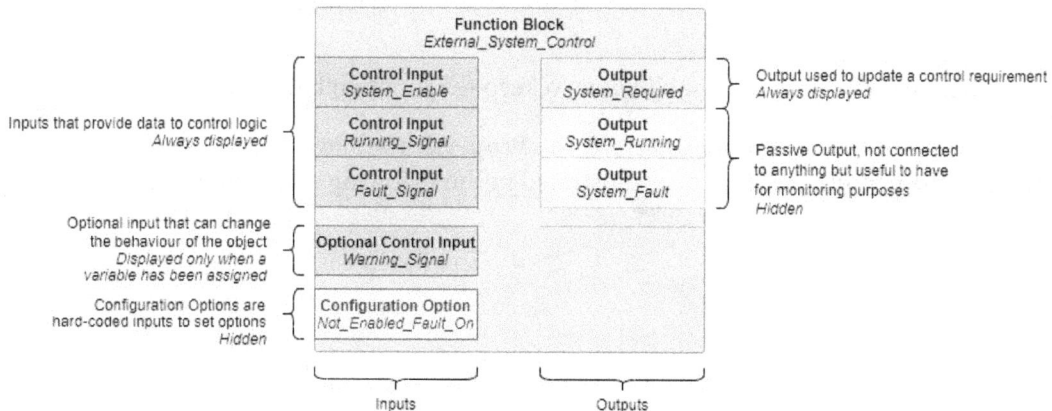

Figure 6.9 – Example layout of a standard control interface

Figure 6.9 shows an example of a standard control object with a well-defined interface. Signal types are categorized by functions on the interface and grouped together. Options are set on each interface element so that less important interface elements are hidden or conditionally hidden.

> **Note**
> TIA Portal refers to interface variables as **parameters**, which is a common term for data passed in through an interface.

The function block represented in *Figure 6.9* has the following types of interface parameters:

- **Control Input**

  - These are parameters that are directly affecting the control of the function block.

  - These parameters will always be required, therefore are set to be *always displayed.*

- **Optional Control Input**

  - These are parameters that may be omitted if not required.

  - These parameters are hidden unless a variable is connected, then they are displayed.

- **Configuration Option**

  - These are parameters that are hardcoded with a value such as `True` or `1`.

  - They change the behavior of the function block by enabling or toggling internal functionality.

  - These parameters are *always hidden* to keep the function block interface clean.

In the function or function block interface, the **Properties** window can be used to set the visibility of the parameter selected, as illustrated in the following screenshot:

| 6 | ⊲⫿ ▪ | Not_Enabled_Fault_On Bool | | 🔳 false | Non-ret... ▾ | ☑ | ☑ | ☑ |

Not_Enabled_Fault_On   🔍 **Properties**   ⁺ᵢ Info ⓘ 📊 Diagnostics   ◻ ▭

| **General** | Texts | Supervisions |

General
Attributes

Attributes _____

                              Retain [Non-retain   ▾]
          ☐ Setpoint

Usage

          ☑ Accessible from HMI/OPC UA/Web API
          ☑ Writable from HMI/OPC UA/Web API
          ☑ Visible in HMI engineering

Visiblity in block calls in LAD/FBD

          ◯ Show
          ⦿ Hide
          ◯ Hide if no parameter is assigned

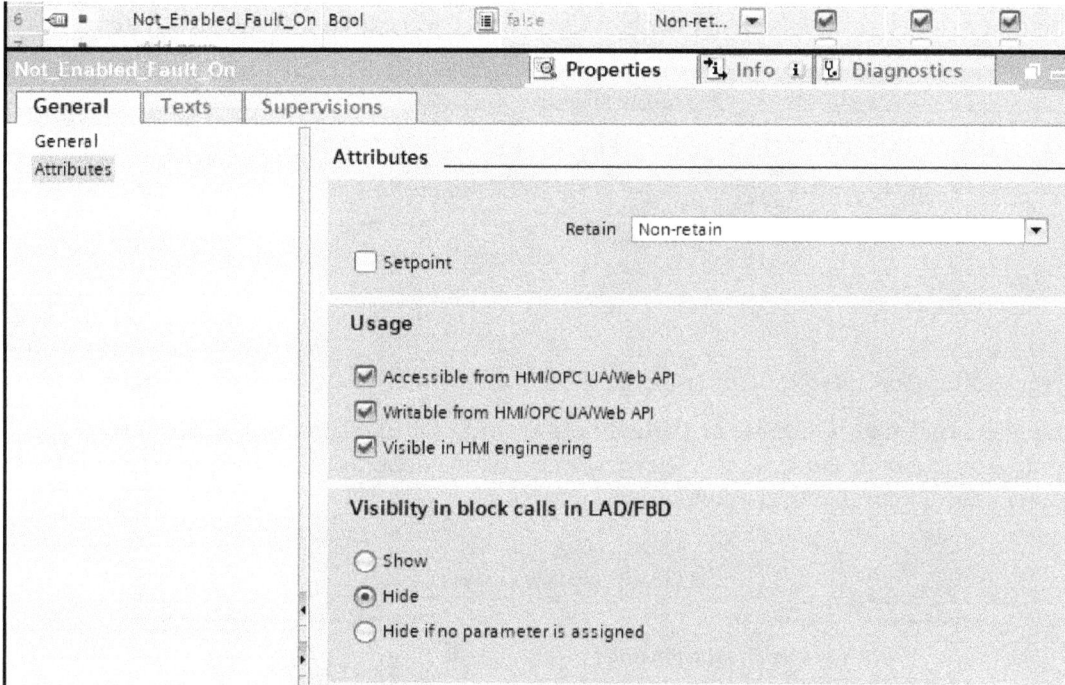

Figure 6.10 – Setting visibility in block calls for interface variables

By setting **Visibility in block calls in LAD/FBD**, the parameters can be hidden under certain conditions.

If a function/function block contains hidden interface parameters, a small drop-down arrow is displayed at the bottom of the block, as illustrated in the following screenshot:

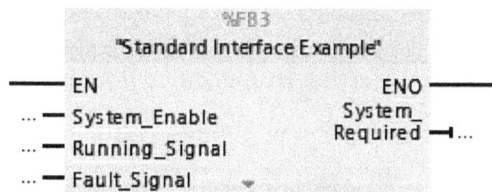

```
                    %FB3
           "Standard Interface Example"

      ——— EN                    ENO ———
      ··· — System_Enable       System_
      ··· — Running_Signal      Required —◀ ···
      ··· — Fault_Signal    ▾
```

Figure 6.11 – Function block with hidden interface parameters (collapsed view)

When the arrow has been pressed, the block's interface will expand, as illustrated here:

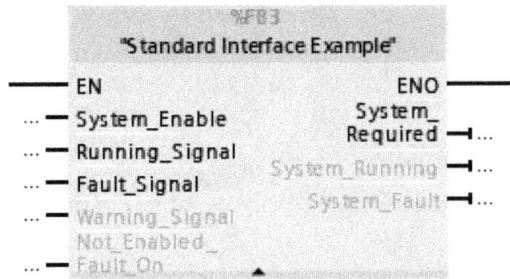

Figure 6.12 – Function block with an expanded interface, showing previously hidden parameters

Parameters that are set to **Hide** or **Hide if no parameter is assigned** will be displayed as gray. If parameters are assigned, the view may change depending on the parameter's properties, as illustrated in the following screenshot:

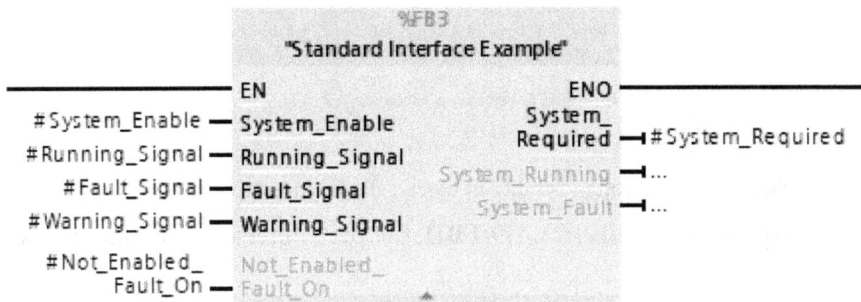

Figure 6.13 – Function block with populated interface

Figure 6.13 demonstrates that the `Warning_Signal` parameter is no longer a hidden parameter (no longer gray). This is because its properties are set to **Hide if no parameter is assigned**, and it now has a parameter (variable) assigned to it.

Notice that the `Not_Enabled_Fault_On` parameter is still gray in the interface. This is because the properties for this parameter are set to **Hide**, so even if a variable is connected to this interface at this point, it will still be hidden on the collapse of the block.

> **Note**
>
> A parameter with the **Hide** or **Hide if no parameter is assigned** property will still be displayed if it must be connected to a variable. For example, functions will ignore the properties assigned until a variable is connected, at which point the parameter will be hidden if required.

Creating control data

By creating control data, programmers create a specific set of criteria or variables that all standard controls should follow. This is a dataset that should follow a requirement or process, such as the following:

- All asset controlling function blocks should return one of the following status types:

  - `Healthy` status

  - `Running` status

  - `Not available` status

  - `Inhibit` status

The preceding status types can be added to a Struct data type called `Control_Data`, as illustrated in the following screenshot:

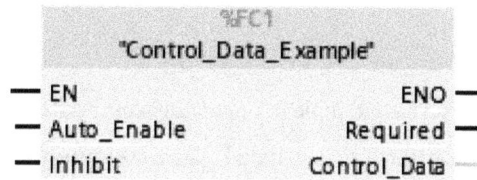

Figure 6.14 – Example of a Control_Data output

Figure 6.14 demonstrates a function with an output for the `Control_Data` elements. The output is a struct containing elements that go on to be used in a program for control and interlocks with other equipment.

The contents of this `Control_Data` struct are shown here:

Control_Data_Example

	Name	Data type	Default value	Comment
▼	Output			
■	Required	Bool		
■ ▼	Control_Data	Struct		
■	Healthy	Bool		Any Fault Active
■	Not_Available	Bool		Pump Not Available For Any Reason
■	Auto_Enable	Bool		Pump Is Enabled To Run In Auto
■	Inhibited	Bool		Pump Inhibited From Running In Auto
■	Running	Bool		
■	Running_In_Auto	Bool		

Figure 6.15 – Control data output

The `Control_Data_Example` asset control function block is taking in inputs, processing the control requirements, and then populating the `Control_Data` struct with data that can be stored and used elsewhere in the project.

This would allow larger systems to access information about an asset directly from `Control_Data` without having to create the desired status elsewhere.

Have a look at the following screenshot:

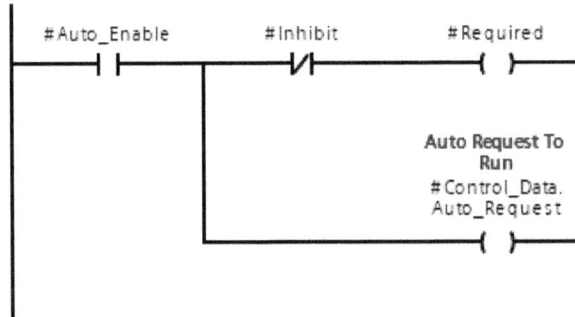

Figure 6.16 – Example of populating control data

Figure 6.16 is an example of populating `Control_Data` elements with a value. `Auto_Enable` is an input into the `Control_Data_Example` function and has multiple conditions that need to be `True`, as illustrated in the following screenshot:

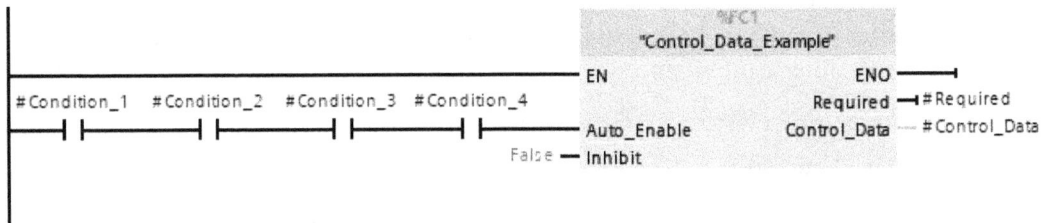

Figure 6.17 – Control_Data_Example Auto_Enable conditions

If a programmer wanted to create interlocks for other devices that ensured that `Auto_Enable` was in a `True` state for the `Control_Data_Example` function, this could now be checked via the `Control_Data` structure, as illustrated here:

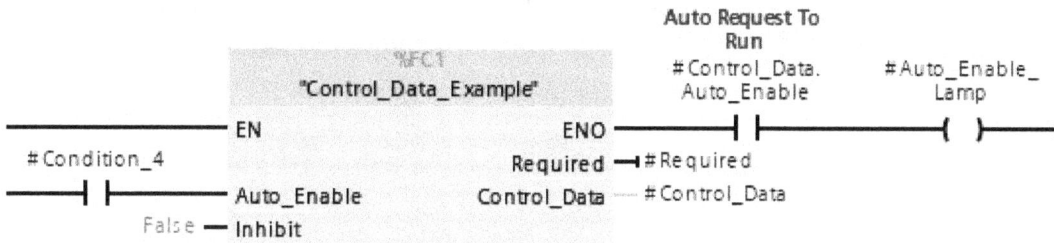

Figure 6.18 – Control_Data usage example

Figure 6.18 demonstrates how `Control_Data` is consolidating all requirements of the `Auto_Enable` input into a single-point `Control_Data.Auto_Enable` signal that can then be used elsewhere in the logic without having to recreate (or understand) the original requirements.

> **Note**
> Consolidating signals into single signals where possible helps keep the repetition of code to a minimum. It also means that should conditions need to change, logic downstream of the consolidation does not need to change.

Improving control data accessibility with UDTs

Instead of using a Struct as an output from the interface, a UDT can be used as an InOut interface instead.

The benefits of this approach are listed here:

- **Less memory consumption**

  - The structure does not exist inside and outside the block as separate anonymous structures.

  - The data is not copied between the interface and the variable connected to the interface (InOut is passed by reference).

- **Consolidated control data**

  - The *control data* can exist within a wider dataset. This can then be accessed by the standard control object and updated. Because the data is referenced, the correct data will be updated without any copying.

  - This allows for large datasets to be used, without impacting performance.

Using the InOut interface instead of output does not change anything about writing control data, but it does mean that control data can also be read and interacted with.

This can allow the opportunity to share control data as a whole UDT structure with other standard control objects, streamlining the logic.

Have a look at the following screenshot:

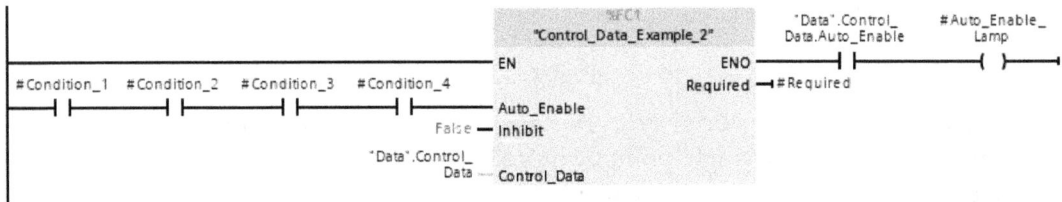

Figure 6.19 – Example of using InOut with a UDT

Figure 6.19 shows that the same logic can still be implemented with InOut usage. The only difference is that the data is part of a UDT and passed as InOut instead of being copied to the output interface.

Example

The benefits of this approach are best demonstrated with an example. A system contains two digital output cards that control various outputs to assets and equipment. The specification states that in the event of an **emergency stop** (**E-Stop**) or other safety-critical signals, all outputs must be switched off and made safe.

The control data approach, shared between multiple areas of the code, allows this to happen easily, leaving room for easy expansion in the future.

The system contains a structure that contains various signals involved in the safety control of the system. This is a single instance of a UDT called `Safety_Control_Data`, stored in a DB called `Data`. The collective data can be accessed via `Data.Safety_Control_Data`, as illustrated in the following screenshot:

Network 2: Write To Safety Control Data

Comment

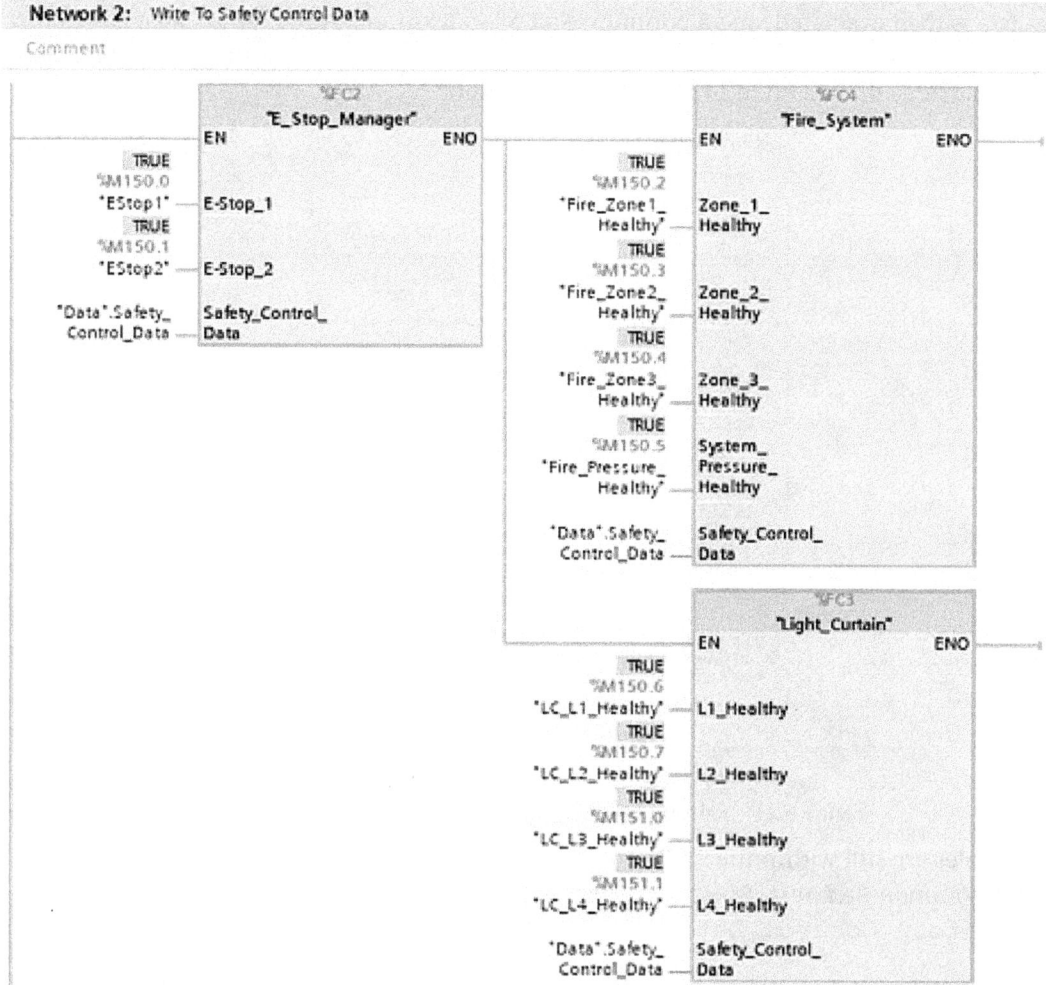

Figure 6.20 – Writing to the Safety_Control_Data structure

Figure 6.20 demonstrates how three separate areas of safety control all use the same control data structure, passed as an InOut variable.

This data is then evaluated, and a common `Safety_System_Active` variable is updated. This consolidates all safety requirements to a single signal for the rest of the system to use, as illustrated in the following screenshot:

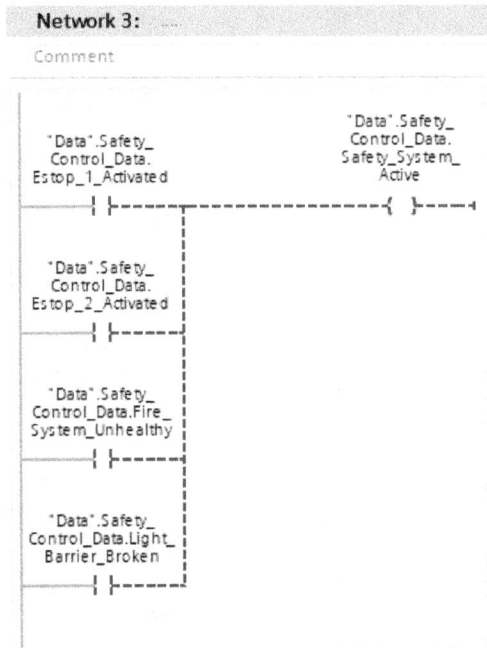

Figure 6.21 – Consolidation of control data to a single variable

All variables are still within the `Safety_Control_Data` structure. This means that the common `Safety_System_Active` signal can still be accessed via the `Data.Safety_Control_Data` variable.

The outputs are managed via a function that maps required outputs to physical output addresses, as illustrated in the following screenshot:

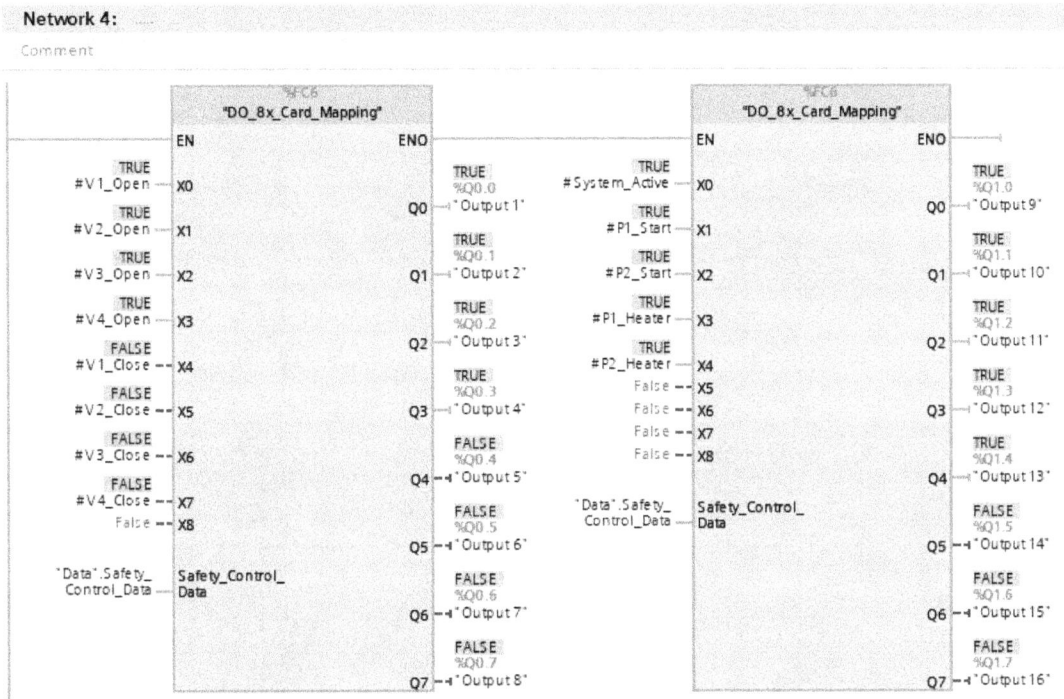

Figure 6.22 – Output mapping

Figure 6.22 shows how the DO_8x_Card_Mapping function contains the Safety_ Control_Data InOut interface. This is being used to turn off all outputs in the case of a safety-related incident, as illustrated in the following screenshot:

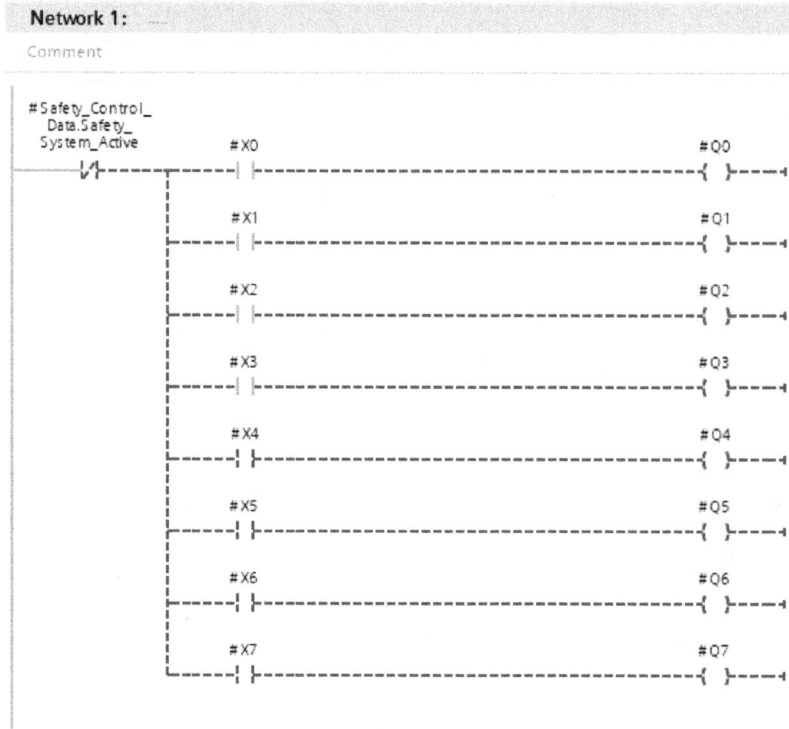

Figure 6.23 – Control data being used to switch off all outputs

The DO_8x_Card_Mapping function reads the control data and checks if the Safety_System_Active variable is False; if so, outputs can be evaluated as normal. If the Safety_System_Active variable is True, all outputs are set to False.

The control data in this example can be passed to every function or function block that needs to respond to a safety-related incident. Should the requirements change over the lifetime of the application, the UDT can be updated and new requirements added with ease. When the UDT is changed, all interfaces that use the UDT will automatically be updated too. This means that making changes to data that is utilized frequently does not result in many hours needed to change interfaces.

Note

If `Safety_Control_Data` were a struct instead of a UDT, every
interface would need to manually be updated if the struct were to be modified.
Remember that struct instances are not connected to each other.

Creating HMI data

Just as with control data, it's usually a good idea to have a general concept of how **human-machine interface** (**HMI**) data will come together.

This is the information that is used to display statuses, control graphics, and any other specific HMI requirements. Segregating these variables from other variables helps keep the vast amount of data that HMI/**supervisory control and data acquisition** (**SCADA**) systems require clean and easy to use.

Creating a UDT to house this information makes it easy to add it into datasets for assets later on, as illustrated in the following screenshot:

	Name	Data type	Default value	Setpoint	Comment
1	Mimic	USInt	0		0 = Off, 1 = Running In Manual, 2 = Running In Auto
2	Hours_Run	USInt	0		Hours Of Active Service
3	Mode	USInt	0		0 = Off, 1 = Manual, 2 = Auto

UDT_Pump_HMI_Data

Figure 6.24 – Example of HMI data for a pump asset

Each instance of a pump asset would then have its own `UDT_Pump_HMI_Data` instance that the HMI would interact with.

This approach, much the same as for control data, allows for each standard control object in the project to have a known and designated area for HMI interaction. This helps unrelated blocks feel familiar, even when the programmer has not used them before.

Setpoints/parameters

Most data transferred between a PLC and an HMI/SCADA system will be setpoints/parameters. These are the settings that change the application's behavior to suit the end user's needs.

Most assets will have some form of common setpoint structure. Whether all of the setpoints are used or not depends upon the application use case, but providing as many as possible simplifies the design process, and the setpoints are already available in UDTs.

Have a look at the following screenshot:

UDT_Level_Controller_HMI_Data			
Name	Data type	Default value	Comment
Mimic	USInt	0	0 = Healthy, 1 = Faulty (Unaccepted), 2 = Faulty (Acc...
Percentage	Real	0.0	Level As %
Level_Status	USInt	0	0 = Healthy, 1 = Low Low, 2 = Low, 3 = High, 4 = Hig...
▼ Setpoints	Struct		Standard Setpoints
■ Normal_Level	Real	50.0	
■ LowLow_Level	Real	10.0	
■ Low_Level	Real	15.0	
■ High_Level	Real	80.0	
■ HighHigh_Level	Real	95.0	
■ Hysteresis	Real	5.0	

Figure 6.25 – Example of a level controller's UDT for HMI data

Figure 6.25 is an example of a level controller's UDT for HMI data. Inside the UDT is a structure for setpoints. This means that every asset that makes use of `UDT_Level_Controller_HMI_Data` will have these base setpoints to work with.

This also means that HMI/SCADA interfaces can guarantee that the data will be available to work with if a type of asset that utilizes this structure is in use.

> **Note**
>
> Offering too much data will result in programmers finding it difficult to filter out what is needed for the current application. Offering too little will result in programmers having to modify HMI data too often.
>
> Filling HMI structures with setpoints and variables that won't be used can also overload tag requirements between the PLC and HMI.

Structuring logic

For standard control objects to feel familiar with each other, the logic should be laid out in roughly the same approach. By following a generic ruleset for each standard control object, code (written in any language) should be easy to pick up and read by anyone that uses the standard controls.

General layout

Control object logic can be easily planned for with a generic template, as illustrated here:

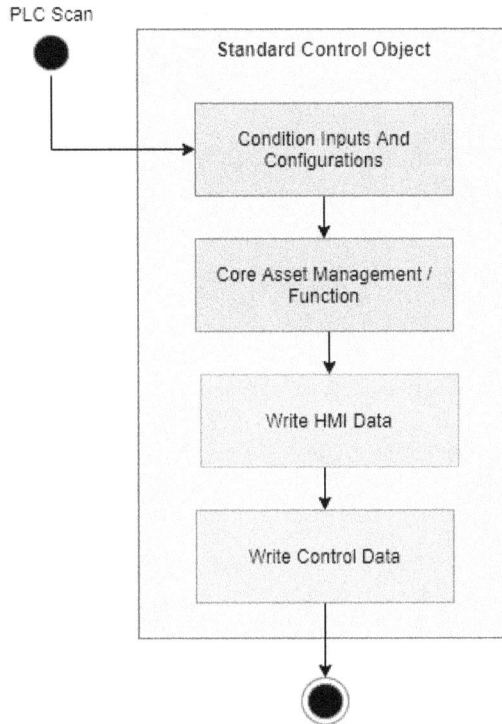

Figure 6.26 – Example of a generic template for standard control block logic

Figure 6.26 is an example of a layout for generic standard control objects. If all objects in the project follow this general paradigm, then programmers and maintainers of the project will have a much easier time reading the logic within the standard control.

Supportive methods

In large standard control objects, each area of the template (*Figure 6.26*) could be created as a function or function block. Each supportive object could then follow the same template internally, keeping the same approach throughout the standard control object.

Have a look at the following diagram:

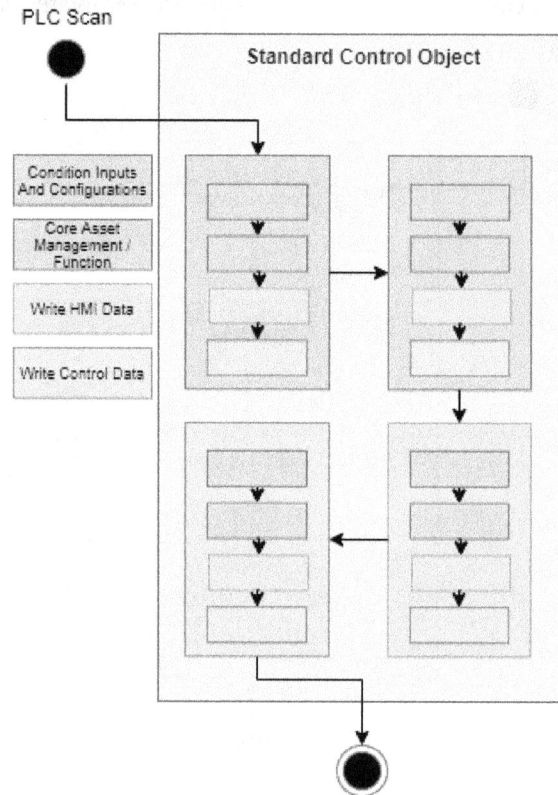

Figure 6.27 – Example of each standard control method following the template

Figure 6.27 demonstrates a standard control object's supportive method also following the control object template.

This means that each step of the template follows all steps of the template again (where possible). Following this approach then means that every standard control object has a generic feel/layout to it, no matter the language or function.

> **Note**
>
> If standard control objects are placed together in a standard control library, this template approach means that third parties, new starters, or infrequent users of the library can get to grips with how the logic is written quickly. It also means that different writing styles of different programmers are still easy to follow as they are governed by an overall template.

Considerations that have an impact on usability

When creating standard control objects, there are many different aspects to consider. Key items are listed here:

- How flexible does the control object need to be?
- How likely is it that the control object will need to be modified?
- What does the control object interact with?

These sorts of questions can alter the approach that is taken for the control object being developed.

How flexible does the control object need to be?

A good example of a flexible control object would be an analog scaling standard control object, as illustrated in the following screenshot:

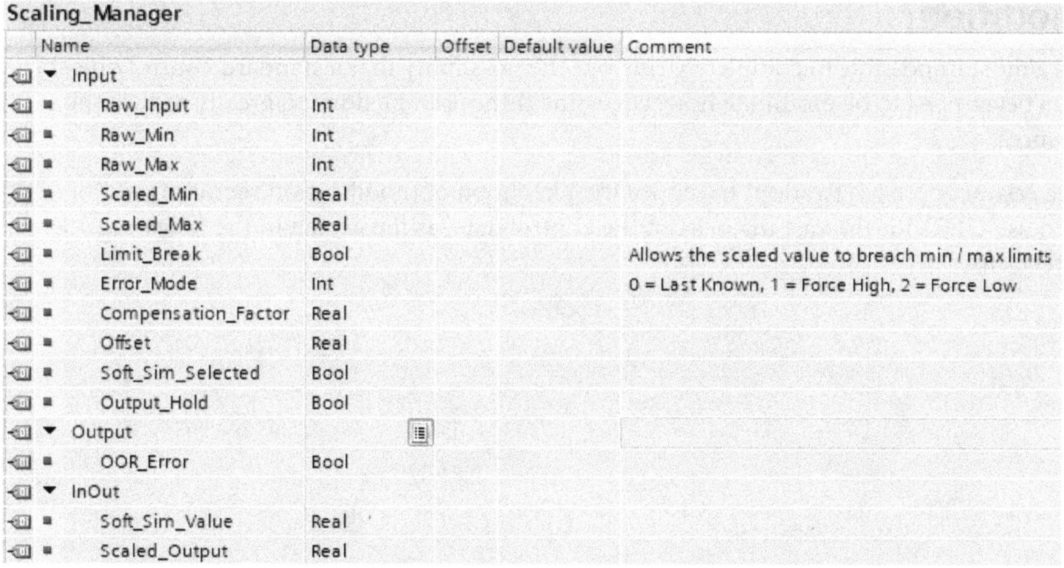

Scaling_Manager

		Name	Data type	Offset	Default value	Comment
⬛	▼	Input				
⬛	▪	Raw_Input	Int			
⬛	▪	Raw_Min	Int			
⬛	▪	Raw_Max	Int			
⬛	▪	Scaled_Min	Real			
⬛	▪	Scaled_Max	Real			
⬛	▪	Limit_Break	Bool			Allows the scaled value to breach min / max limits
⬛	▪	Error_Mode	Int			0 = Last Known, 1 = Force High, 2 = Force Low
⬛	▪	Compensation_Factor	Real			
⬛	▪	Offset	Real			
⬛	▪	Soft_Sim_Selected	Bool			
⬛	▪	Output_Hold	Bool			
⬛	▼	Output				
⬛	▪	OOR_Error	Bool			
⬛	▼	InOut				
⬛	▪	Soft_Sim_Value	Real			
⬛	▪	Scaled_Output	Real			

Figure 6.28 – Interface for a standard control object that controls analog scaling

The `Scaling_Manager` interface pictured in *Figure 6.28* has many different inputs that exceed the basic requirements to scale a value. This particular control method is capable of performing the following additional methods:

- Scaling beyond the minimum/maximum range of the scaled value (`Limit_Break`)
- In the event of an error (raw out of range), driving the scaled value high, low, or holding the last value that was not an error (`Error_Mode`)
- Compensation and offset of the final scaled value (`Compensation Factor/Offset`)
- Ability to hold the output, ignoring any raw value movement (`Output_Hold`)

These additional methods all extend the basic functionality of simply scaling the raw value. This allows for much more flexibility and reduces the likelihood of a new block needing to be written for a project, without introducing high overhead if it is not used.

How likely is it that the control object will need to be modified?

It's almost impossible to completely rule out the possibility that a standard control object won't ever need to be modified; however, some things can be done to greatly reduce the chances.

The easiest approach to adopt to reduce the likelihood of modification requirements is to use UDTs for the vast majority of the control data, as illustrated in the following screenshot:

	▼ InOut				
	▼ Generic_Analog_Data	"UDT_Generic_Analog_HMI_Values"	32.0		Generic Analog Data Structure
	Raw_Value	DInt	0.0		Analog Raw Value
	Scaled_Value	Real	4.0		Analog Scaled Value
	Trend_Value	Real	8.0		Manipulated Trend Value
	Soft_Sim_Value	Real	12.0		Soft Sim Value From SCADA
	Raw_Min	Int	16.0		Minimum Raw Value
	Raw_Max	Int	18.0		Maximum Raw Value
	Instrument_Min	Real	20.0		Minimum Displayed Instrument Value
	Instrument_Max	Real	24.0		Maximum Displayed Instrument Value
	Scale_Min	Real	28.0		Minum Scaled Value
	Scale_Max	Real	32.0		Maximum Scaled Value
	Offset	Real	36.0		Offset Applied To Scaled Value
	Compenstation_Fa..	Real	40.0		Multiplication Factor Applied To Scaled Value
	HiHi	Real	44.0		HiHi Alarm Setpoint
	Hi	Real	48.0		Hi Alarm Setpoint
	Lo	Real	52.0		Lo Alarm Setpoint
	LoLo	Real	56.0		LoLo Alarm Setpoint
	Hysteresis	Real	60.0		Hysteresis
	HiHi_Release	Real	64.0		HiHi Alarm Release
	Hi_Release	Real	68.0		Hi Alarm Release
	Lo_Release	Real	72.0		Lo Alarm Release
	LoLo_Release	Real	76.0		LoLo Alarm Release
	Alarm_IND	Byte	80.0		Instrument Alarm Indication For SCADA
	Fault_IND	Byte	81.0		Fault Indication For SCADA
	Alarm_Trigger_Del...	Time	82.0		Alarm Trigger Delay
	Alarm_Release_Del..	Time	86.0		Alarm Release Delay
	Fail_Delay	Time	90.0		Fail Delay (OOR)
	OOR_Fault	Bool	94.0		OOR Fault Status (Pre Delay)
	HMI_Tag_Check	Bool	94.1		RESERVED FOR HMI/SCADA
	▶ Alarm_Status	Struct	96.0		

Figure 6.29 – Example of a Generic_Analog standard control object interface

Figure 6.29 shows a control object that is used to manage generic analog signals and provide alarms and other functions. The vast majority of the information is stored within the UDT_Generic_Analog_HMI_Values UDT. This means that if the UDT is ever modified, the control objects associated with the UDT will update automatically.

If 10 objects used this UDT but an update was required to provide additional functionality to 1 of the 10 control objects, the other 9 would *not* require an update. TIA Portal would re-initialize the interfaces, and the corresponding memory would be assigned to the new-sized UDT.

This approach is the easiest way to minimize the effects of data causing changes in a standard control object.

What does the control object interact with?

If the control object being developed is to interact with hardware, considerations as to how information is passed to the control object should be made.

Creating interface data using `Any` or `Variant` data types can help create flexible areas that allow for more than one configuration.

Have a look at the following screenshot:

▼ InOut		
▪ ▶ VSD_Data	"UDT_VSD_Drive"	54.0 i4.0
▪ Hardware_Data	Variant	
▪ Control_Data	Variant	

Figure 6.30 – Example of Variant usage

Figure 6.30 shows how an InOut interface can be used to pass a `Variant` data type to a standard control object. This type of interface allows the control object to detect a UDT/struct and check what the type is. The control object can then change its behavior based on the type passed, as illustrated in the following screenshot:

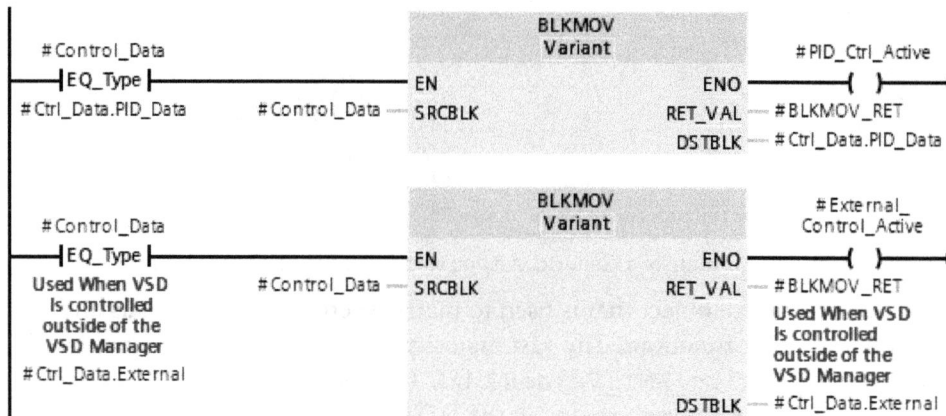

Figure 6.31 – Example of Variant being checked for data type

The EQ_Type instruction is used to check the data type of the Control_Data variable against a temporary instance of another type (Ctrl_Data.PID_Data in this case), as illustrated in the following screenshot:

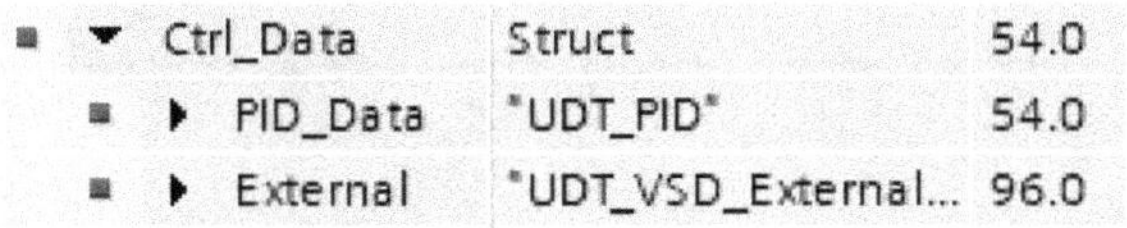

	Ctrl_Data	Struct	54.0
▶	PID_Data	"UDT_PID"	54.0
▶	External	"UDT_VSD_External...	96.0

Figure 6.32 – Temporary UDT instances

If the Control_Data data type passed into the control block matches the UDT_PID data type, then the BLKMOV command (*Figure 6.31*) moves the control data into the temporary instance of the UDT_PID data type.

If the Control_Data data type passed into the control block matches the UDT_VSD_External_Control data type, then the BLKMOV command (*Figure 6.31*) moves the control data into the temporary instance of the UDT_VSD_External_Control data type.

The control logic of this block then goes on to handle the operation of the variable speed drive differently, according to the UDT that has been passed to it.

This approach can help common use cases where a standard control object may interact with more than one type of equipment and control different equipment in different ways.

Have a look at the following screenshot:

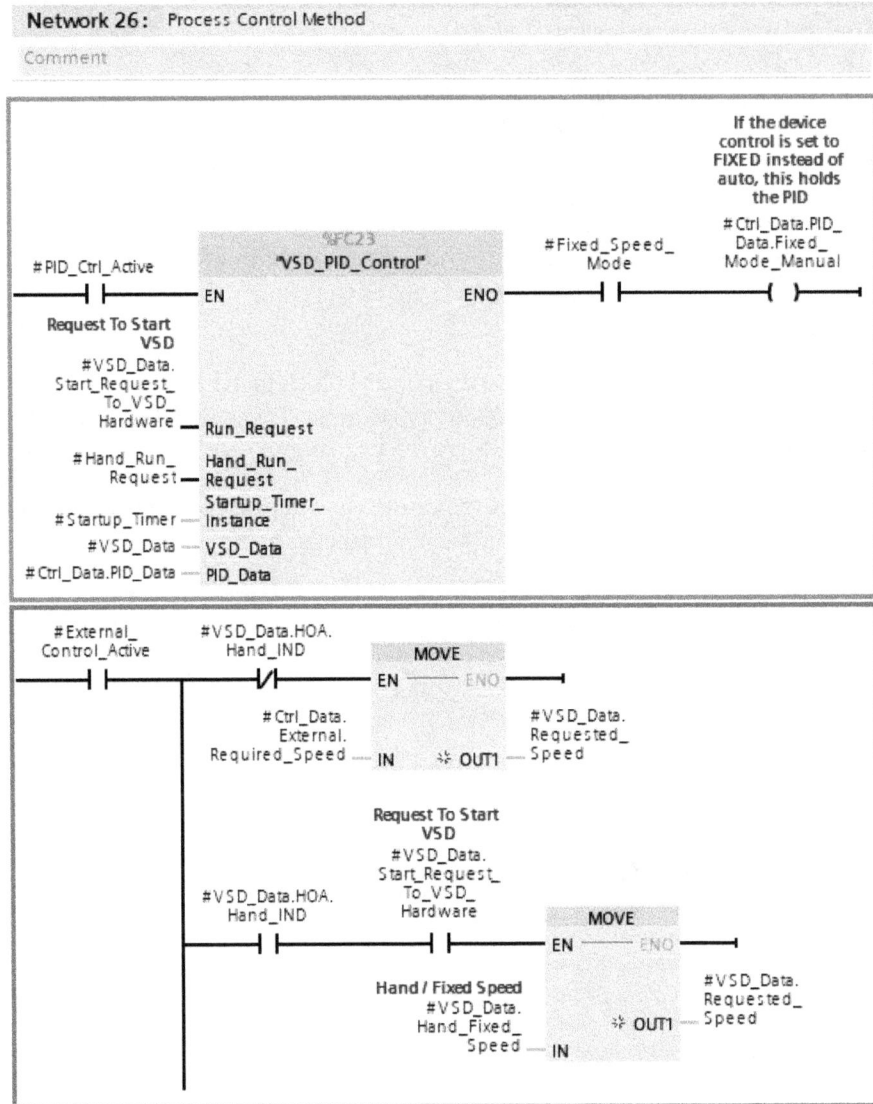

Network 26: Process Control Method

Comment

Figure 6.33 – Example of different control methods based on the data provided

Figure 6.33 shows that when the UDT_PID data is passed to the control object, the VSD_PID_Control function is used. When the UDT_VSD_External_Control data is passed to the control object, loose-ladder logic is used to control the **variable speed drive** (**VSD**) in the required way.

At the end of the control object, the information is moved back to the `Control_Data` variable, as illustrated in the following screenshot:

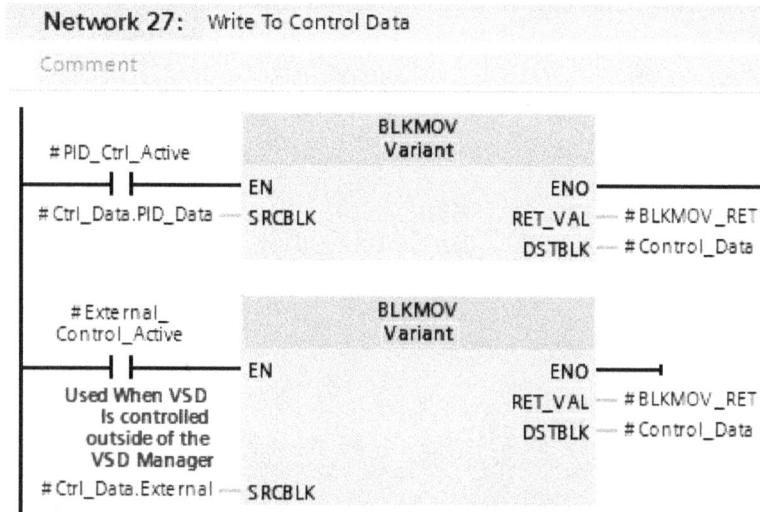

Figure 6.34 – Moving temporary data back to Control_Data

Figure 6.34 shows how the information is passed back to the `Control_Data` variable, ready to be passed back out of the control object (as InOut).

> **Note**
>
> This step is necessary. Without it, the information being acted upon in the `Ctrl_Data` structure would be left in temporary memory and the `Control_Data` variable would never be updated.

Summary

This chapter covered creating standard control objects, with a viewpoint of structuring data and logic layouts, as well as the management of interface variables.

The knowledge gained from this chapter should help programmers to write standard control objects that are easy to maintain, expand, and modify while retaining a standard approach to the development.

The next chapter focuses on the simulation of code so that programmers can test the inputs and outputs of the project without any hardware requirements. This includes using standard control objects to manage the mapping of information through *mapping layers*.

7
Simulating Signals in the PLC

In this chapter, we'll look at different methods we can use to simulate devices and logic.

Siemens TIA Portal has a built-in simulation package that lets you simulate a variety of different CPUs and the logic within them.

In this chapter, we're going to cover the following topics:

- Running PLC/HMI in simulate mode
- Managing simulated inputs
- Creating a simulation interface
- Safeguarding outputs when in simulation mode

Running PLC/HMI in simulate mode

In TIA Portal, both PLCs and HMIs can be run in soft simulate mode. When this occurs, the hardware runtime is run from TIA Portal, instead of the hardware. This runtime then behaves in the same way it would as if it were running in the hardware.

Starting a PLC simulation

To start a PLC simulation, select the device that's required from the **Project tree** area and click the **Start Simulation** button via the toolbar at the top of TIA Portal:

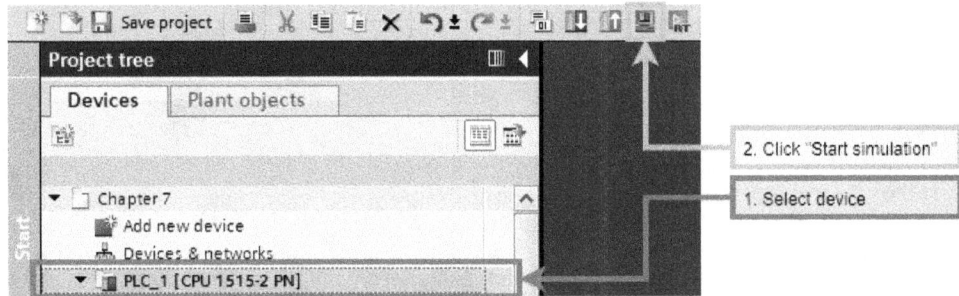

Figure 7.1 – Starting a simulation PLC

Clicking the **Start Simulation** button will automatically switch the network connection interface to **PLCSIM**.

> **Note**
>
> When you're running the PLCSIM interface, it is not possible to connect to another physical device, even with a different instance of TIA Portal (or even SIMATIC Manager, if it is installed).

Enabling simulation support

TIA Portal automatically disables the ability to simulate new blocks in a new project. If you've clicked the **Start Simulation** button, the following message may appear:

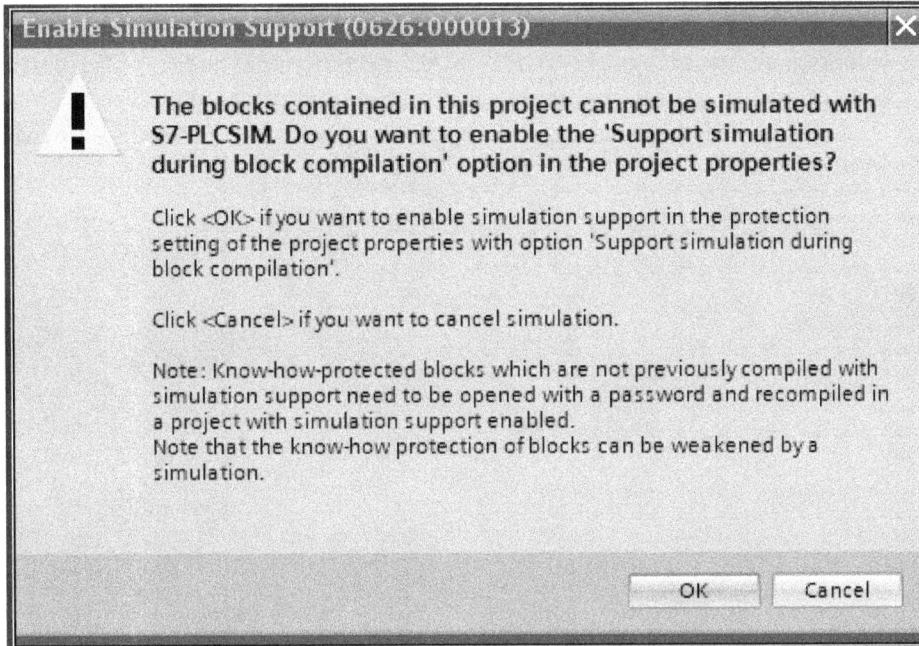

Enable Simulation Support (0626:000013)   ☒

! **The blocks contained in this project cannot be simulated with S7-PLCSIM. Do you want to enable the 'Support simulation during block compilation' option in the project properties?**

Click <OK> if you want to enable simulation support in the protection setting of the project properties with option 'Support simulation during block compilation'.

Click <Cancel> if you want to cancel simulation.

Note: Know-how-protected blocks which are not previously compiled with simulation support need to be opened with a password and recompiled in a project with simulation support enabled.
Note that the know-how protection of blocks can be weakened by a simulation.

| OK | Cancel |

Figure 7.2 – The Enable Simulation Support message

By clicking **OK**, the project and program blocks will have the **S7-PLCSIM** simulation option of **Can be simulated with SIMATIC S7-PLCSIM** enabled. This option can be found in the **Compilation** section of the program block's properties.

Note

Project library blocks that are **types** and have *not* had the **Can be simulation with SIMATIC S7-PLCSIM** option enabled will *not* have that option enabled. The block must be edited (**In Test**) to change that option. Once the option has been changed, the block can be **released** under a new version with that option enabled.

The S7-PLCSIM interface

TIA Portal will open an instance of the **S7-PLCSIM** interface once the **Start Simulation** button has been pressed (and the **Enable Simulation Support** message has been accepted). While **S7-PLCSIM** is loading, TIA Portal may display the **Extended download to device** window:

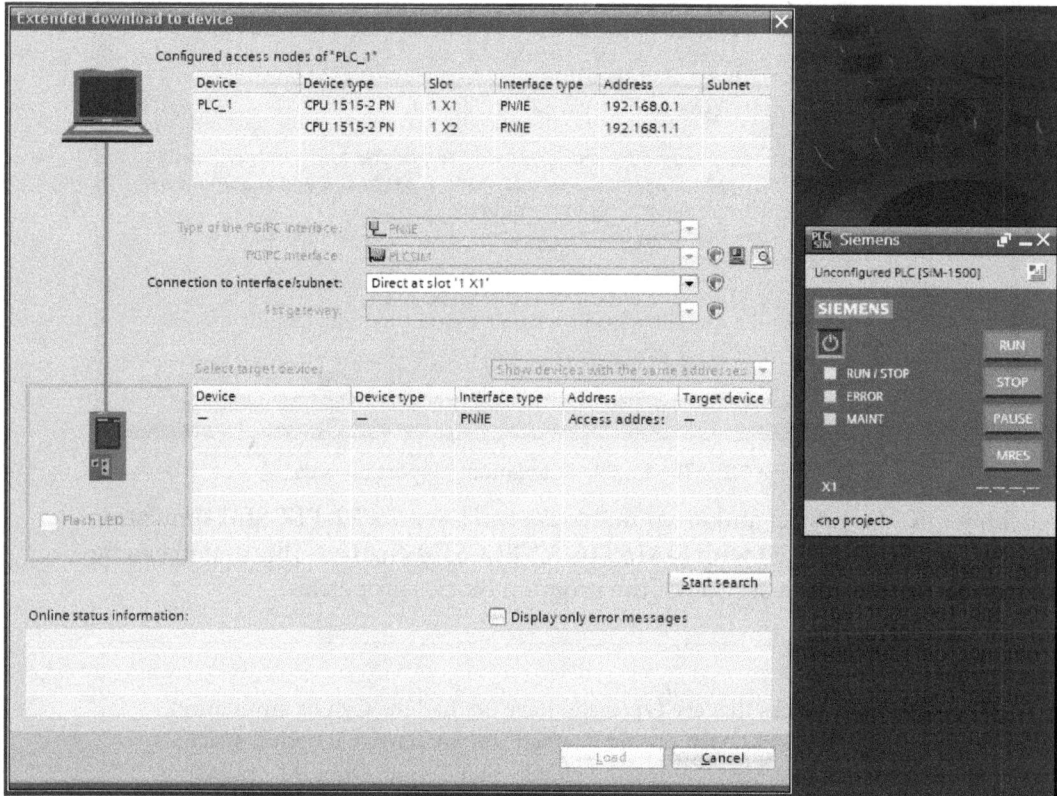

Figure 7.3 – The S7-PLCSIM window (right) and the Extended download to device window (left)

Notice that the *PG/PC interface* that's in use is *PLCSIM* and cannot be changed. By clicking **Start search**, the simulation PLC should be identified:

Select target device:				Show devices with the same addresses	▼
Device	Device type	Interface type	Address	Target device	
CPUcommon	CPU-1500 Simulation	PN/IE	192.168.0.1	CPUcommon	
—	—	PN/IE	Access address	—	

☐ Flash LED

Start search

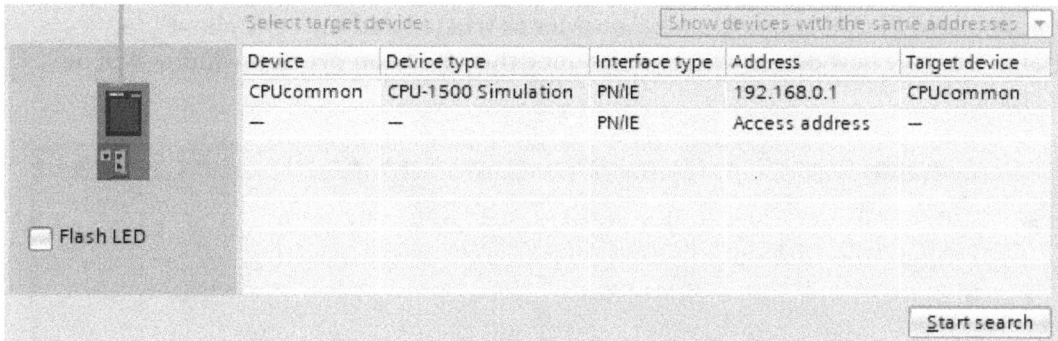

Figure 7.4 – Successfully discovered CPU-1500 simulation PLC

The simulated PLC will be displayed at *CPUcommon* for the device's name and the IP address will be 192.168.0.1. By selecting this device and clicking **Load**, the TIA Portal project will be prepared to be downloaded to the simulated PLC.

Trustworthy devices

Even though TIA Portal started creating and connecting to the simulation PLC, it will still warn you that the simulated PLC may not be a trustworthy device:

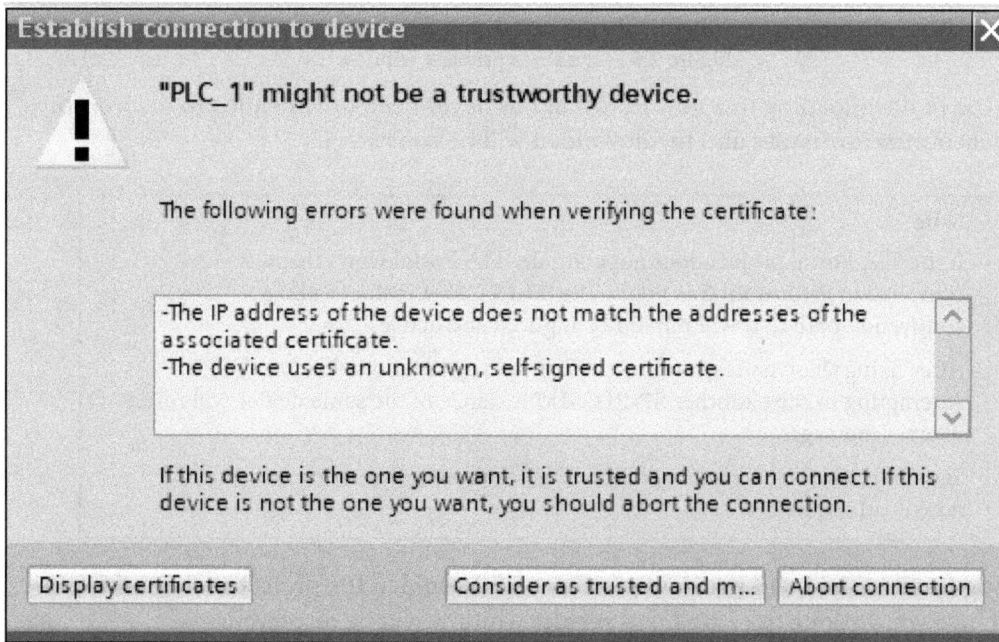

Establish connection to device   ✕

⚠ **"PLC_1" might not be a trustworthy device.**

The following errors were found when verifying the certificate:

-The IP address of the device does not match the addresses of the associated certificate.
-The device uses an unknown, self-signed certificate.

If this device is the one you want, it is trusted and you can connect. If this device is not the one you want, you should abort the connection.

Display certificates   Consider as trusted and m...   Abort connection

Figure 7.5 – The "PLC_1" might not be a trustworthy device message

For a simulated PLC, it is safe to click **Consider as trusted and m…** and allow TIA Portal to trust the new device. Once you've done this, the **Load preview** window will be displayed:

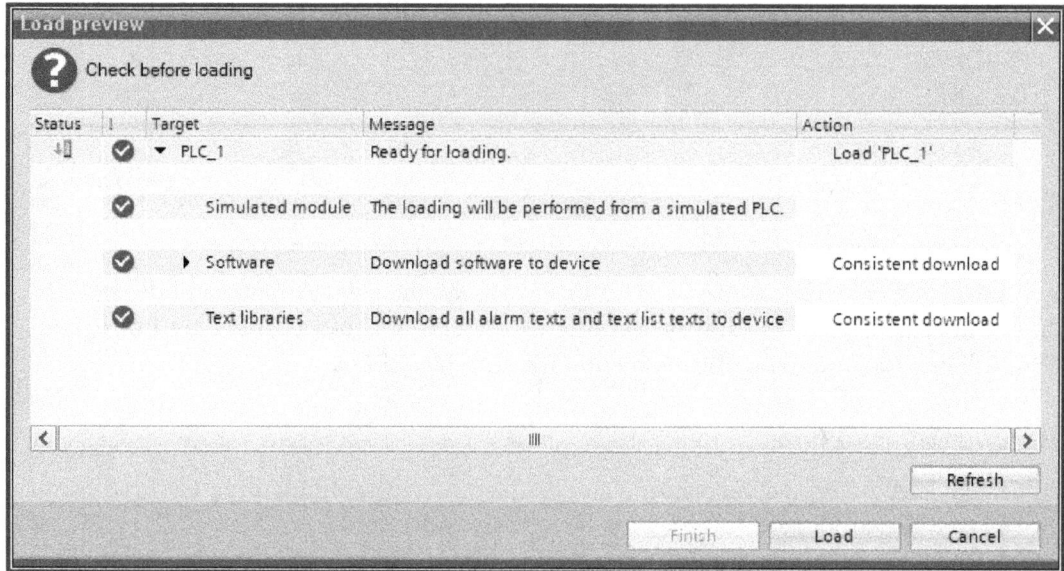

Figure 7.6 – The Load preview window

Since you're downloading to a new PLC that has never been downloaded to before, you won't encounter any issues and the download will be consistent.

> **Note**
>
> If the TIA Portal project does not compile, TIA Portal won't throw a compilation error until this point. The **S7-PLCSIM** instance and the connection path to it will remain configured and active.
>
> After fixing these issues, you can perform a standard download to the PLC. Attempting to start another **S7-PLCSIM** instance of the same device will result in error messages.
>
> See *Chapter 13, Downloading to the PLC*, for more information on standard download sequences.

Clicking **Load** in the **Load preview** window will download the project to the simulated PLC. Once you've done this, the **Load results** window will be displayed:

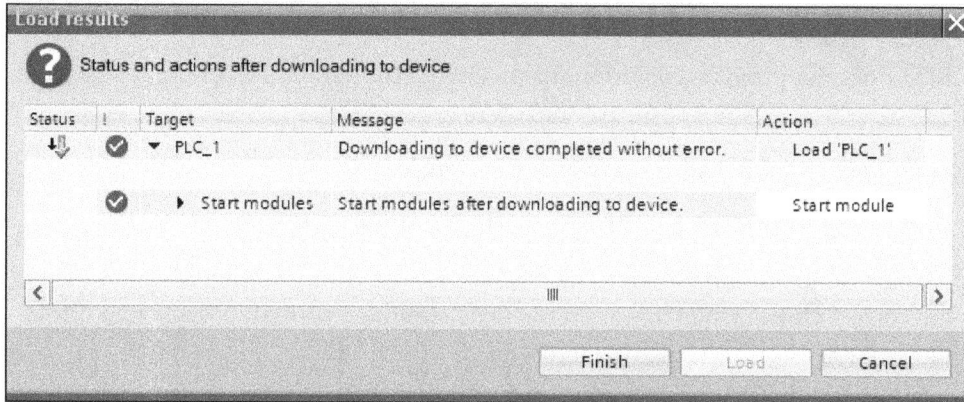

Figure 7.7 – The Load results window

Ensure that **Action** is set to **Start module** and click **Finish**. The simulated PLC will then be set to **RUN** mode and the TIA Portal project will be loaded so that it can be executed:

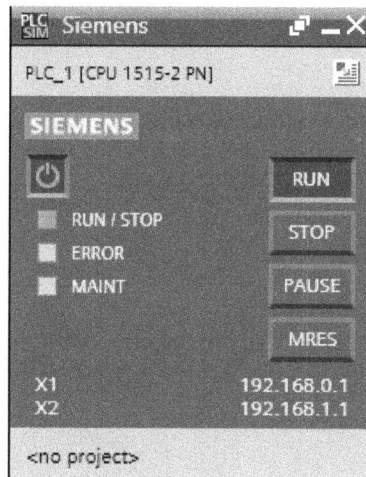

Figure 7.8 – The S7-PLCSIM instance running

The simulated PLC will now execute. At this point, the PLC will behave as a physical CPU would behave. I/O, networking, and other connectivity features will be unavailable.

Note that the **X1** and **X2** network interfaces are now populated with the IP addresses that have been configured for the PLC. However, they can't be accessed via TCP connections.

Managing simulated inputs

S7-PLCSIM does not provide a solution for managing the simulation of input signals. It is up to the programmer to choose one of the following solutions:

- Create a **watch table** and modify its input signals
- Create an **input mapping layer** with dedicated simulation data

These two approaches both have strengths and weaknesses, but creating an input mapping layer is the correct choice when you wish to create **standard control objects** and **UDTs**.

Using watch tables to change inputs

In the **Project tree** area, open the **Watch and force tables** folder. Double-click the **Add new watch table** item; a new watch table called **Watch table_1** will be created:

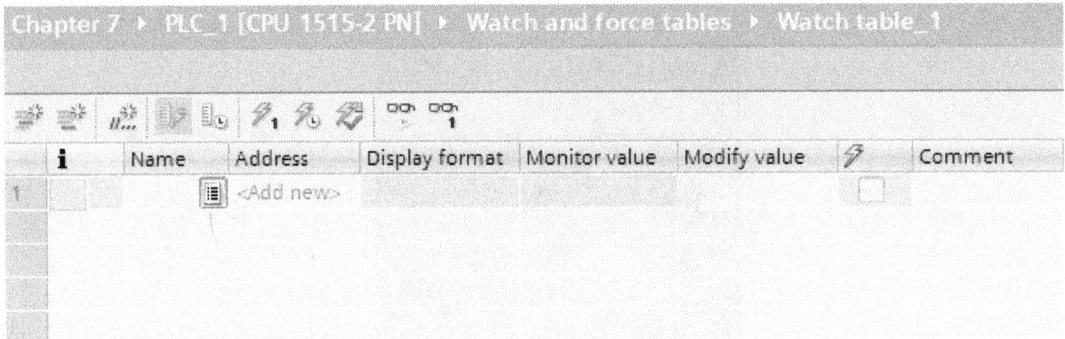

i	Name	Address	Display format	Monitor value	Modify value	🕭	Comment
1		📋 <Add new>				☐	

Figure 7.9 – The watch table view

The new watch table will open with no tags to be monitored. In the **Address** column, an **Input address** can be added that you can monitor.

> **Note**
>
> When input cards and remote I/O are added to a CPU in the **Device configuration** window, input memory addresses (in the form of tags) are required to map physical signals to a symbolic reference. These memory areas are prefixed with %I. Therefore, a 16-channel digital input card may use input addresses %I0.0 to %I1.7. These addresses can be added to the watch table and monitored.

Once tags/addresses have been added to the watch table, the value at the address can be monitored by clicking the **Monitor all** button from the toolbar above the watch table:

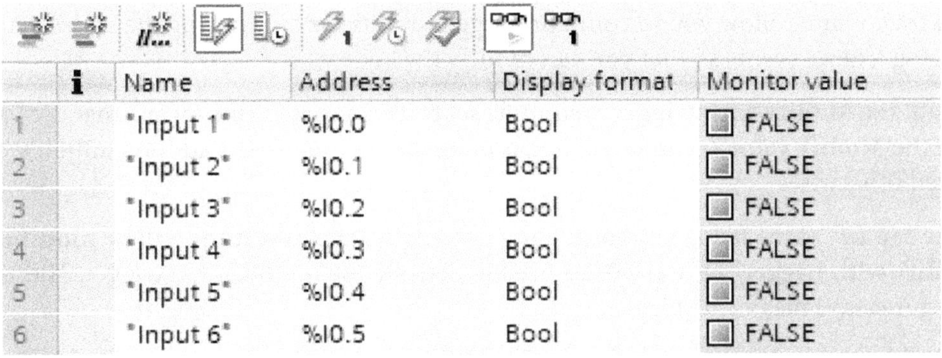

Figure 7.10 – Watch table in monitor mode

The current values will be displayed in the **Monitor value** column. As with other areas in TIA Portal (such as **Data blocks**), the value in the **Monitor value** column can be modified by right-clicking and selecting **Modify** and then selecting or entering the value that is required.

With input addresses, this approach does not work. This is because every time the PLC finishes a scan cycle, the input memory is written to by the physical hardware layer. Because only the CPU is being simulated, the value that's written back is always the default value for the associated data type. In the case of a 16-channel digital input card, this would mean all the values would return `False` at the start of the scan. If the modified value is changed to `True`, it may only be `True` for one scan.

To change this behavior, we must change the type of modification process we are using. You can do this by selecting the **Show/hide expanded mode columns** button from the toolbar above the watch table:

Figure 7.11 – Expanded mode columns button enabled

The icon of a list with a small clock to the right of it depicts the **Show/hide expanded mode columns** button. When this is enabled, some additional columns will appear in the watch table:

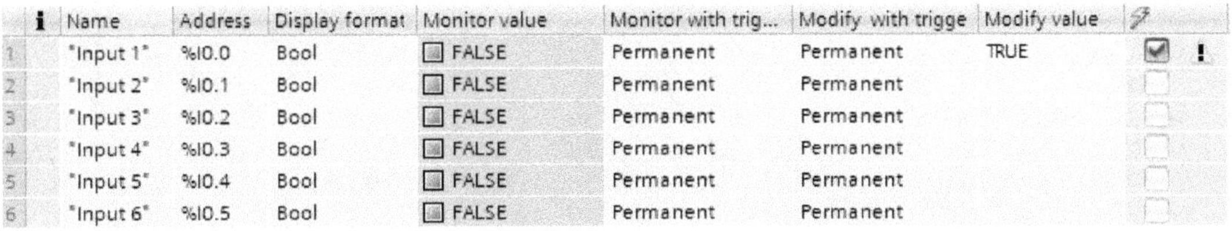

Figure 7.12 – Watch table view with expanded mode columns enabled

These extra columns allow you to configure trigger events for monitoring the address and configuring trigger events so that you can modify the address.

By default, the **Modify with trigger** column is set to **Permanent**. This means that the value in the **Modify value** column will be permanently written to the address until it's instructed otherwise.

Once the **Modify value** field contains the required data, the **All actions will be modified by "modify with trigger"** button can be used to activate the modification trigger. The following message may be displayed:

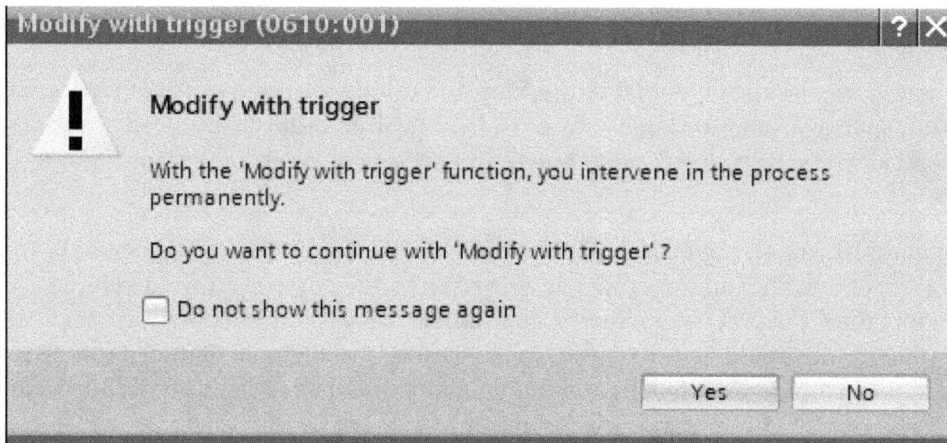

Figure 7.13 – The Modify with trigger dialog

This dialogue explains that the watch table will intervene in the process, permanently setting all the values that have been checked for modification to the **Modify value** column.

Click **Yes** to start modifying the values. Those values will now show the requested modify value.

Once the modification by trigger process starts, it will need to be deactivated with the **All actions will be modified by "modify with trigger"** button again before the updating will stop.

> **Note**
>
> Forcing values via a force table works similarly, but a warning will appear on the CPU's MAINT indicator while a force is active. The difference between the **watch table** and the **force table** is that going offline from the CPU in a force table will not stop the variables from being forced into a value.
>
> With a watch table, a programmer must remain connected and online to the CPU to continue writing the modified value.

Using an input mapping layer to change inputs

A more scalable solution, compared to using a watch table, is to use a specified layer to bring inputs into the program:

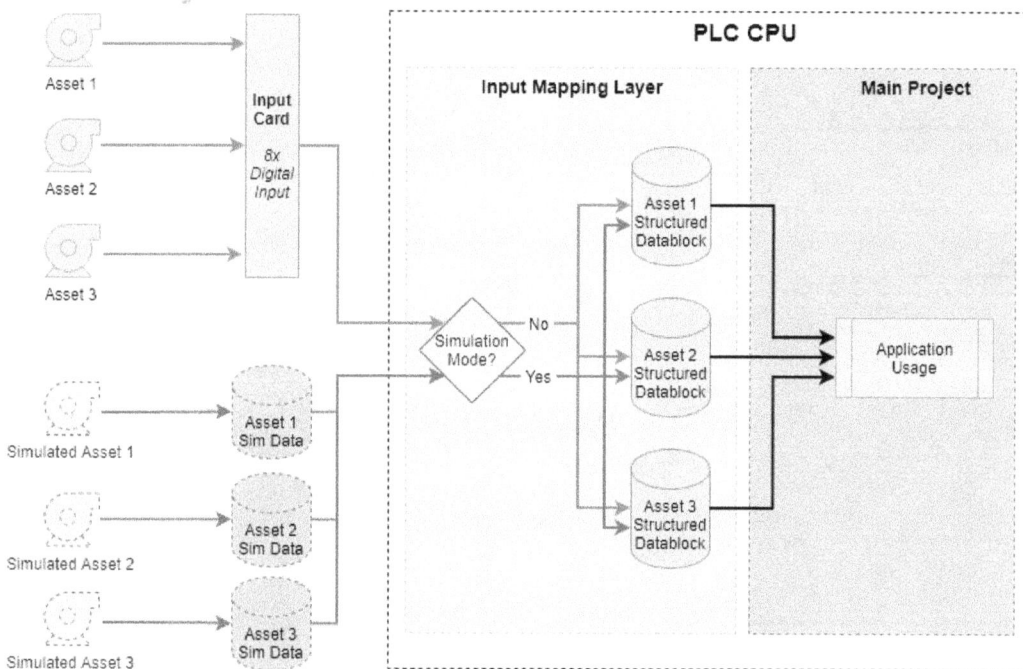

Figure 7.14 – The input mapping layer and its simulation

The preceding diagram shows an example of using an **input mapping layer** and **simulation data**. This method segregates data that comes from the physical input layer and simulation data that comes from asset-dependent data blocks.

This approach means that simulation data can be used in place of the real input data that will be used when it's not in simulation mode:

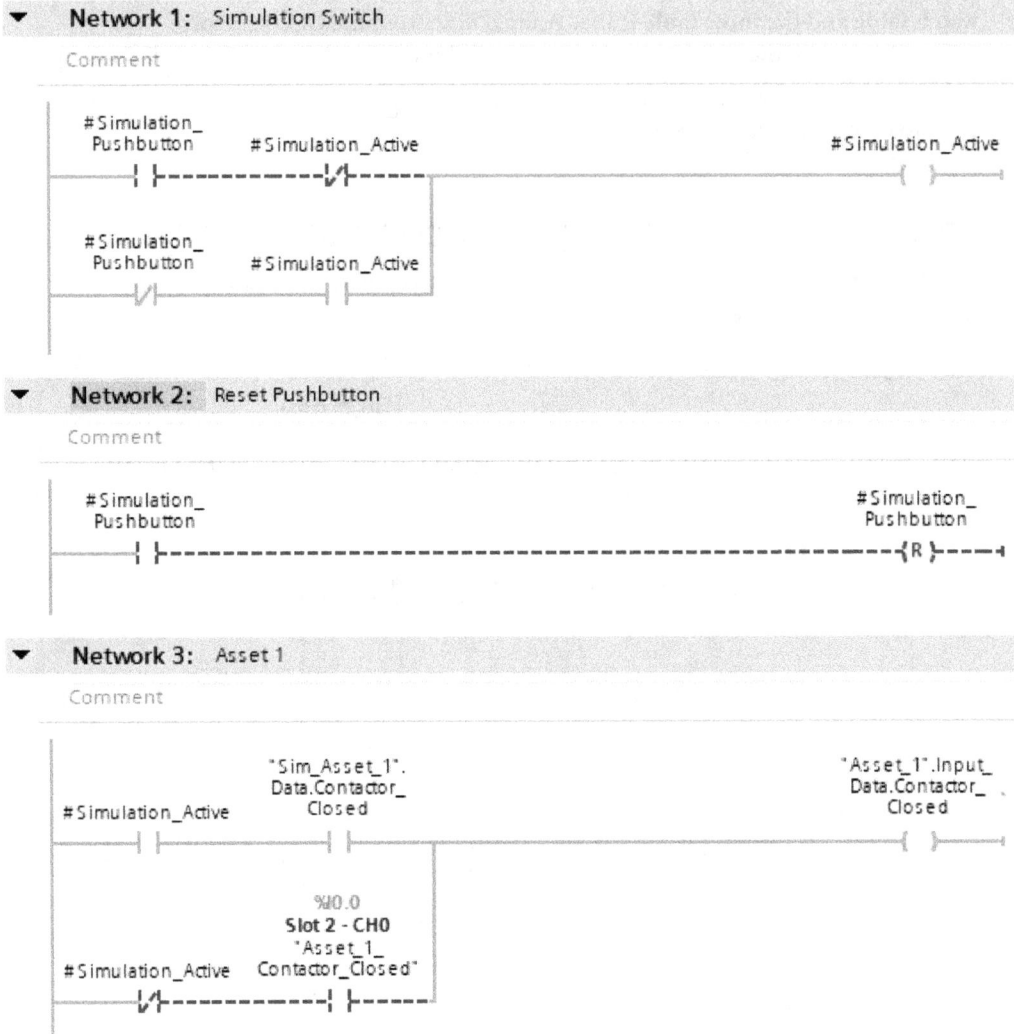

Network 1: Simulation Switch

Comment

```
  #Simulation_
   Pushbutton      #Simulation_Active                                    #Simulation_Active
  ──┤ ├──────────────┤/├──────                                            ──( )──

  #Simulation_
   Pushbutton      #Simulation_Active
  ──┤/├─────────────┤ ├──
```

Network 2: Reset Pushbutton

Comment

```
  #Simulation_                                                          #Simulation_
   Pushbutton                                                            Pushbutton
  ──┤ ├───────────────────────────────────────────────────────────────────(R)──
```

Network 3: Asset 1

Comment

```
                  "Sim_Asset_1".                                      "Asset_1".Input_
                  Data.Contactor_                                      Data.Contactor_
  #Simulation_Active    Closed                                              Closed
  ──┤ ├──────────────┤ ├──                                               ──( )──

                     %I0.0
                   Slot 2 - CH0
                   "Asset_1_
  #Simulation_Active  Contactor_Closed"
  ──┤/├──────────────┤ ├──
```

Figure 7.15 – The input mapping layer with active simulation

The preceding diagram shows how to use the input mapping layer and an active simulation:

- **Network 1**: Manages the simulation's status:

  - This network checks whether the `Simulation_Active` variable is `True` or `False`.

- **Network 2**: Resets `Simulation_Pushbutton`:

  - This network resets `Simulation_Pushbutton` if it's active. This ensures that when the next scan is done, the simulation won't be turned off again.

- **Network 3**: Maps simulation data or *real* data:

  - Based on the status of `Simulation_Pushbutton`, either the data from the physical I/O layer is used or the simulation data is used.

  - Regardless of the status of `Simulation_Active`, the data that's used is written to the `Asset_1` data block. This means that the `Asset_1` data block does not distinguish between *real I/O* data and *simulation* data.

> **Note**
> It is important to recognize that the data that's used for the project is the same data that's used when the simulation is active and not active. By moving the mapping of the I/O outside of the main project process, the process logic cannot be changed by changing the mapping. This has the advantage of guaranteeing that the logic behaves and reacts to input changes in the same way for *real I/O* data and simulation data.

There are some key benefits to implementing the input mapping layer approach when it's coupled with simulation data:

- Simulation data can be connected to a **Human Machine Interface (HMI)**:

  - You can use an HMI to make interacting with the system even easier.

  - This can help facilitate commissioning/testing, especially with numerical inputs for analog input data.

- Simulation data can be turned on and off easily.

  - The entire simulation system hinges on the `Simulation_Active` variable. If it is set to `False`, *real I/O* data is used. If it is set to `True`, simulation data is used.

- Standard control objects can be created for the mapping layers:

  - Asset mapping can be condensed into standard control objects, where *real I/O* data and simulation data are passed as structures, and mapping is performed inside the standard control object.

  - This has the benefit of speeding up writing time and ensuring that assets are handled the same way in all instances.

Advanced simulation using standard control objects

An input mapping layer, with a simulation system and structured data, is a recipe for advanced simulation. By creating standard input mapping control objects (blocks), structures for simulation can be used alongside the standard structures for asset data:

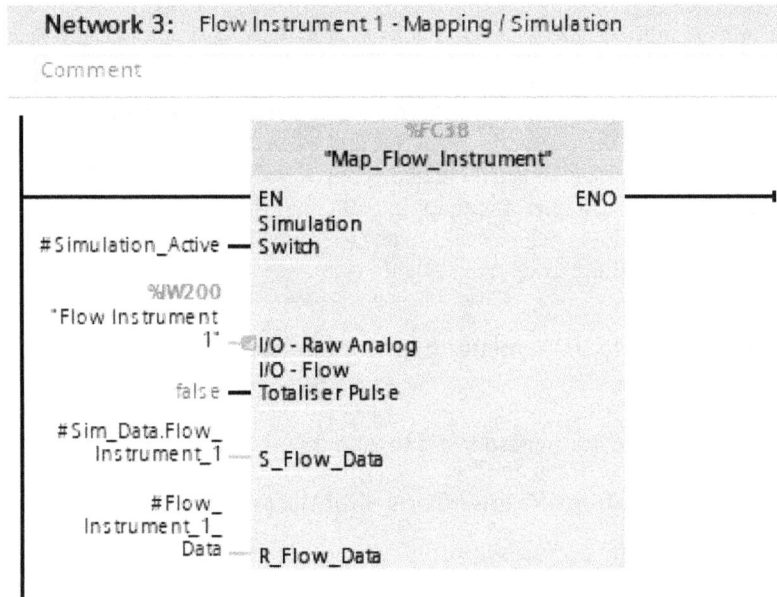

Figure 7.16 – Example of a standard mapping block for a flow instrument

This type of standard control object can ensure that all the assets of the same type follow the same conventions:

- Assets are mapped in the same manner:

  - This is controlled by the Map_Flow_Instrument block in the preceding diagram.

- Assets all contain the same datasets:

  - This is controlled by the fact that the mapping block that's being used requires both simulation data (S_Flow_Data) and real asset data (R_Flow_Data).

When the Simulation_Active variable is set to True, Map_Flow_Instrument moves the data contained in Sim_Data.Flow_Instrument_1 into the relative locations in Flow_Instrument_1_Data. When Simulation_Active is set to False, Flow_Instrument_1_Data is updated with the I/O values from the interface.

> **Note**
>
> The S_Flow_Data and R_Flow_Data interface variables are both **InOut**, which means that pointers are used to reduce the overhead of copying data.

Creating a simulation interface

Having an interface for the simulation system is vital if you wish to manipulate your simulation signals with ease. When the TIA Portal project is tested, simulation signals will need to be set to particular values to test how the production code reacts to the simulated input values. While this can be done by creating a watch table, it would be far more beneficial to the users of the simulation system to create an HMI in TIA Portal that interacts directly with the simulation code:

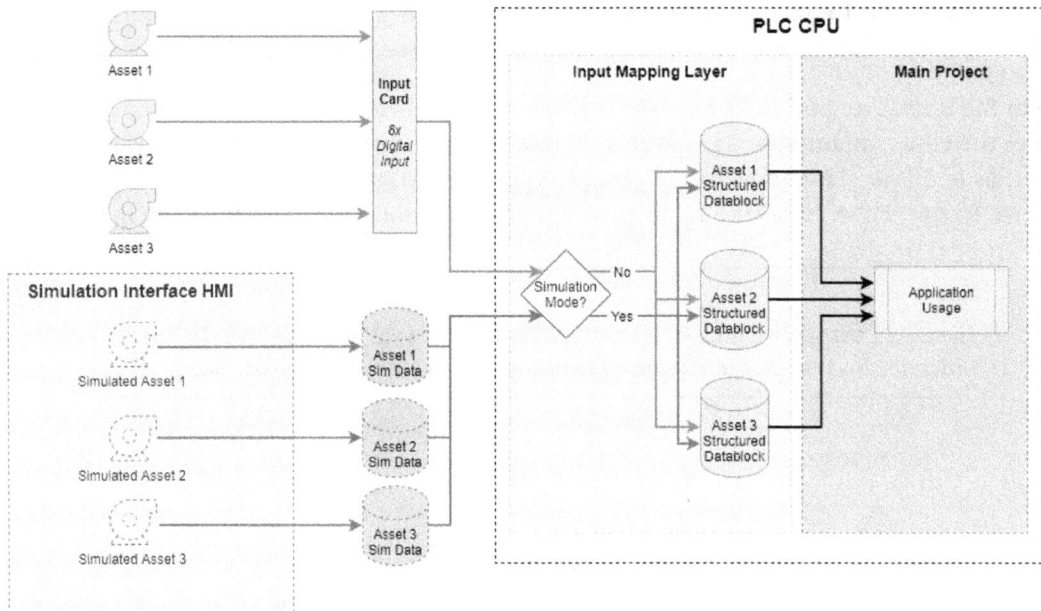

Figure 7.17 – The input mapping layer with simulation and an associated HMI

The concept behind this approach is that each asset has a dataset defined as a UDT that's either stored in a *data block* or as a variable in a wider simulation data block.

Simulation Interface HMI is connected to **Asset Sim Data** via a standard Siemens connection interface. This enables the HMI to change values and specify what is currently being sent to the PLC's **Input Mapping Layer**.

Configuring a simulation HMI in TIA Portal

To use an HMI with the simulation data, one needs to be added to the project. This can be done in the **Project tree** area by double-clicking **Add new device**. Select an HMI to use and add it to the project:

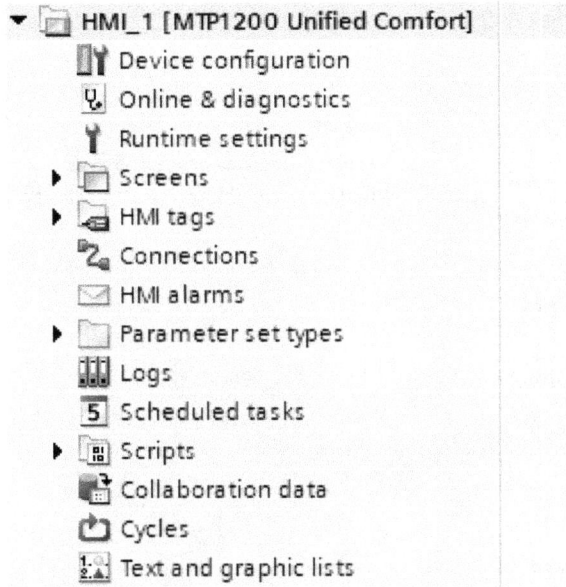

Figure 7.18 – Unified Comfort HMI added to a project

Once the HMI has been added, HMI tags that point to the simulation data can be created:

Default tag table

	Name ▲	Data type	Connection	PLC name	PLC tag
▶	Asset_1_Sim	UDT_Pump_Input_Data	HMI_Connection_1	PLC_1	Sim_Asset_1.Data
▶	Asset_2_Sim	UDT_Pump_Input_Data	HMI_Connection_1	PLC_1	Sim_Asset_2.Data
▶	Asset_3_Sim	UDT_Pump_Input_Data	HMI_Connection_1	PLC_1	Sim_Asset_3.Data

Figure 7.19 – HMI tags mapped to PLC tags

The PLC tags that are being used should point to the structure of the asset's simulation data. The PLC simulation structure should consist of the same I/O data that is to be used in the application.

In the HMI, create a **Faceplate** by opening the **Libraries** tab to the right of TIA Portal, expanding the **Project Library** > **Type** folder, and double-clicking **Add new type**:

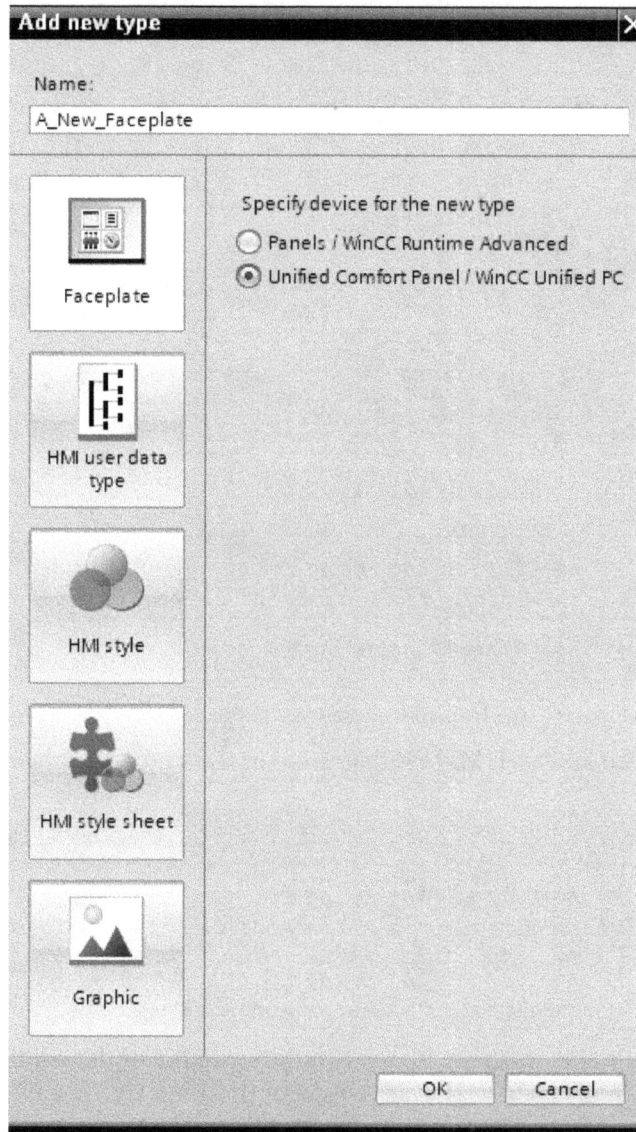

Figure 7.20 – Adding a new faceplate

Choose **Faceplate** from the dialog that opens, ensuring that the correct HMI type is selected. A new **Faceplate** will be created, and the **Visualization** tab will be displayed. The controls for your **Faceplate** can be placed here, just as a normal screen would have them placed, although not all the controls will be available:

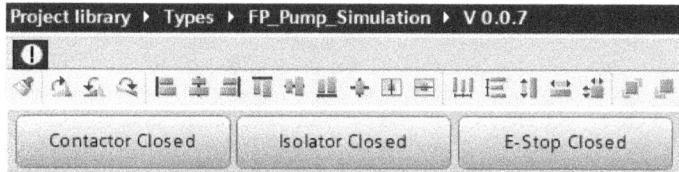

Figure 7.21 – Faceplate visualization design

Once the faceplate's controls have been placed and the size of the faceplate has been defined, you can create its interface. Because the simulation data that's being used is in the form of a UDT structure, only one tag interface is required. By clicking the **Tag interface** tab, the interface element can be assigned to the faceplate:

Name	Data type	User data type structure
Sim_Data	Struct	UDT_Pump_Input_Data V 0.0.1

Figure 7.22 – Tag interface assignment in the Tag interface tab

This interface data type is a **struct**, which is then further defined in the **User data type structure** column as UDT_Pump_Input_Data. This is the same UDT that is being used in the PLC for the simulation data.

> **Note**
>
> To use a UDT in the HMI, it must exist in the same project library. This means that all the UDTs that are being used in the HMI must be version-controlled. If the UDT changes, the faceplate must be updated manually as the old UDT version will be used until it's updated.

Once the tag interface has been created, data from the interface can be assigned to various controls:

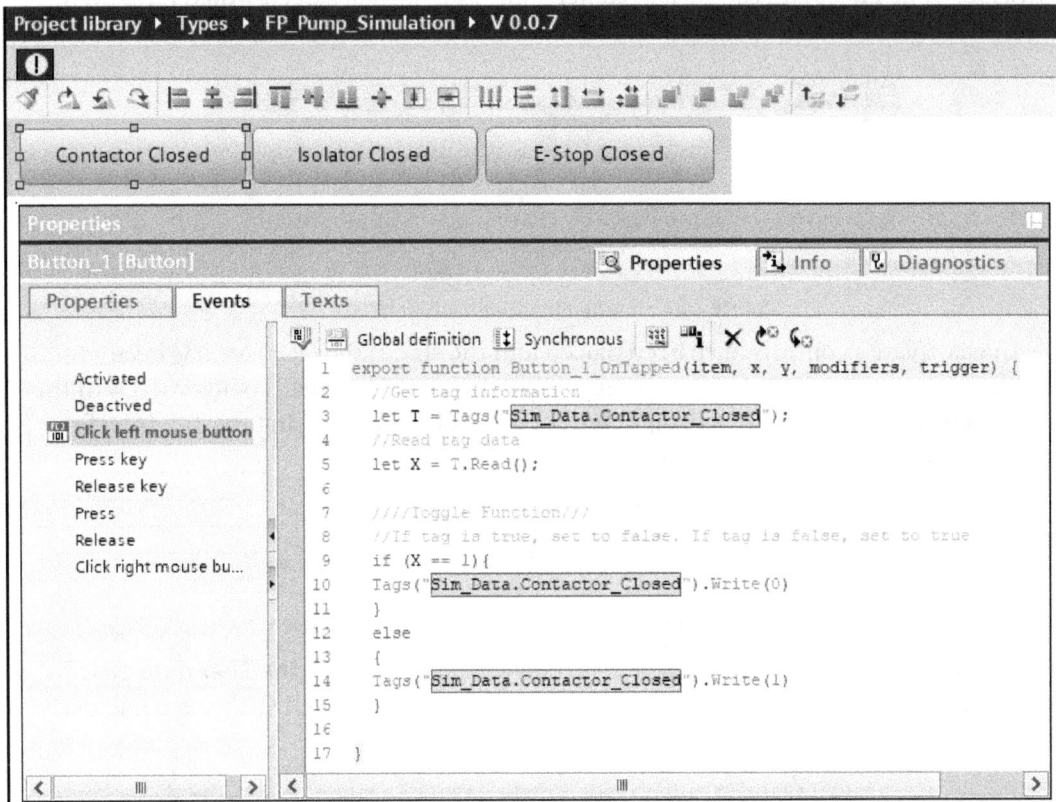

Figure 7.23 – Assigning JavaScript to the Contactor Closed toggle button

By accessing the `Sim_Data` property of **Tag Interface Element**, the faceplate will have access to the outside instance of the UDT and can access the variables within it.

On the screen where the faceplate will be used, **Interface** must be set with the appropriate **HMI tag**:

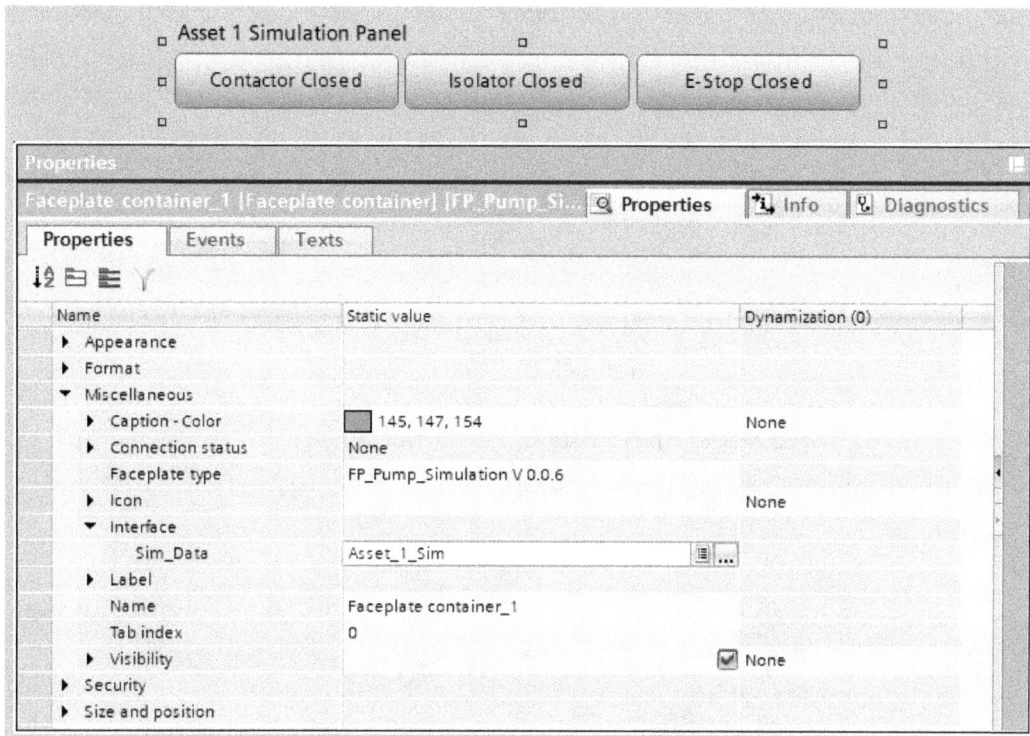

Figure 7.24 – The Tag Sim_Data interface has been assigned the Asset_1_Sim HMI tag

With that, we have connected the PLC simulation data to a simulation HMI that can interact with that data directly.

The following diagram shows the connections that have been made here:

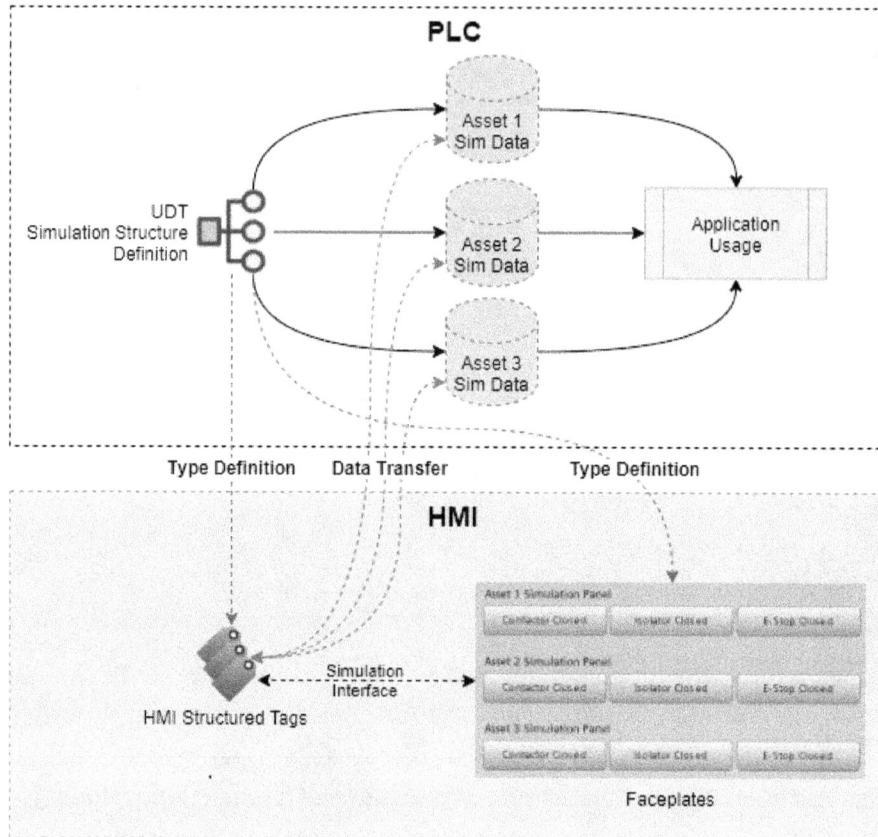

Figure 7.25 – The simulation HMI concept

The preceding diagram helps visualize how connections are being made between the PLC and the HMI and how the UDT is the structure definition that ties the two systems together.

> **Note**
>
> Physical devices are not required for testing. Remember that both the PLC and the HMI can be simulated by clicking the **Simulation** button at the top of TIA Portal.
>
> Unified HMIs can be accessed in a standard web browser using the `127.0.0.1` IP address.

Safeguarding outputs when in simulation mode

When any automation system is running in simulated mode, **outputs** should be protected and made safe so that when the simulation is running, they aren't processed.

This can be achieved with an **output mapping layer**. Much like the input mapping layer, if the simulation system is active, the output memory addresses are filled with data that ensures they are in a *safe* state:

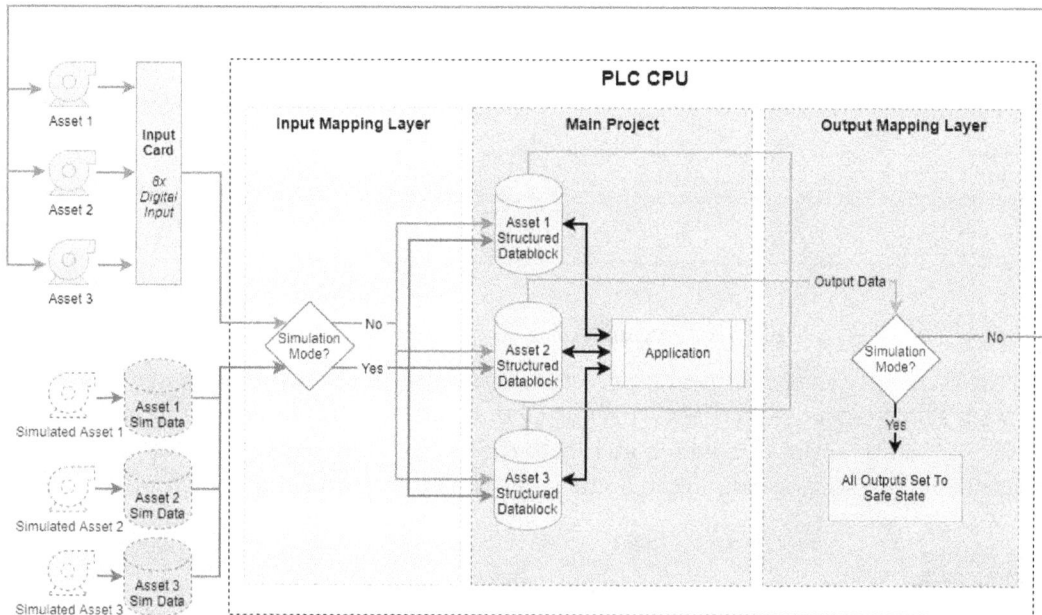

Figure 7.26 – Setting outputs to a safe state when in simulation mode

> **Note**
>
> The preceding diagram shows how all the physical data, both **input** and **output**, can be stored in the asset's dedicated structured data block. Preparing output data in the asset's dataset is a good way of encapsulating the process data and segregating the input and output layers of the project.
>
> This means that our **Input Mapping Layer** could be changed to a network protocol such as **Modbus** and our **Output Mapping Layer** could be changed to a remote I/O solution. This can be done without changing any core logic in the **Main Project** section.

With outputs set to a `safe` state, simulation data cannot damage any of the equipment involved in the project. The following screenshot shows an example of an output mapping layer:

```
1   #Zero := 0;
2   IF NOT "H1_HSim_Select".Master_Sim_Sel THEN
3
4       "H1_M1_Node26A_O-1" := "E501".Typed.UV_System.Run_Output_To_UV_System;
5       "H1_M1_Node11_AO" := "AV302".Typed.PQW_Word;
6       "H1_M1_Node15_AO" := "IF501".Typed.Raw_Value;
7
8       //P501
9       #Temp_INT := BLKMOV(SRCBLK := "P501_G120".Typed.Control_Telegram, DSTBLK => P#Q1840.0 Byte 20);
10
11      //RO Mapping
12      "Output_Mapping_R1A"();
13      "Output_Mapping_R1AN"();
14
15  ELSE
16
17      #Temp_INT := FILL(BVAL := #Zero, BLK => P#Q0.0 Byte 3000);
18
19
20  END_IF;
```

Figure 7.27 – Example of an output mapping layer

When the `Master_Sim_Sel` variable is set to `True`, the `IF` statement becomes `False`, and the `ELSE` statement is executed. This fills all the Q addresses (physical outputs) with the value in the `Zero` variable, which was set to the value of 0 in the first line of the logic. This ensures that all the outputs are in a safe state when they're in simulation mode.

Summary

This chapter explored the concepts behind using standardized structures in simulations, as well as how to get starting with the built-in simulation platforms that TIA Portal offers.

Whether programmers are building large-scale bespoke projects or small-scale standard projects, the principles behind structuring data ensure that simulating any project is easier. It also means that no additional software is required.

In the next chapter, we will highlight some of the difficulties of standardizing code and data and what can be done to ensure that flexible options are available when projects deviate from the standards. We will also look at the dangers of using InOut interfaces with HMI/SCADA data and how to mitigate those risks.

8
Options to Consider When Creating PLC Blocks

This chapter explores additional options to consider when creating **programmable logic controller** (**PLC**) blocks. There are many different approaches that programmers can take when designing, creating, and then implementing logic solutions. This chapter explores how to retain standardization while offering a flexible approach.

In this chapter, the following topics are covered:

- Extending standard functions
- Extending standard data
- Managing data through instance parameters
- Asynchronous data access considerations

Extending standard functions

A well-built paradigm or design pattern should allow for the extension of standard objects without affecting the object itself. For example, standard objects should consist of other standard objects that can also be used outside of their parent object.

Have a look at the following screenshot:

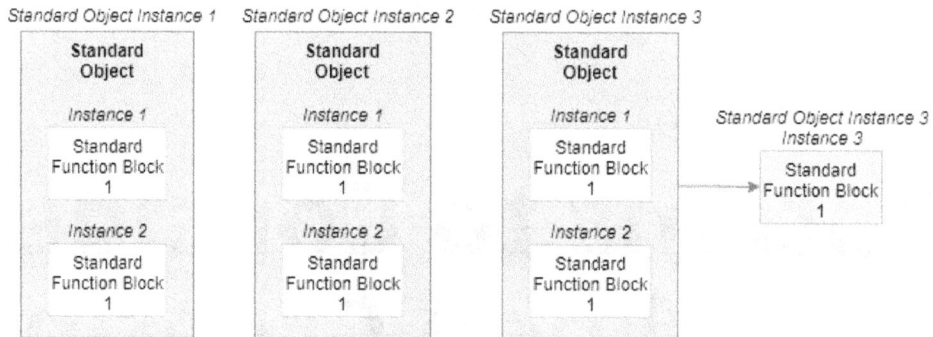

Figure 8.1 – Example of an extension

Figure 8.1 shows an example of an extension whereby **Standard Object Instance 3** has an additional instance of **Standard Function Block 1** that is used outside of the normal parent. Because all instances of **Standard Object** use **Standard Function Block 1**, and **Standard Function Block 1** is in itself a standard, it can be used outside of **Standard Object**, and data that would normally be passed to **Standard Function Block 1** from the **Standard Object** interface can be passed directly instead.

If the example in *Figure 8.1* were to be replicated in **Totally Integrated Automation Portal (TIA Portal)**, the **Standard Object** interface might look something like this:

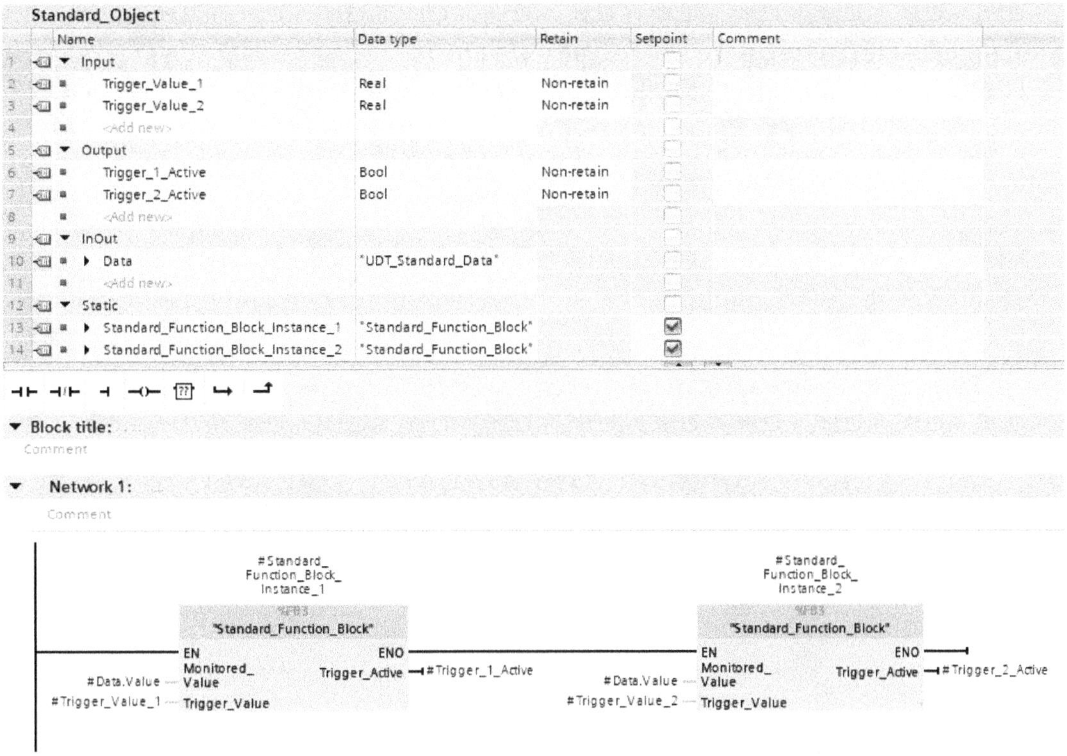

Standard_Object					
	Name	Data type	Retain	Setpoint	Comment
1	▼ Input				
2	Trigger_Value_1	Real	Non-retain		
3	Trigger_Value_2	Real	Non-retain		
4	<Add new>				
5	▼ Output				
6	Trigger_1_Active	Bool	Non-retain		
7	Trigger_2_Active	Bool	Non-retain		
8	<Add new>				
9	▼ InOut				
10	▶ Data	"UDT_Standard_Data"			
11	<Add new>				
12	▼ Static				
13	▶ Standard_Function_Block_Instance_1	"Standard_Function_Block"		✓	
14	▶ Standard_Function_Block_Instance_2	"Standard_Function_Block"		✓	

▼ **Block title:**
Comment

▼ **Network 1:**
Comment

Figure 8.2 – TIA Portal example

Figure 8.2 shows the **Standard Object** interface from *Figure 8.1* realized in TIA Portal. This object can then be called in a parent object, and the two `Standard_Function_Block` instances will be executed with data from the parent interface.

If a particular instance of the **Standard Object** interface needs extending, this is done in the **parent object**. This is the object in which the **Standard Object** instance exists.

The extension also uses the same data that the block it is extending uses. When using **user-defined types** (**UDTs**), it may be possible that all of the variables required to extend functionality reside within the UDT; this may mean that only a single InOut interface is required.

Have a look at the following screenshot:

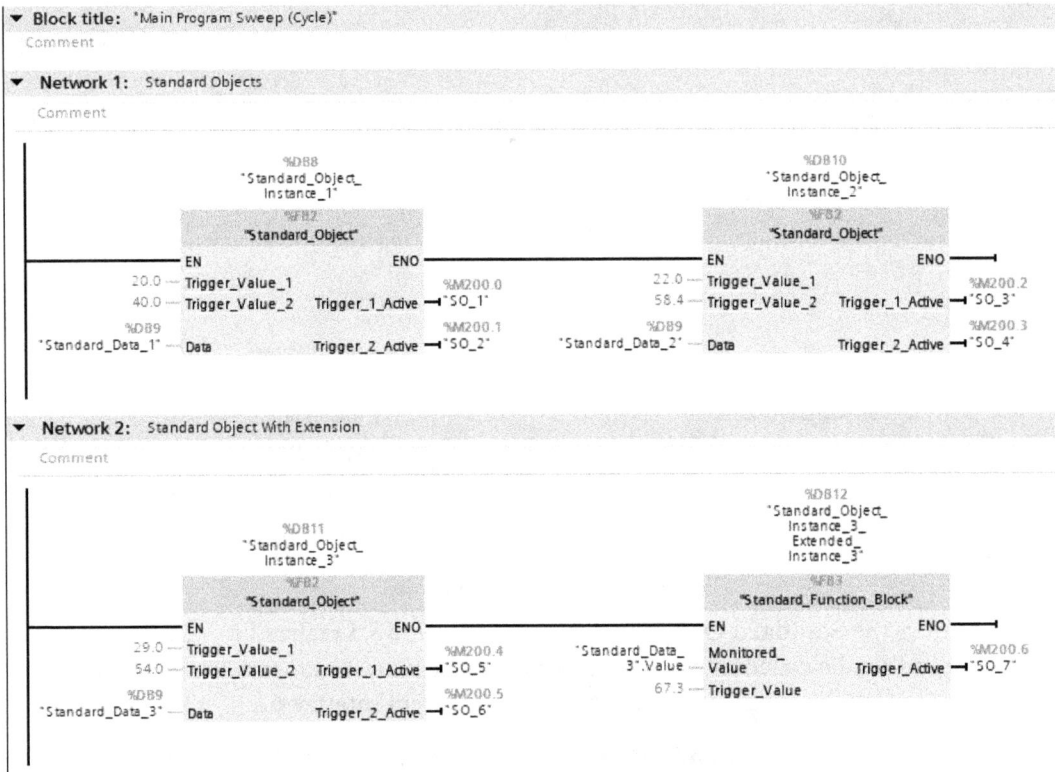

Figure 8.3 – Extension in TIA Portal

Figure 8.3 demonstrates a typical extension requirement. **Network 1** contains two Standard_Object instances that are not extended and function as intended when developed. **Network 2** contains a single instance of Standard_Object and an additional Standard_Function_Block object (which *Figure 8.2* reveals is called twice in Standard_Object).

The Standard_Object_Instance_3_Extended_Instance_3 instance of the Standard_Function_Block object is the extension of Standard_Object_Instance_3. Both blocks in **Network 2** are using data from Standard_Data_3.

While this is only a small example of an extension, the principle is the same no matter the size of the extension. By taking functions that a **Standard Object** interface utilizes and using them alongside the standard object—with data from the same source—the **Standard Object** interface effectively has access to an additional instance of a function.

Extending standard data

Similar to program blocks, standard **data blocks** that use UDTs or structs can also be easily extended without breaking standardization. For this to work, a data block needs to be structured in a particular way.

Have a look at the following screenshot:

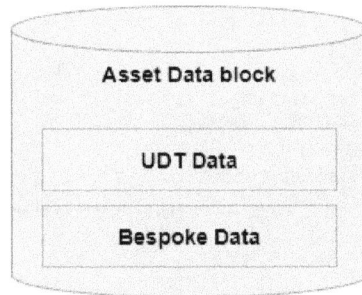

Figure 8.4 – Example of a data block structured with UDT and bespoke data

Figure 8.4 is an example of a data block that consists of the following two areas:

- **Asset UDT Data**—This is the data that standard objects will access.
- **Asset Bespoke Data**—This is an *extension* of the standard data.

Now, have a look at the following screenshot:

Figure 8.5 – Example in TIA Portal data blocks

Figure 8.5 demonstrates this pattern in data blocks. Asset_1 has the UDT defined as a variable named Typed, and that is the only variable that appears in the data block. This means that any standard object that requires data from Asset_1 can access the UDT under the Asset_1.Typed path.

Asset_2 has an additional Struct extension that holds data that is not available in the UDT associated with the asset but is still related to the asset. The Bespoke struct contains an additional variable, External_Interlock. This extended data can be used as part of the extension of a standard object, to change or enhance behavior.

Next, have a look at the following screenshot:

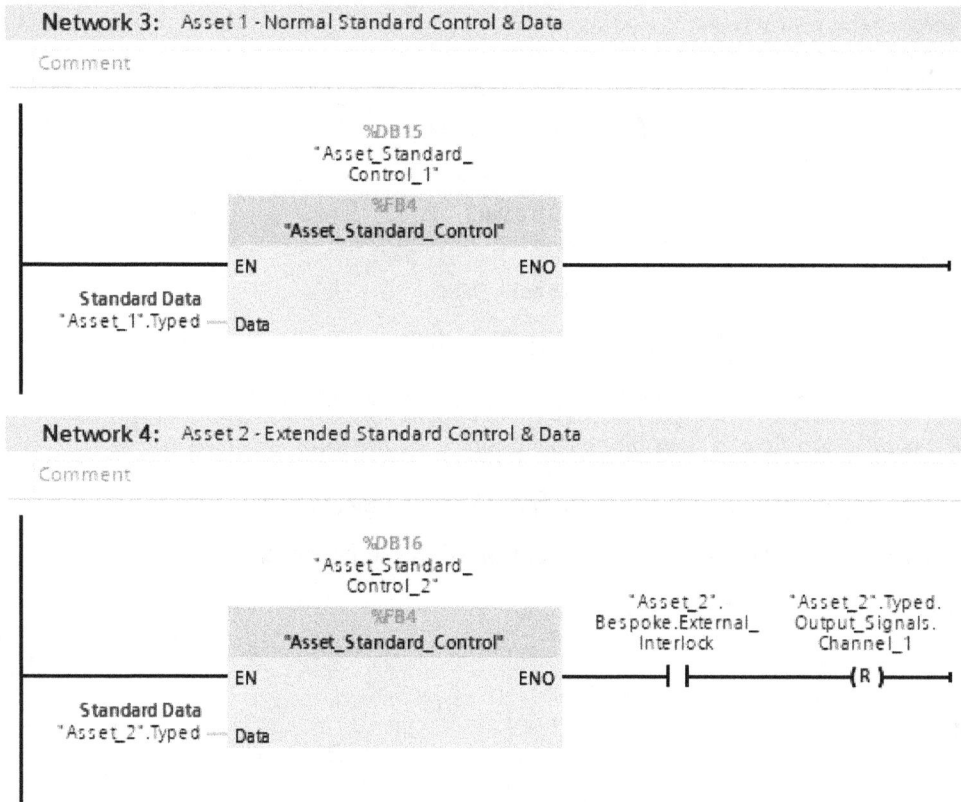

Figure 8.6 – Example of both extended control and data

Figure 8.6 demonstrates this approach in a practical example. **Asset 1** on **Network 3** passes the standard UDT data from the `Asset_1` data block, through the InOut interface of the `Asset_Standard_Control` function block, and the block then performs its logic on the data and passes it back through the interface.

Asset 2 on **Network 4** performs the same actions that are performed for **Asset 1**, as the same `Asset_Standard_Control` function block is called, only this time the data passed to it is `Asset_2` UDT data. Once the standard object has been executed, the extension (which is a loose ladder after the `Asset_Standard_Control` function block) is then processed. The **contact** checks if `Asset_2.Bespoke.External_Interlock` is True and, if so, sets `Asset_2.Typed.Output_Signals.Channel_1` to False.

This means that the code has changed the standard UDT data, using an extended method. If `Asset_2.Typed.Output_Signals.Channel_1` had been set to True in the `Asset_Standard_Control` function block, the extension logic would have overridden that to False before the rest of the logic continued.

The use of this approach means that no new version of `Asset_Standard_Control` needed to be created, and no structural changes to the UDT were required either. This protects and simplifies managing large libraries with potentially thousands of blocks and structures.

Managing data through instance parameters

When creating standard objects that have multiple methods of operation, it is worth considering which resources may be shared between the methods that an object can use. Sharing resources has a positive impact on memory allocation and simplification of an object.

Principle to this approach

Consider a standard control object that can be run in one of three modes. Only one mode can be run at any one time. You can see an example of this in the following screenshot:

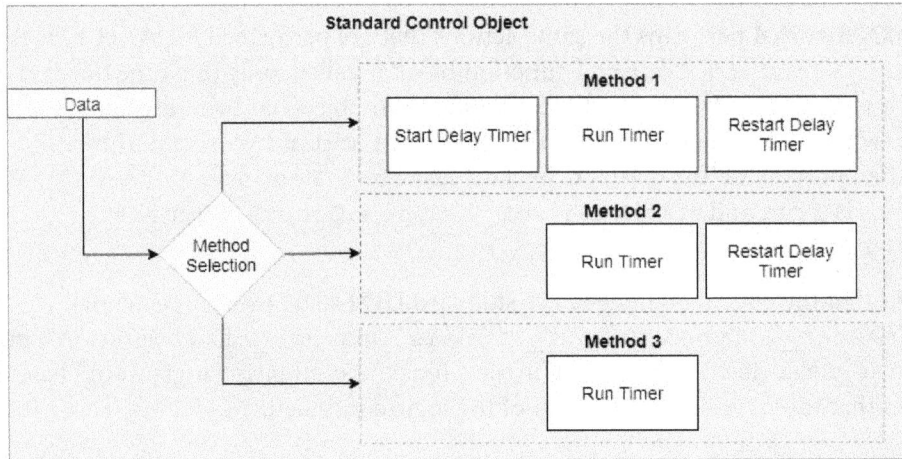

Figure 8.7 – Example layout of a standard control object with multiple control methods

The standard control object accepts a common data structure (UDT, for example) and contains logic to read the selected method. In *Figure 8.7*, the three methods contain different timers that are in use. All methods require at least one timer, with the largest method (**Method 1**) requiring three timers.

If a function block were written in this manner, it would contain six timers, even though it is not possible to run more than three at any one time.

It would be better to declare the timers as an external resource to the methods, as depicted here:

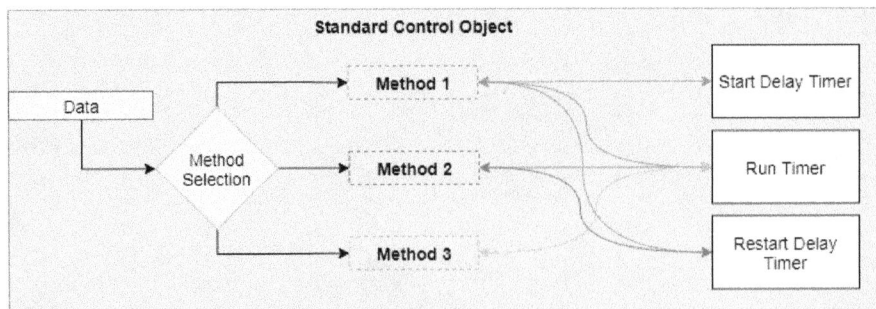

Figure 8.8 – Methods with common resource access

Figure 8.8 demonstrates methods 1, 2, and 3 all having access to timers that they require to function. This means that only three timers are required, no matter which method is in use.

This approach optimizes data and helps reduce the amount of complexity when maintaining the standard control object in the future.

TIA Portal example

This example consists of a `Standard_Control_Object` instance that contains an interface for a UDT called `Data` and three timers, listed here:

- `Start_Delay_Timer`
- `Run_Timer`
- `Restart_Delay_Timer`

The timers are all declared in the static scope of `Standard_Control_Object`, as illustrated in the following screenshot:

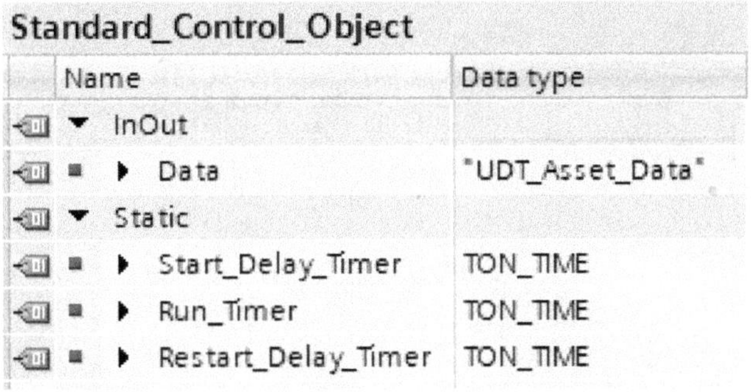

Standard_Control_Object	
Name	Data type
▼ InOut	
▶ Data	"UDT_Asset_Data"
▼ Static	
▶ Start_Delay_Timer	TON_TIME
▶ Run_Timer	TON_TIME
▶ Restart_Delay_Timer	TON_TIME

Figure 8.9 – Timers declared in static data

The methods are then called as **functions** in the logic of Standard_Control_Object, as illustrated in the following screenshot:

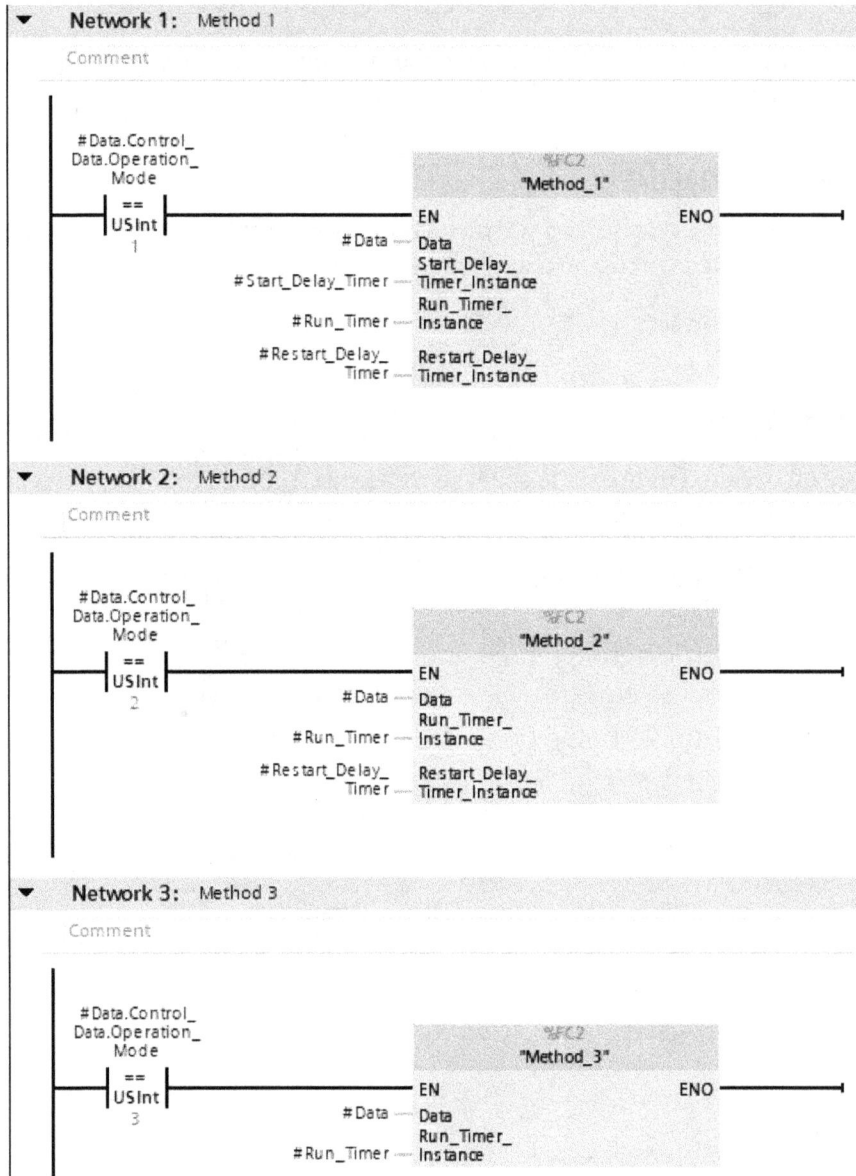

Figure 8.10 – Method 1 call

Figure 8.10 shows that Method_1 has three interface parameters, once for each timer required. The **static** data from the parent object (Standard_Control_Object) is passed through this interface, such that Method_1 has access to the timers.

This means that, as a function cannot hold instance data, the function's parent holds the instance data on behalf of the child object.

Now, have a look at the following screenshot:

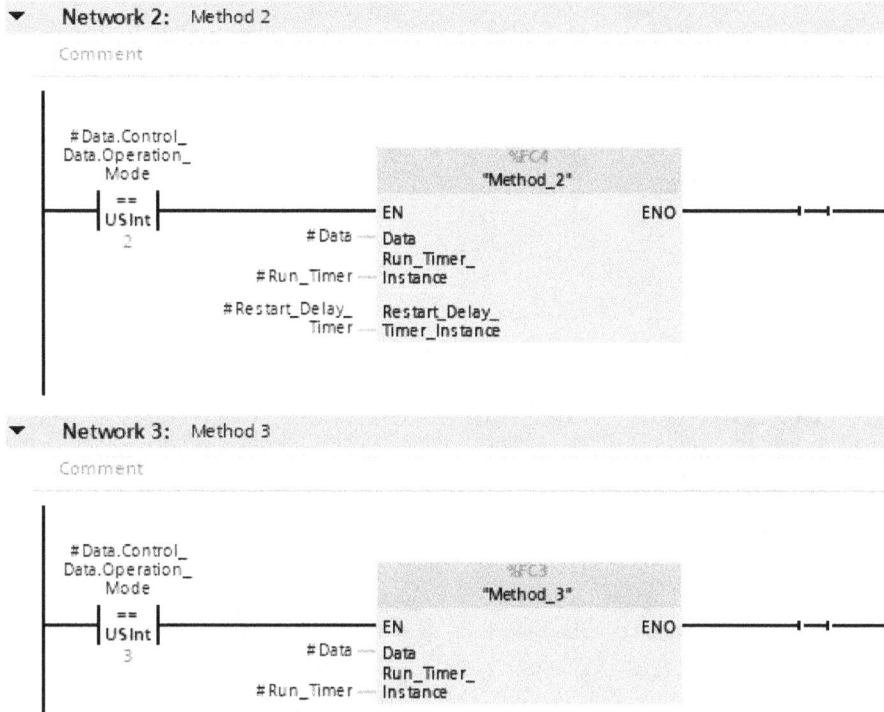

Figure 8.11 – Additional methods share the same timer instance data

Figure 8.11 shows that Method_2 and Method_3 also use the same instance data, from the same source. Because the methods cannot be run at the same time, it is safe to share the timers across different methods.

> **Note**
>
> Careful consideration should be taken when using instance parameters to ensure that it is safe to use the data more than once. It is also just as important to remember that data held by the parent object will not reset or update if no method is utilizing it anymore.

In this example, if the method being executed were to change while a timer was in use, the timer instance data would still contain the same values when the next method first accessed it. This could cause adverse effects depending on the required logic.

Function block interfaces

When creating instance data for a function block, the following window is displayed by TIA Portal to determine the *call* options:

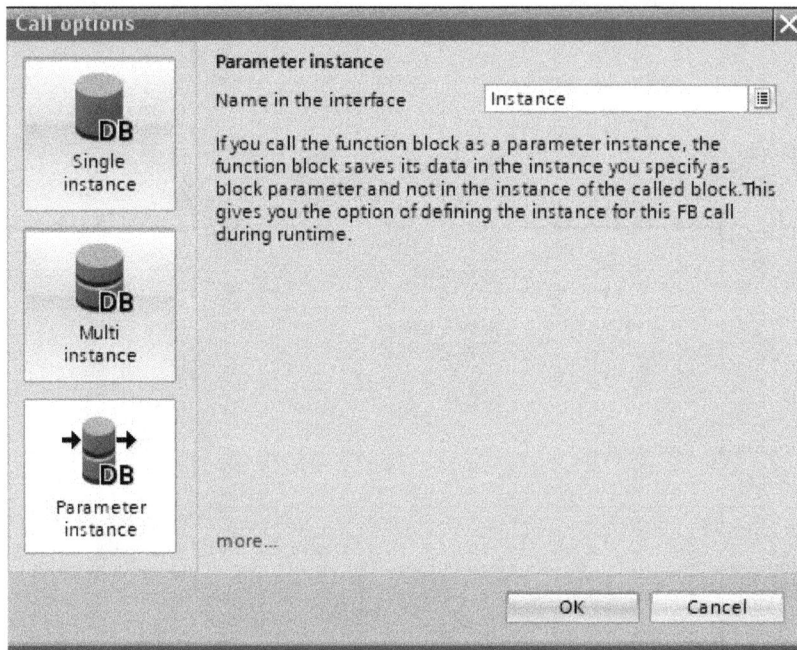

Figure 8.12 – Call options for database instance data

If an instance is to be stored in the **parent** object, as in the example in this chapter, then **Parameter instance** should be selected. This will create an InOut interface with the data type set as that of the instance data of the function block that is being called.

> **Note**
>
> **Parameter instance** is only available to be selected when a function block is called inside a function or another function block. When called in an organization block, this is not an available option.

Asynchronous data access considerations

It is important to remember that many modern **human-machine interfaces (HMIs)/ supervisory control and data acquisition (SCADA)** systems will access PLC data asynchronously. This means that instead of waiting till the end of the PLC scan to obtain or write data, the HMI/SCADA system will interrupt the scan to update or obtain data.

Normally, this is not an issue. However, when references via InOut interfaces are in use, it can cause unexpected behavior. This can be very difficult to diagnose because the next time data is updated, it may not be updated in the same place in the scan.

Have a look at the following screenshot:

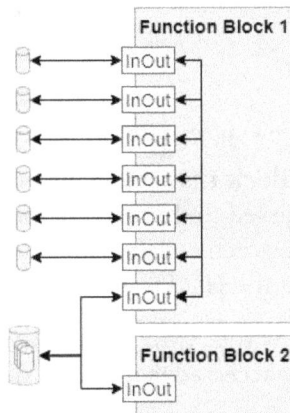

Figure 8.13 – Example of an erroneous configuration with InOut interfaces

Figure 8.13 is an example of a configuration that is valid by the compiler but introduces issues if data blocks are interfaced with an HMI.

The blue data blocks represent *asset data*—they are connected to InOut interfaces of **Function Block 1**. **Function Block 1** is moving the data contained in each *asset data block* to a consolidated data block (represented by the green data block in *Figure 8.13*) so that **Function Block 2** can utilize all assets on a single interface element.

The issue with this particular setup is that the InOut interface between the consolidated data block and **Function Block 1** is not actually InOut in the true sense.

Now, let's see what's happening here:

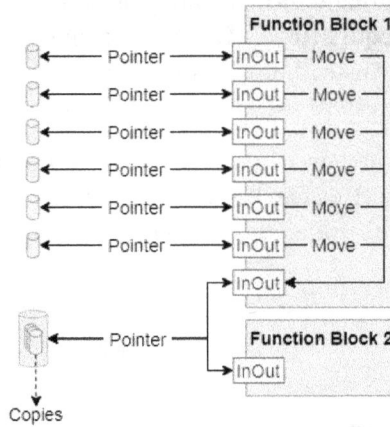

Figure 8.14 – Example of how copies of InOut-referenced data are created

The Move instruction in **Function Block 1** is actually dereferencing the data from **Pointer** and creating a copy, which is then passed to the relevant location in the consolidated data block. This means that all data in the green consolidated data block is a *copy* of the asset data and not the original dataset that the HMI is interfaced with.

Because of how the data is accessed by the HMI, the following example of a configuration cannot guarantee that the data being accessed is the latest data:

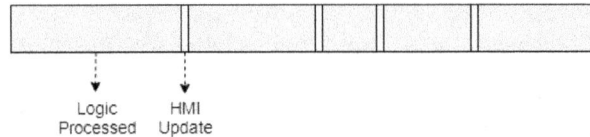

Figure 8.15 – Example of HMI asynchronous update

If **Function Block 1** performed the copy of the asset data after the first HMI update, the rest of the HMI updates would be missed by the PLC until the next update was performed, as illustrated here:

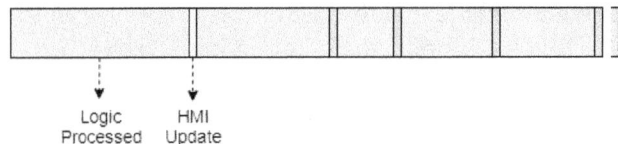

Figure 8.16 – Example of missed updates

This can result in strange behavior, especially if later on in the program, data is written back to the original data blocks. This would mean that data that has been updated by the HMI would revert to original values before the HMI update occurred.

The correct method

Instead of trying to consolidate large datasets into a common data block, it is always better to interface data blocks separately. This means that there may be many interfaces on a function block; however, it is safer from issues brought on by unexpected copy events. You can see an example of this in the following screenshot:

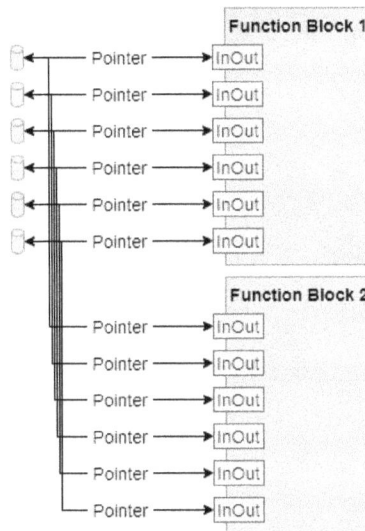

Figure 8.17 – Example of full interface

This approach safeguards against accidental copy events that dereference the pointer data. All function block interface elements point directly back to the original source data.

In this scenario, it does not matter when the HMI or SCADA system asynchronously updates data as all references would update at the same time.

Summary

This chapter has covered the topic of **extension**. This is when standard control objects' standard methods are extended outside of the standard control object, usually by using a child object that already exists.

This form of extension is a solution to adding methods to a project without changing the standardization of objects.

This chapter has also covered the extension of standard data. This allows asset-oriented data to consist of standard UDT data that is extended by bespoke data. This further supports the method extension by allowing non-standard setpoints and other data.

The next chapter moves away from PLC and over to the TIA Portal WinCC environment, in which HMIs and visualizations can be created to support projects.

Section 3 – TIA Portal – HMI Development

Learn how to develop HMI screens that interact with PLC data. Learn how to enhance this further with faceplates and structures.

This part of the book comprises the following chapters:

- *Chapter 9, TIA Portal HMI Development Environment*
- *Chapter 10, Placing Objects, Settings Properties, and Events*
- *Chapter 11, Structures and HMI Faceplates*
- *Chapter 12, Managing Navigation and Alarms*

9

TIA Portal HMI Development Environment

This and the next few chapters explore the visualization side of **Totally Integrated Automation Portal** (**TIA Portal**), creating **human-machine interfaces** (**HMIs**) and connecting them to data in associated programmable logic controllers (**PLCs**). HMIs are important to the overall feel of a project as they are the mechanism by which people interact with the project.

Programmers that can both develop a PLC application and the associated HMI applications are expected in today's working environment. TIA Portal allows the two different program environments to fall under one application, making it much easier for programmers to write both the PLC and HMI with ease.

In this chapter, the following topics are covered:

- Adding an HMI to a project
- HMI development environment overview
- Screen objects
- Special objects

TIA Portal Comfort Panel

In TIA Portal **Version 17 (V17)**, there are now two different types of Comfort Panel HMIs, as outlined here:

- SIMATIC Comfort Panel
- SIMATIC Unified Comfort Panel (new)

This chapter (and others that reference HMIs) will focus on the newest *Unified* Comfort Panel.

Adding an HMI to a project

Just as with a PLC, an HMI needs to be added as a device in a TIA Portal project. Once an HMI device has been added, different objects appear under the HMI object in the **Project tree** pane.

To add an HMI, the following steps will be used:

1. Double-click **Add new device** in the **Project tree** pane. This will open the **Add new device** window, as illustrated in the following screenshot:

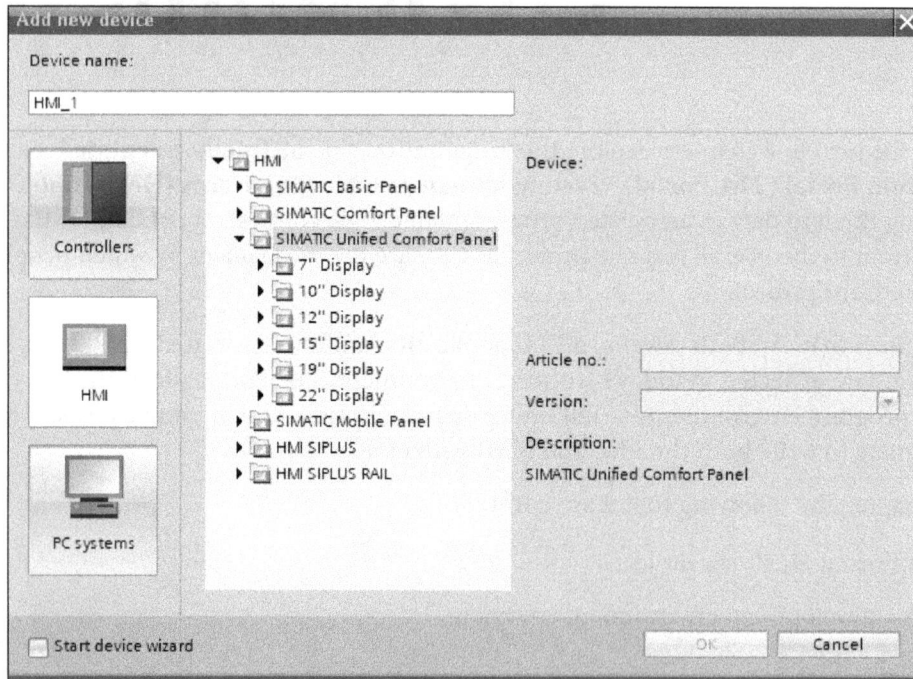

Figure 9.1 – Add new device window

2. Selecting **HMI** from the options in the left column will display the available HMIs
    in TIA Portal V17. **SIMATIC Unified Comfort Panel** is the latest version of
    Siemens' HMIs.

3. Select a display size and open the corresponding folder. Inside will be the device
    that is to be added to the project, as illustrated in the following screenshot:

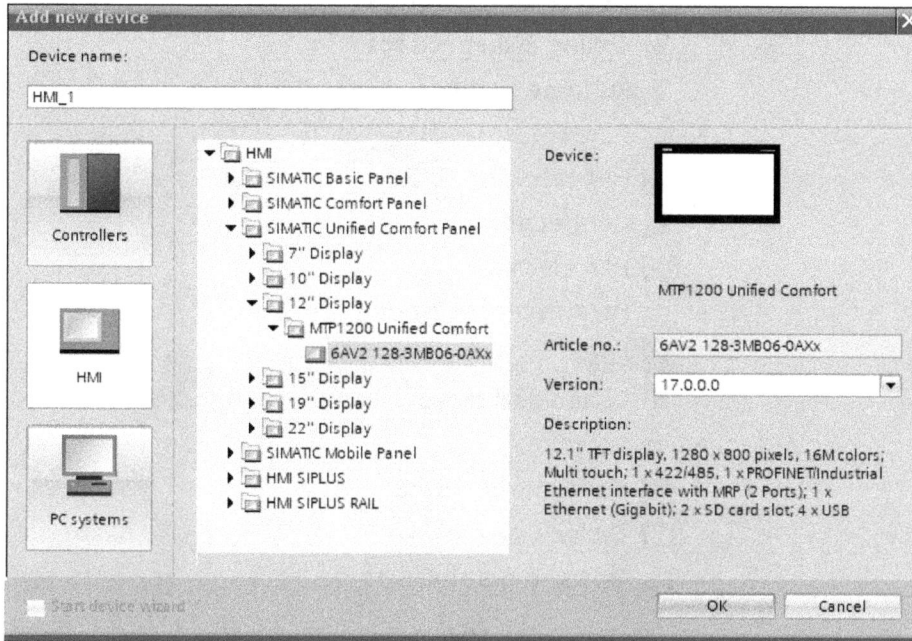

Figure 9.2 – Adding a 12" Unified Comfort Panel

At this point, different versions of the Unified Comfort Panel can be selected. If
the panel is to be paired with a version 16 PLC, select version 16.0.0.0 from the
drop-down list; otherwise, leave it as version 17.0.0.0.

4. Clicking **OK** will then add the HMI to the project and **Project tree** pane, as
    illustrated in the following screenshot:

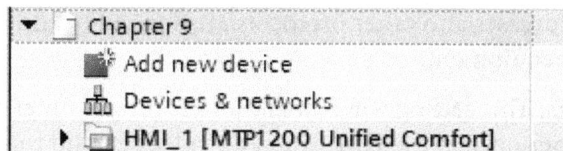

Figure 9.3 – Project tree pane with the new Unified Comfort Panel added

5. Once the HMI has been added to the **Project tree** pane, it can be expanded with the arrow to the left of the HMI object. This reveals additional objects, most of which are different from those of a PLC, as illustrated in the following screenshot:

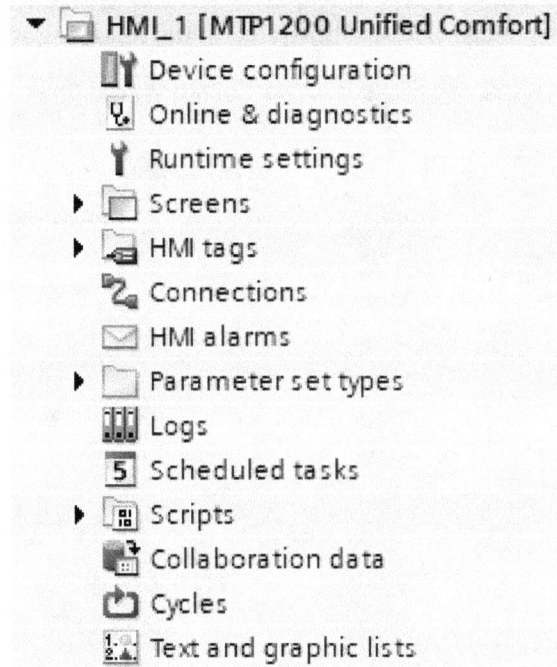

Figure 9.4 – HMI objects

The most important of these new types are listed here:

- **Screens**: This is where visual elements are built so that the HMI can display them to the end user.

- **HMI tags**: These interface with the PLC so that data can be written and read between the HMI and PLC.

- **Connections**: These are the interface connection profiles, where **Internet Protocol (IP)** addresses and other protocols are defined to allow the HMI to connect to other equipment.

- **Runtime settings**: This setting window allows the programmer to change the behavior and experience of the HMI when downloaded and the project's runtime is active.

These are collectively used to configure the HMI.

HMI development environment overview

HMI is made up of many different aspects and areas, and hence it is important to understand these key aspects and the tools that are offered to build an HMI.

Runtime settings

Runtime settings configures how the HMI behaves once downloaded to the hardware. It is necessary to access this and change the configuration once the HMI has been added to the project.

In TIA Portal V17, Unified Comfort Panels fail to compile when they are added to a project due to invalid security settings, and an invalid start screen configured, as illustrated in the following screenshot:

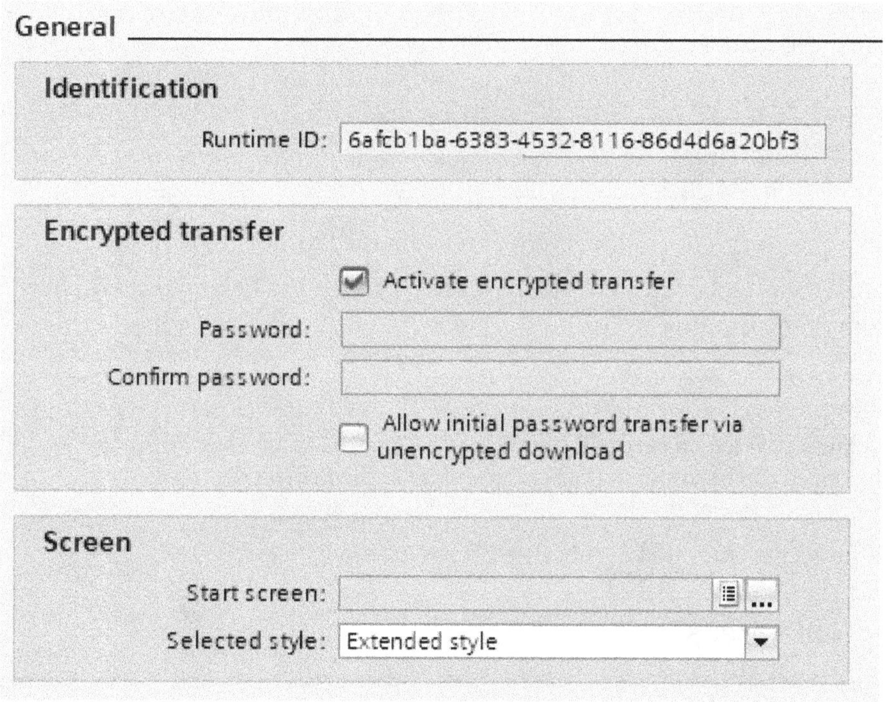

Figure 9.5 – Invalid runtime configuration

A password can be set for **Encrypted transfer,** or the activation of the option can be turned off by unchecking the **Activate encrypted transfer** option.

The start screen can be defined by clicking the button with the three dots (**...**), which will open the following window:

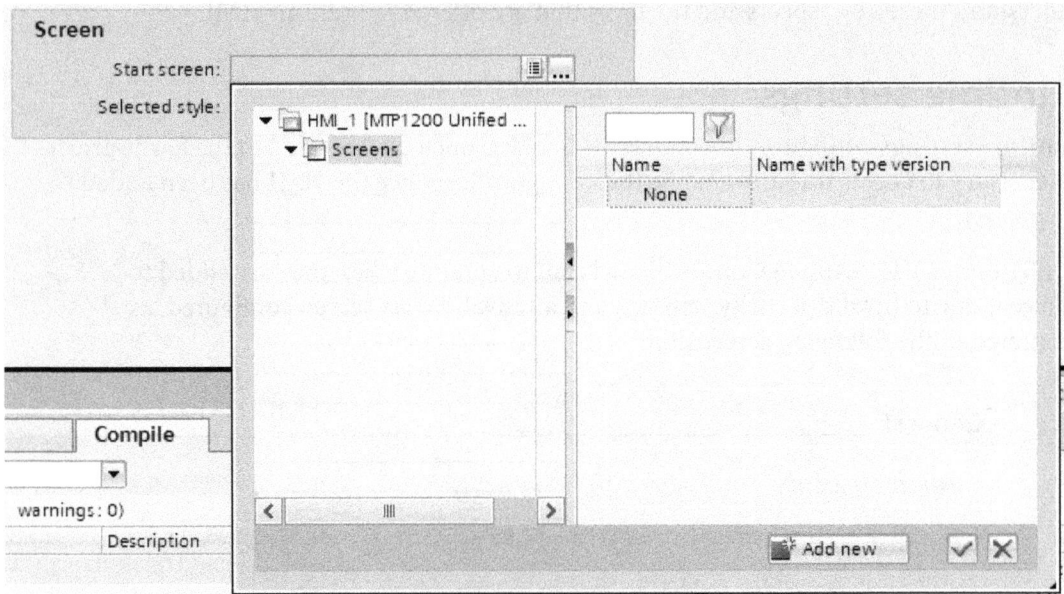

Figure 9.6 – Start screen window

Here, a start screen can be selected, or a new screen added by clicking the **Add new** button at the bottom of the window.

> **Note**
>
> Adding a new screen here will simply create Screen_1 or a similar name.
> The screen can be found in the **Screens** folder in the **Project tree** pane.

If the HMI project is compiled at this stage, it should now be error-free.

Other notable areas

The **Runtime settings** option also contains some other areas that can be of use, depending on how the project is to be used and configured. Let's look at a few of these here:

- **Alarms**: This allows the editing of static texts for alarms and other options around handling alarms.

- **Services**: Allows the HMI to operate as an **Open Platform Communications** (**OPC**) server and allows the designation of a **Simple Mail Transfer Protocol** (**SMTP**) port.

- **Language & font**: Allows more than one runtime language to be defined. Remember that these are actually set in the **Language & resources** object in the **Project tree** pane and are shared across the entire project.

- **Remote access**: These options configure how the HMI may be accessed remotely—via a web client, for example. They include the following options:

  - **Collaboration**: The settings under this heading allow the HMI to share (or collaborate) with other Unified HMIs and Unified PC stations. When configured correctly, this allows HMIs to share resources such as screens, despite them not being part of the same project.

    - This requires an additional runtime license.

  - **Web client**: Allows access to the runtime via a web browser and requires an additional runtime license.

  - **Smart Server**: Allows access to the Unified Panel from the **Siemens Smart Client** application.

- **Storage system**: Allows configuration of where the storage system is located (**Universal Serial Bus** (**USB**) or **Secure Digital** (**SD**) card) and which medium is being used for tag logging and alarm logging operations.

  Folders can be created in mediums, too, to help organize logging.

- **Tag settings**: Settings here change how the HMI synchronizes tags with any PLCs connected.

  It is recommended to leave the settings as their default settings; however, it may be beneficial to some projects to amend these if required.

- **User administration**: Allows the setting of local user management or **User Management Component** (**UMC**). UMC requires additional resources and licensing to operate.

Screens

A **screen** is a visualization object; it is the area that contains objects that will be presented to the end user on a physical HMI device. An HMI can consist of many different screens that hold objects. These screens collectively form different areas of a project and require navigation methods to switch between screens.

At least one screen must be created for an HMI to compile correctly. The **start screen** (the screen the HMI will display on startup) is denoted by a small green arrow in the **Project tree** pane, as illustrated in the following screenshot:

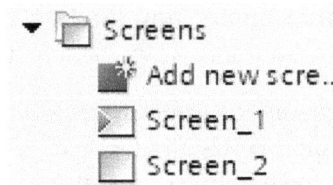

Figure 9.7 – Start screen with green arrow indication

The start screen can also be referred to as the **base screen**. This screen may simply navigate to other screens or contain screen windows that display other screens.

Screen toolbox

Opening a screen (or creating a new one by double-clicking **Add new screen** from the **Screens** folder in the **Project tree** pane) will open the screen and update the **Toolbox** section to the right of the main TIA Portal window, as illustrated in the following screenshot:

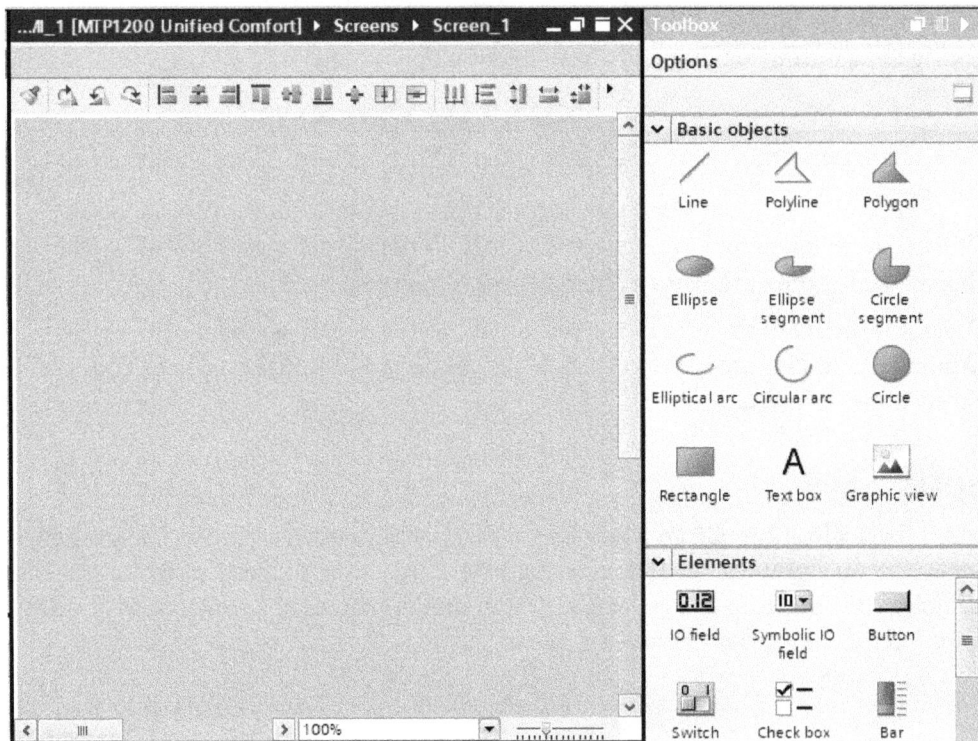

Figure 9.8 – Screen development toolbox

Toolbox contains all of the necessary objects required to build a visual interface for the screen selected. **Toolbox** objects are split into different categories, ranging from basic objects to smart elements and controls, and Siemens-supplied graphics and widgets.

> **Note**
> The **Graphics** and **Dynamic Widgets** sections of **Toolbox** are best viewed with all other areas collapsed. This is because they contain another graphics window below that displays graphic contents. Items can be dragged out and dropped on the screen from this additional window.

Screen layout

The screen **Layout** tab, located to the right of the main TIA Portal window, allows the configuration of the **layer** the screen object resides on. Layers are used to place screen objects on top of each other, splitting them into areas that can be hidden if required. You can see an overview of this in the following screenshot:

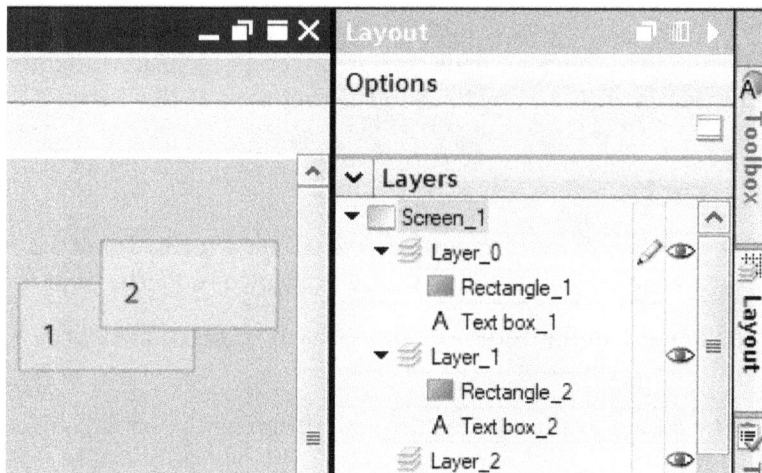

Figure 9.9 – Screen objects and layers example

For objects to sit on top of another object, they must be configured to be part of a higher layer number, or if in the same layer, further toward the next layer in the list. *Figure 9.9* demonstrates that Rectangle_2 is positioned on top of Rectangle_1 as Rectangle_2 is in Layer_1, which is a different layer that has a higher number. Text box_1 is positioned on top of Rectangle_1 because it is further toward the next layer in the Layer_0 list.

> **Note**
>
> Moving objects by dragging them around in the **Layers** window is the same as right-clicking on an object and changing the arrangement by choosing an option in the **Arrange** menu. Remember that arranging objects only affects their position against objects in the same layer. An object can be moved to a different layer by dragging it there in the **Layers** window.

Layers can be hidden in the editor by turning off the layer with the eye icon, as shown in *Figure 9.9*.

> **Note**
>
> At least one layer must be active at all times. An active layer is represented by the pencil icon. Active layers cannot be hidden. Hidden layers are still visible in **Runtime** settings.

TIA Portal V17 bug

When toggling the visibility of layers, sometimes the layers do not come back on in the correct order. Unfortunately, this is an issue that has no easy resolution other than to re-organize the layer again. This may be fixed in later versions of TIA Portal V17.

Screen objects

Screen objects are items that are used to build up visuals on screens. There are many different types of screen objects, and each comes with its own properties and events.

Screen objects can be found in the **Toolbox** window to the right of TIA Portal when a screen object is open in the editor.

Screen objects can be placed by simply dragging and dropping the screen object into the **Editor** window.

Object properties

When an object is selected, the **Properties** tab at the bottom of TIA Portal contains the relevant properties of the object, as illustrated in the following screenshot:

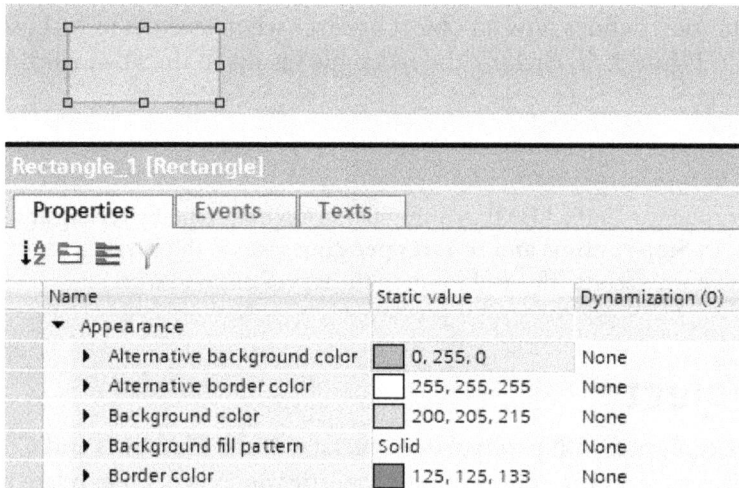

Figure 9.10 – Example of object property configuration

These properties define how an object looks. Most can also be made dynamic by changing the **Dynamization** option to the right.

> **Note**
>
> When **Dynamization** is selected, additional options appear to the right of the **Properties** tab.

Object events

When an object is selected, the **Properties** window at the bottom of TIA Portal contains relevant events for the object on the **Events** tab, as illustrated in the following screenshot:

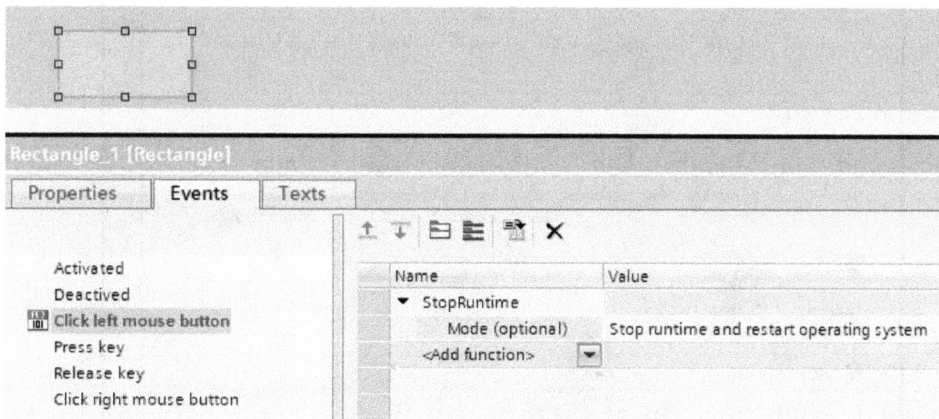

Figure 9.11 – Example of object event configuration

The event configuration defines how an object behaves when interacted with, such as a left mouse click. In *Figure 9.11*, clicking the rectangle results in the HMI runtime shutting down and restarting.

> **Note**
>
> If we are running a Unified HMI in a simulation environment, by raising an event against **Stop runtime and restart operating system**, the development environment will restart. This will result in the loss of unsaved work.

Special objects

TIA Portal comes equipped with two types of special objects: **Elements** and **Controls**. Both types of object have a range of pre-built properties and events that relate to their designated purpose.

Both **Elements** and **Controls** can be found in the **Toolbox** window to the right in TIA Portal when a screen is open, as illustrated in the following screenshot:

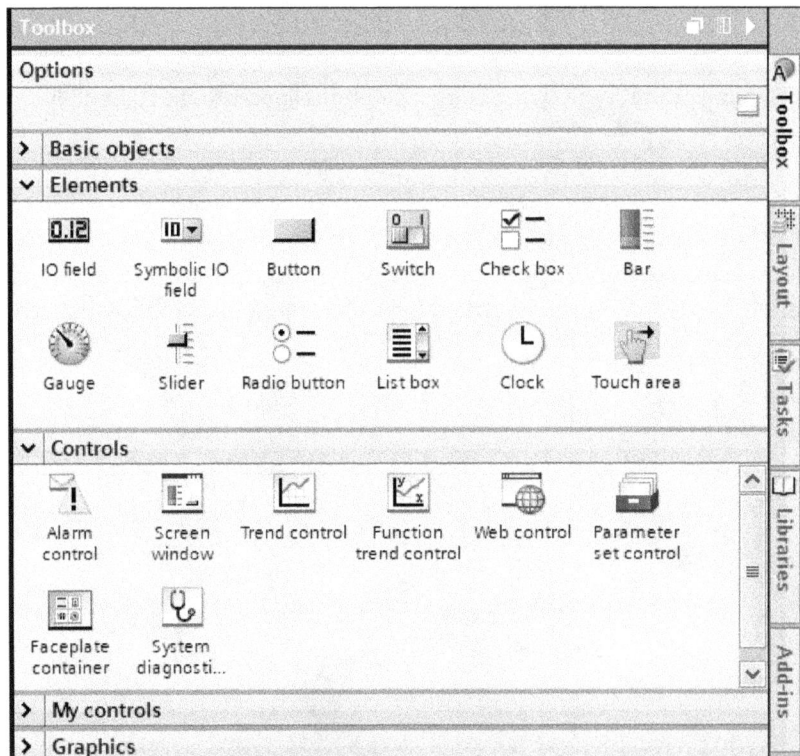

Figure 9.12 – Elements and Controls in the Toolbox section

These objects have more specific roles than **Basic objects**. **Elements** are typically objects that a user may interact with, such as a *button* or *slider*. **Controls** are objects that allow a TIA Portal's Unified environment to make use of a particular function, such as displaying a *faceplate* or managing *alarms*.

Elements

Elements are essentially ready-made faceplates that programmers can drop into a project to enable quick visualization of data. You can see an example of an element in the following screenshot:

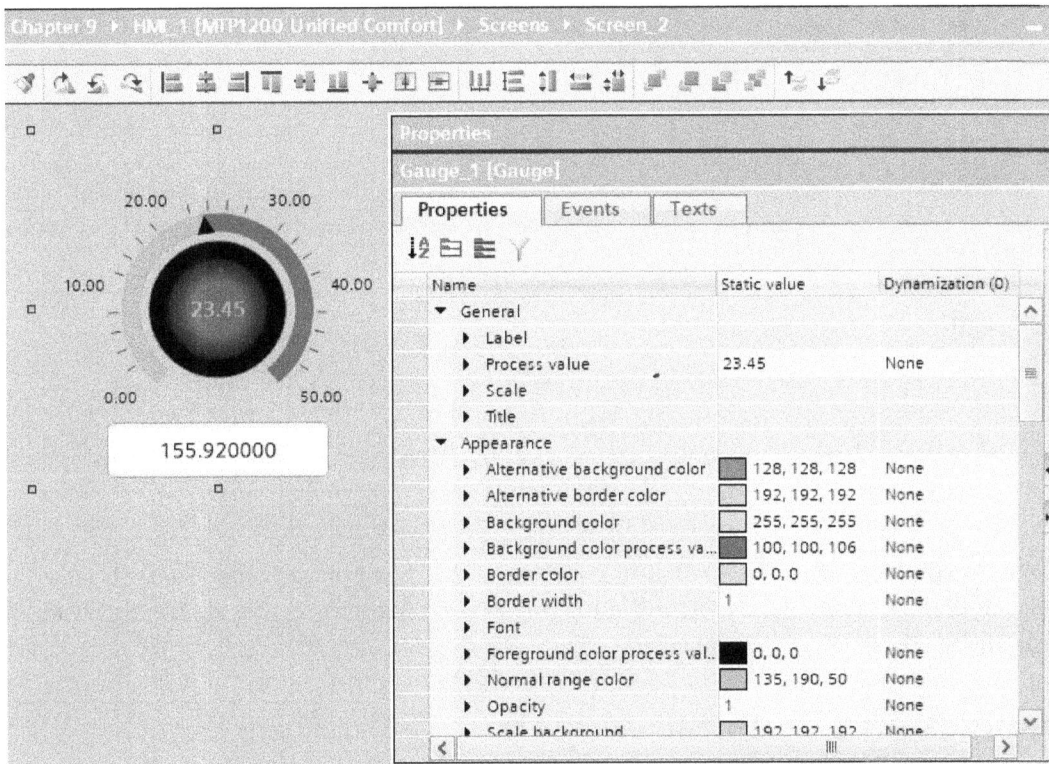

Figure 9.13 – Example of a Gauge element and its properties

Elements are highly configurable, allowing easy configuration and *dynamization* of options and signals. Building these controls from scratch can be done; however, these types of controls and indicators are well known in HMI development, and it makes sense to use a highly configurable object instead of starting from scratch.

Controls

Controls are more involved than **Elements**, from a configuration point of view. **Controls** are objects that directly interact with the background systems in the TIA Portal and Unified environment, allowing access to data stored in alarm databases and tag logging. You can see an example of a control in the following screenshot:

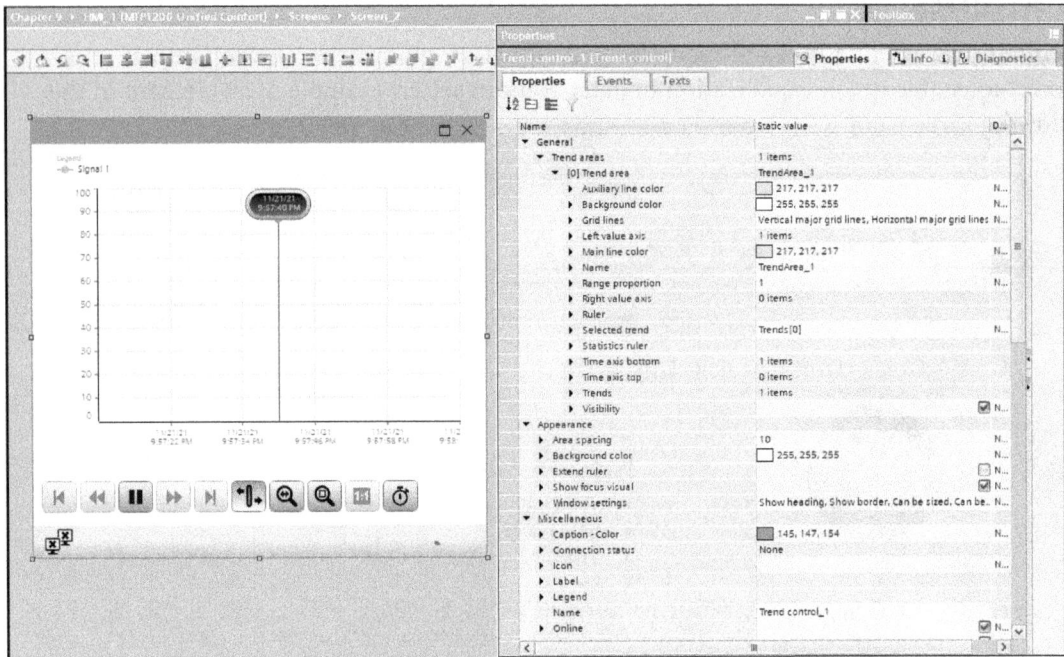

Figure 9.14 – Example of a Trend control, with many more properties

Figure 9.14 shows an example of a **Trend** control that has been placed onto a screen. This particular control allows access to the tag logging system, something that is not otherwise accessible by other, more basic objects.

Graphics and Dynamic widgets

Two more types of special controls relate to visually displaying graphics objects on the screen. These contain pre-built graphics files that have a number of configurable properties for color so that an indication of status can be achieved. There are many different control types to choose from, categorized by different asset types. You can see an example of these in the following screenshot:

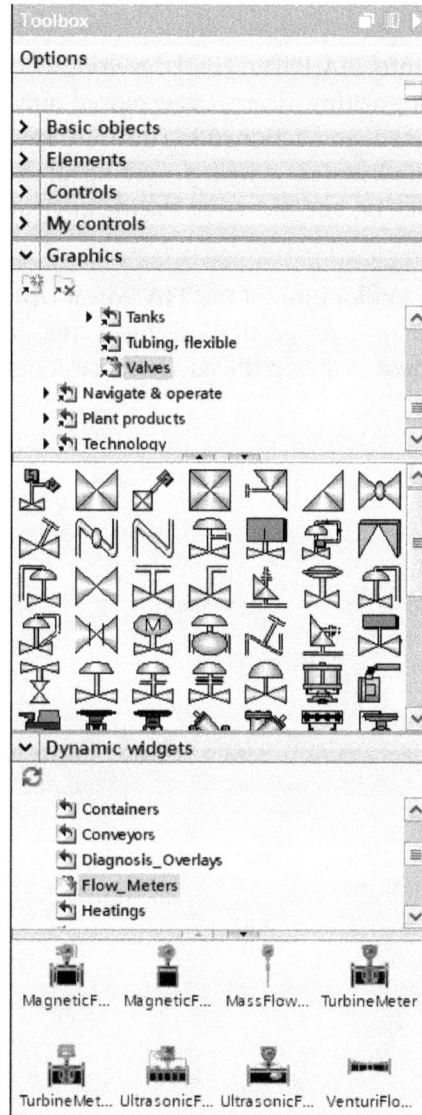

Figure 9.15 – Toolbox with Graphics and Dynamic widgets expanded

These can be found in the **Toolbox** section to the right of TIA Portal when a screen is open, as shown in the preceding screenshot.

Summary

This chapter has given insight into TIA Portal HMI development and what the environment feels like to develop within. The areas explored only touch the surface of what can be configured, visualized, and actioned in runtime. The TIA Portal Unified environment is very powerful and designed well, providing all that is needed to help programmers create smart, efficient, and visually pleasing environments, with minimal work.

The next chapter continues the exploration of the TIA Portal Unified environment. The chapter discusses static and dynamic properties, raising events, and using scripting and dynamization of properties. These make up the basic principles of HMI development in TIA Portal.

10
Placing Objects, Settings Properties, and Events

This chapter explores screen objects further, explaining how to set static properties and how to make them dynamic. **Human-machine interfaces** (**HMIs**) accept inputs from operators too. This chapter explains how to create events and how to use scripts within the Unified environment.

Properties, events, and scripts are all used to change the behavior and feel of the HMI, from how it looks to how it responds to buttons being touched.

In this chapter, the following topics are covered:

- Setting static properties
- Setting dynamic properties
- Using scripts
- Raising events

Setting static properties

Nearly all properties of a screen object have the ability to define a **static value**. This static value can be considered a *hardcoded* value or a **constant**. This means that it does not change during the execution of the **HMI runtime**.

Every screen object that can be placed from the **Toolbox** section has its own list of properties that affect the look and feel of the object. Most screen objects contain similar properties, such as visibility, authorization level, size, and position. You can see an overview of this in the following screenshot:

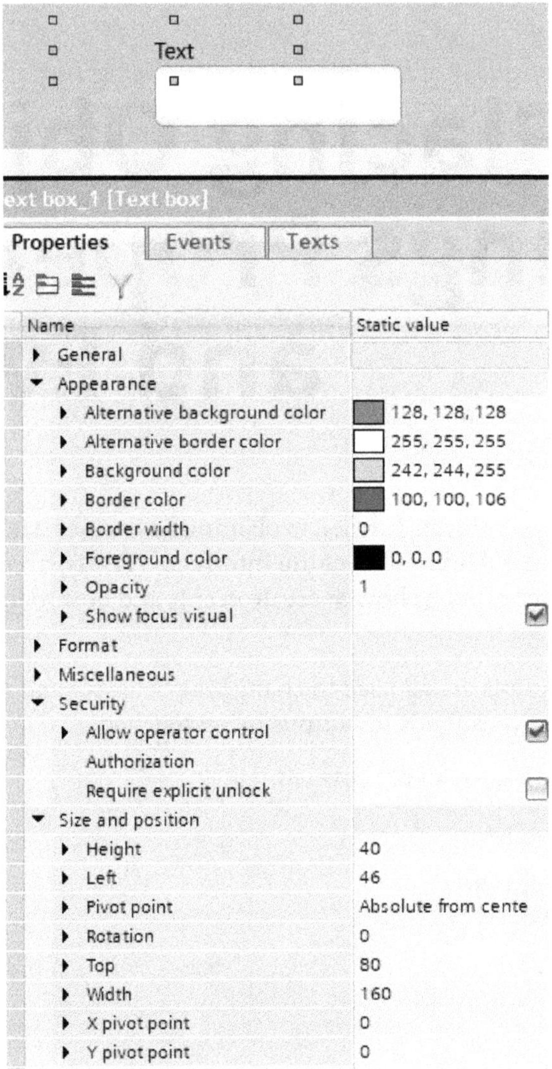

Name	Static value
▶ General	
▼ Appearance	
▶ Alternative background color	128, 128, 128
▶ Alternative border color	255, 255, 255
▶ Background color	242, 244, 255
▶ Border color	100, 100, 106
▶ Border width	0
▶ Foreground color	0, 0, 0
▶ Opacity	1
▶ Show focus visual	☑
▶ Format	
▶ Miscellaneous	
▼ Security	
▶ Allow operator control	☑
Authorization	
Require explicit unlock	☐
▼ Size and position	
▶ Height	40
▶ Left	46
▶ Pivot point	Absolute from cente
▶ Rotation	0
▶ Top	80
▶ Width	160
▶ X pivot point	0
▶ Y pivot point	0

Figure 10.1 – Properties when screen object selected

The **Properties** list contains everything available to be changed for the selected screen object.

The **General** section of the **Properties** list contains properties that relate to the selected object only. This means that other screen objects may not have these properties—for example, a **Text box** object has the **Text** property, but a **Line** object does not.

When static values are changed for a property, **Totally Integrated Automation Portal (TIA Portal)** is able to action the change immediately in a visual format (in most cases). This means that when the property is changed, the object on the screen changes. You can see an example of this in the following screenshot:

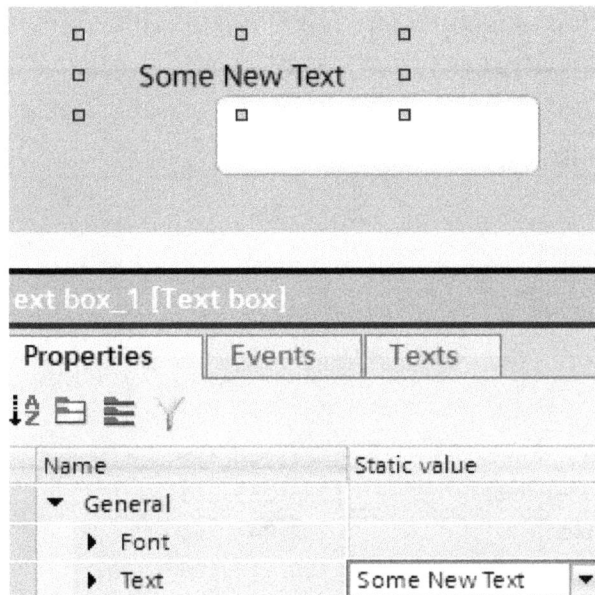

Figure 10.2 – Example of the Text property static value being changed and the object updating to reflect the change

This allows programmers to preview changes as properties' static values are modified.

> **Note**
> Not all static values update an object's look in development time. Properties such as **Alternative background color** are only displayed in the HMI runtime.

Types of static values

There are different types of static values, determined by their functionality and how they are used. For example, the appearance section of a **Text box** object consists of *color properties*, *numerical properties*, and also *Boolean properties*, as illustrated in the following screenshot example:

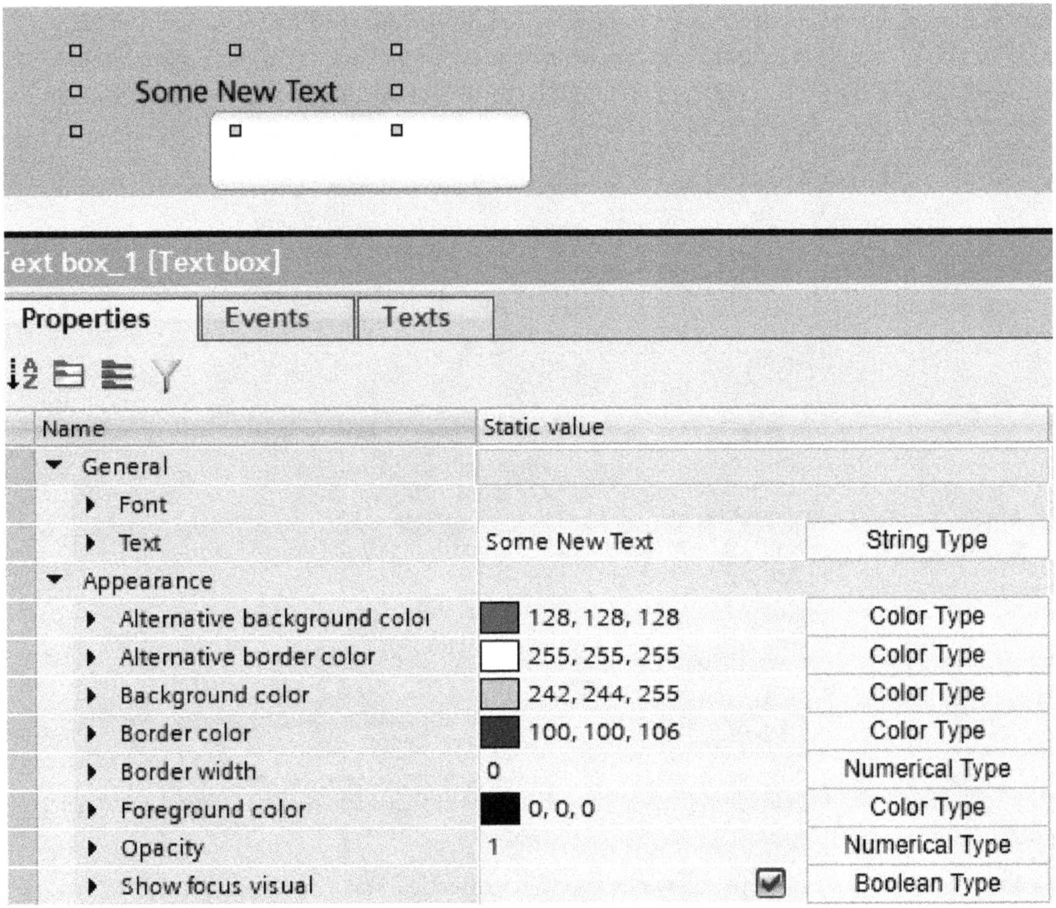

Name	Static value	
▼ General		
▶ Font		
▶ Text	Some New Text	String Type
▼ Appearance		
▶ Alternative background color	▨ 128, 128, 128	Color Type
▶ Alternative border color	☐ 255, 255, 255	Color Type
▶ Background color	▨ 242, 244, 255	Color Type
▶ Border color	▨ 100, 100, 106	Color Type
▶ Border width	0	Numerical Type
▶ Foreground color	▮ 0, 0, 0	Color Type
▶ Opacity	1	Numerical Type
▶ Show focus visual	☑	Boolean Type

Figure 10.3 – Example of different types of static values

These different types of static values change how data is entered. When interacting with **Color** properties, a color chart is displayed. When interacting with a **String** or **Numerical** property, nothing is displayed, and the programmer can type in the required property value. Some properties are simply checkboxes that either enable or disable the property.

Key properties

Most objects have some *key properties* that programmers will interact with often and are relative to nearly all screen objects, including *elements*, *controls*, and *graphics*. Here are some of them:

- **Miscellaneous**

    - **Name**

        - Sets the name of the object in the development environment

        - Used to access the object via **scripting**

        - Automatically assigned a unique name when an object is placed on the screen; however, this can be modified if required

    - **Visibility**

        - Sets if the object is visible in the HMI runtime.

        - The default value is True.

- **Security**

    - **Allow operator control**

        - When enabled, allows operator interaction events to be actioned for the object.

        - When disabled, operator interaction events associated with the object are ignored.

        - The default value is True.

    - **Authorization**

        - A drop-down box that allows the setting of the minimum required login level in order to interact with an object.

        - The default value is unset—this means any level can access an object, even when no user is logged in.

- **Size and position**

    - This section contains many different properties that all relate to where an object appears on the screen and the size of the object.

- **Appearance**

  - This section contains many different properties that all relate to the appearance of an object.

- **General**

  - This section contains many different properties that all relate explicitly to the associated object.

  - This section differs for different object types but is considered a key section as it is nearly always required to be interacted with during the configuration of an object.

> **Note**
>
> The **Properties** tab on the **Properties** window can be filtered or sorted using the buttons that appear between the table of properties and the tab header. See *Figure 10.3*—note the buttons directly beneath the **Properties** tab header.

Setting dynamic properties

Nearly all static properties can be made dynamic through the **Dynamization** feature built into TIA Portal and Unified HMIs. Dynamization allows the modification of property values during runtime. This means that properties such as color, visibility, and more can be modified to react to data that is being passed to a property. You can see an example of this in the following screenshot:

Name	Static value	⚡ Dynamization (1)
▶ Appearance		
▼ Miscellaneous		
▶ Connection status	None	
▼ Interface		
BasicColor	☐ 238, 238, 238	None
Name	DynamicSVG_1	
▶ Show connection quality		☑ None
Tab index	0	
▶ Tooltip		None
▶ Visibility		☑ Tag

Figure 10.4 – Example of Visibility property set to Tag for dynamization

By setting the dynamization of a property, additional information needs to be provided to the property, as illustrated in the following screenshot:

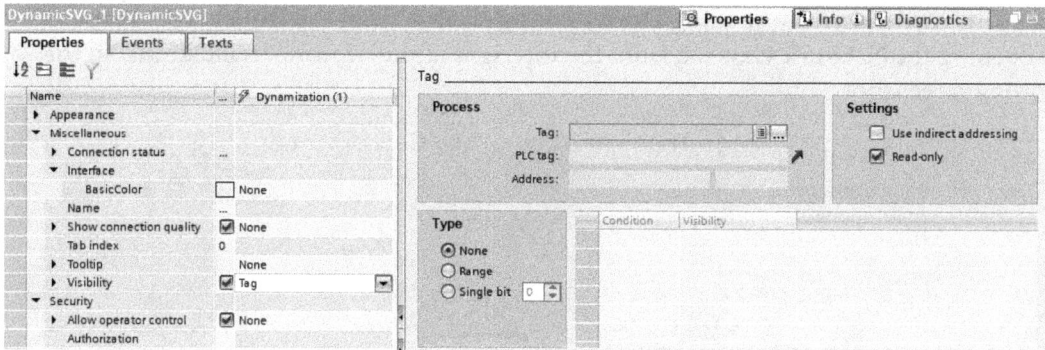

Figure 10.5 – Tag dynamization additional properties

These additional requirements appear to the right of the **Properties** list when a dynamization property is selected.

> **Note**
>
> Depending on the type of dynamization used, the view may look different and have different options.

Assigning tags to dynamization properties

Selecting **Tag** as the dynamization method allows programmers to assign a **programmable logic controller** (**PLC**)- or HMI-based tag to inspect for a condition. Depending on the conditions set, an outcome can be defined that affects the property that is being made dynamic.

Connecting the HMI and PLC

Before tags from a PLC can be used, a connection between the HMI and PLC must be made. This can be done by opening **Device configuration** from the **Project tree** pane and opening the **Network view** tab from the top right of the window, as illustrated in the following screenshot:

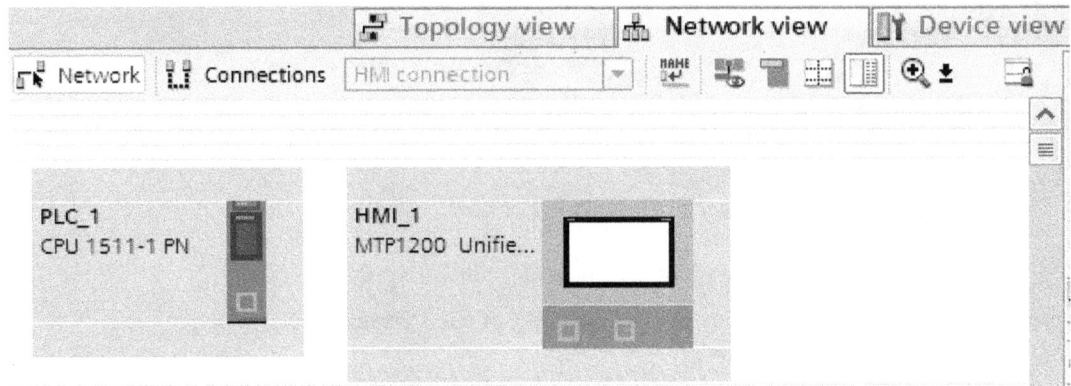

Figure 10.6 – Network view displaying a PLC and HMI with no connection

To connect the two devices together, a *network connection* must be made. This can be done by ensuring the **Network** button in the top left of the window is selected and then dragging between the green network ports on both of the devices.

When this is complete, TIA Portal will create a **Profinet/Industrial Ethernet (PN/IE)** connection between the devices, as illustrated in the following screenshot:

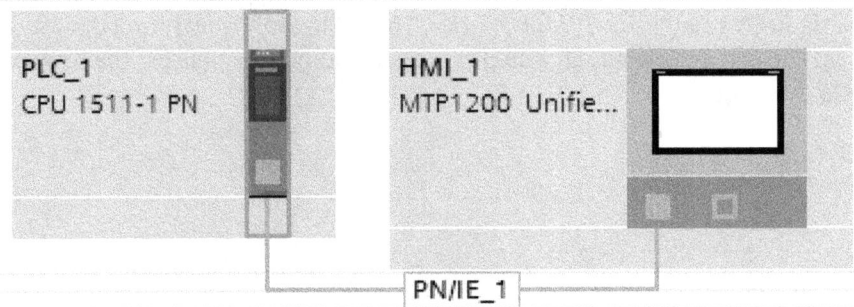

Figure 10.7 – PN/IE network between devices

Once this has been set up, HMIs need an additional step to share tag information between the HMI and PLC. An *HMI connection* is required in order to point internal HMI tag data to the relevant location in a connected PLC.

To make an HMI connection, the **Connections** button in the top left of the window must be selected, and the drop-down menu must be set to **HMI connection**, as illustrated in the following screenshot:

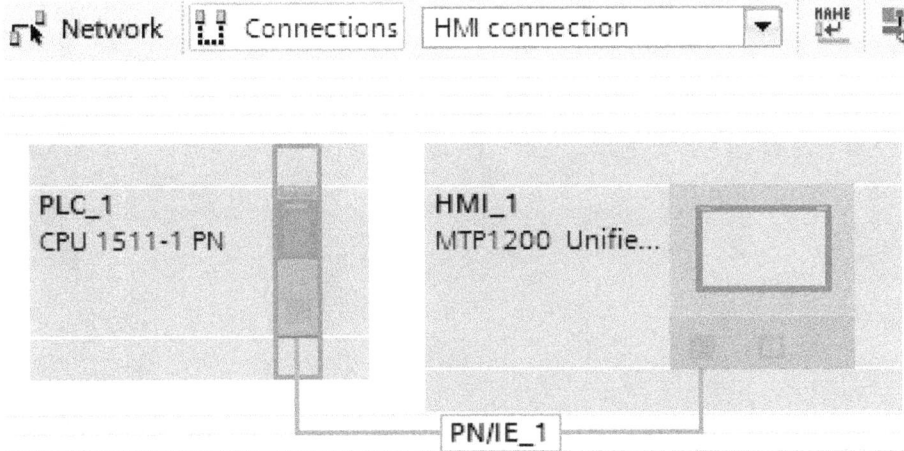

Figure 10.8 – HMI connection selected, but connection not yet made

TIA Portal will highlight devices that are eligible to have an HMI connection established. Dragging a connection between the green network ports on the devices will create a connection, as illustrated in the following screenshot:

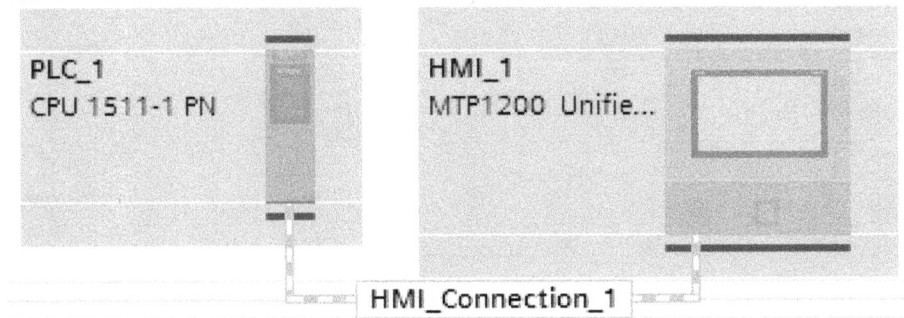

Figure 10.9 – HMI connection between devices

Now that this has been completed, the HMI is able to read tags and data blocks from the PLC without additional configuration.

Assigning tags

Once the PLC and HMI have been connected, assigning a tag to a dynamization property is as simple as selecting a tag from a list. If a tag does not exist yet, TIA Portal allows the selection of data in the PLC data block and will automatically create an associated HMI tag, as illustrated in the following screenshot:

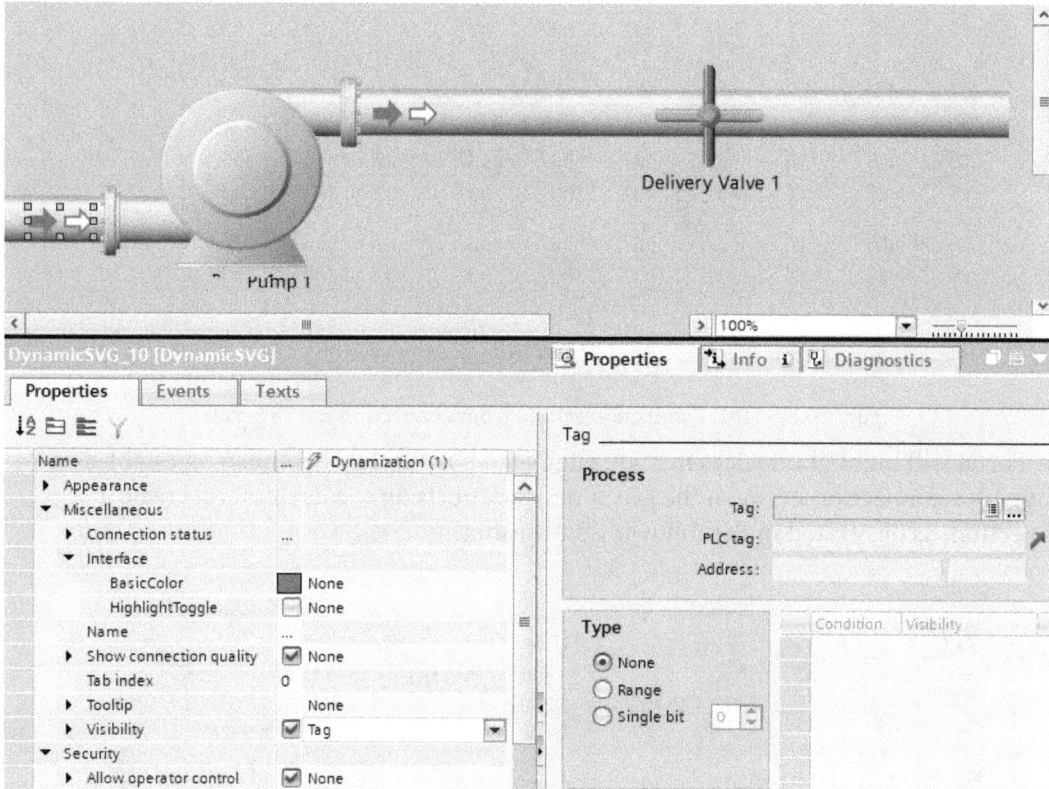

Figure 10.10 – Assigning a tag for dynamization

Figure 10.10 shows an object selected and the **Visibility** dynamization set to **Tag**. The **Dynamization** window appears to the right and highlights the **Tag** field in the **Process** area. The **Tag** field refers to the HMI tag required to set the associated **Visibility** property. The **PLC tag** field is only used if the tag is connected to PLC data.

Tags can be selected by clicking one of the two available buttons in the **Tag** field. The first button displays a list of already defined HMI tags, as illustrated in the following screenshot:

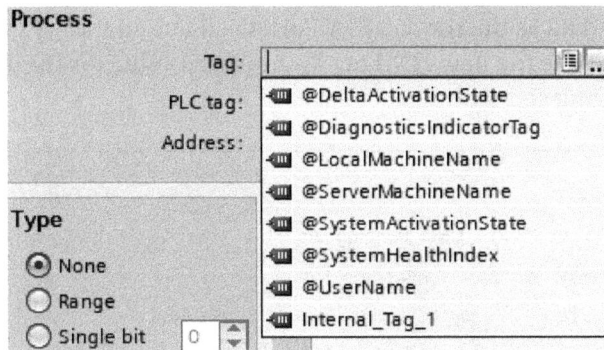

Figure 10.11 – HMI tag list

These tags are identifiable by a purple icon. These tags can be created and edited in the **HMI tags** object in the **Project tree** pane, and connections to PLC data can also be made; however, TIA Portal offers a simple method to create PLC data connections.

Clicking the second button in the **Tag** field loads a project view where the PLC data can be seen and selected, as illustrated in the following screenshot:

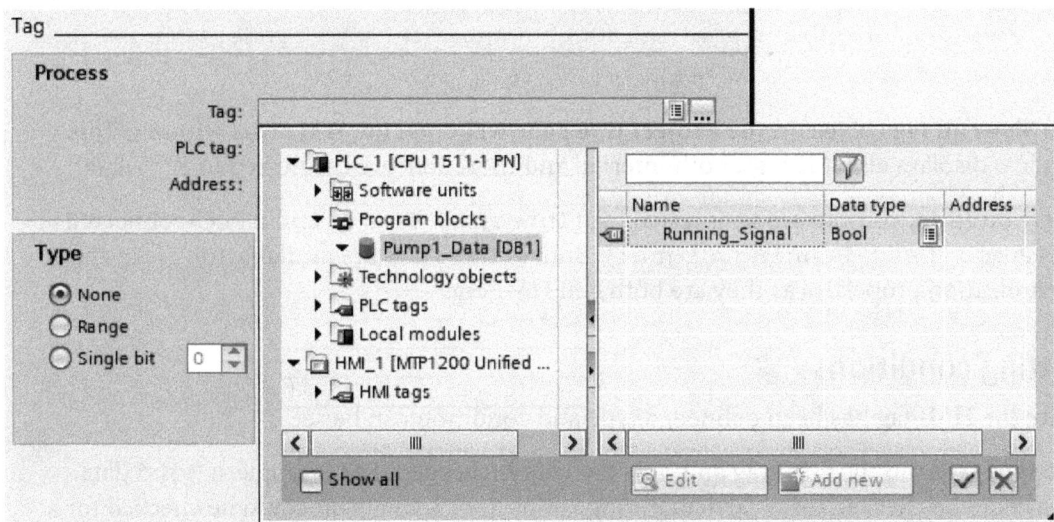

Figure 10.12 – Project browser and selection of PLC data

On acceptance of PLC data as the source, TIA Portal will automatically create an associated HMI tag and use the new HMI tag as the actual value for the **Tag** field, as illustrated in the following screenshot:

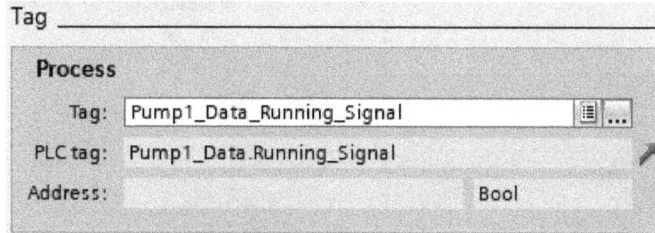

Figure 10.13 – New HMI tag and associated PLC tag location

The declaration of the HMI tag and the connection to the PLC tag can be viewed and modified by opening the **HMI tags** object in the **Project tree** pane and then opening **Show all tags**. A table like this will then appear:

Name ▲	Tag table	Data type	Connection	PLC name	PLC tag
Internal_Tag_1	Default tag table	Bool	<Internal tag>		<Undefined>
Pump1_Data_Running_Signal	Default tag table	Bool	HMI_Connectio...	PLC_1	Pump1_Data.Running_Signal

Figure 10.14 – HMI tag table

The **Show all tags** object in the **Project tree** pane will open the **HMI tags** window. This window displays all HMI tags, both internal and those that are connected to PLC data.

In *Figure 10.14*, you can see an internal tag (`Internal_Tag_1`) and a PLC-connected tag (`Pump1_Data_Running_Signal`). Both of these tags are available for use in the dynamization properties as they are both still HMI tags.

Setting conditions

Once the HMI tag has been defined, a type and condition can be set.

The **type** refers to the HMI tag and how the data is checked. For example, a `Word` data type could be checked for a particular range of `0-2`, or a single bit could be checked for a `1` or `0` value.

Once a type has been set, the condition value and associated action can be set, as illustrated in the following screenshot:

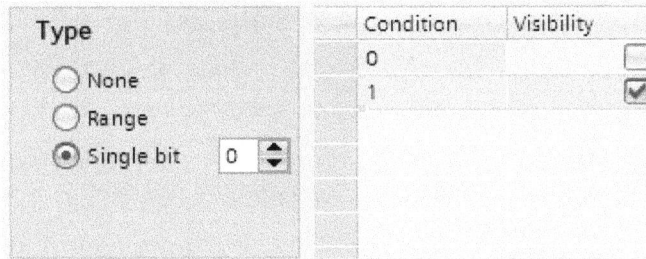

Figure 10.15 – Setting Condition and Visibility properties

Figure 10.15 shows that when the associated tag value's bit 0 is 1 (or True), then the **Visibility** property is set to True. If bit 0 is 0 (or False), then the **Visibility** property is set to False.

In the HMI runtime, the visibility property value causes the screen object to be displayed or hidden, as illustrated in the following screenshot:

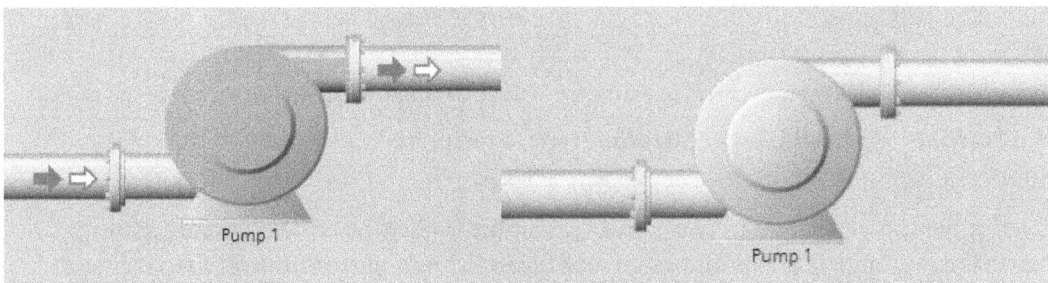

Figure 10.16 – Runtime view of dynamization (left pump is when the running signal is True)

Figure 10.16 shows dynamization in use in **Runtime**. When the tag Pump1_Data_ Running_Signal is set to True, the pump changes color and flow arrows appear. This is all completed with the **Dynamization** properties set to **Tag**.

Using scripts

TIA Portal's Unified system can make use of JavaScript against screen object properties. Historically, Siemens used **Visual Basic Script (VBS)** for dynamization in an HMI environment; however, Unified has moved to a more advanced and commonplace JavaScript language.

Scripting can be used by setting **Dynamization** to **Script**, as illustrated in the following screenshot:

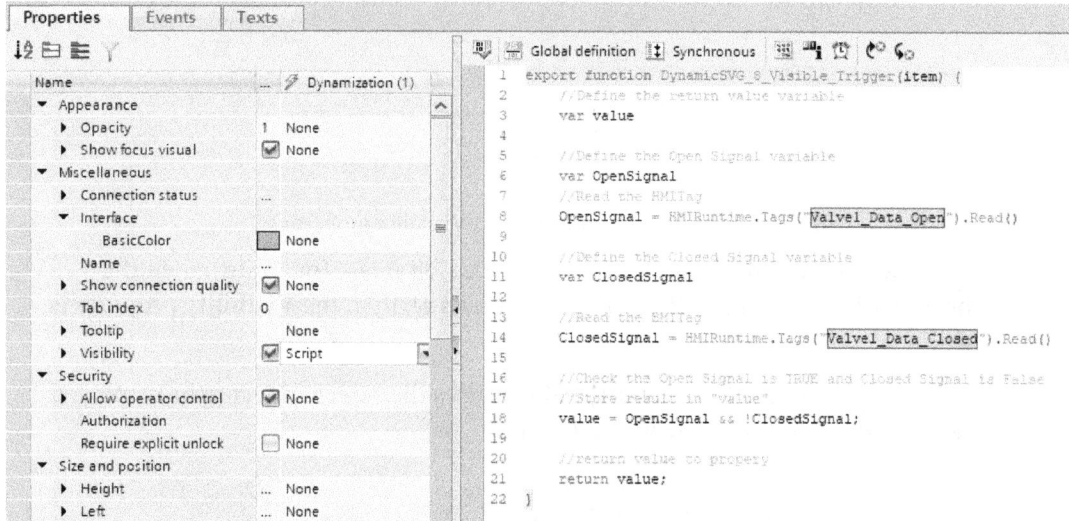

Figure 10.17 – Visibility property set to Script for Dynamization, with the script shown to the right

When **Script** is used as the **Dynamization** type, a script file is displayed to the right of the window.

The script file can be used to perform more complex evaluations of data before setting the associated properties. This allows greater flexibility and customization than other dynamization options.

Construction of script files

Scripts are semi-constructed when the **Script** option is selected from the **Dynamization** list. TIA Portal will provide the necessary function declaration and an associated return variable, which is usually declared as `value`. You can see an example of this in the following screenshot:

```
1  export function DynamicSVG_8_ToolTipText_Trigger(item) {
2      var value;
3      return value;
4  }
```

Figure 10.18 – Example of TIA Portal-provided script

It is up to the programmer to fill in the desired function between `var value;` and `return value;`. You can see an example of this here:

Line	Code	Description
1	`export function DynamicSVG_8_Visible_Trigger(item) {`	Function Declaration - Automatically added
2	`//Define the return value variable`	Declare Return Value
3	`var value`	
4		
5	`//Define the Open Signal variable`	Declare Tag Signal
6	`var OpenSignal`	
7	`//Read the HMITag`	Read Tag Signal
8	`OpenSignal = HMIRuntime.Tags("Valve1_Data_Open").Read()`	
9		
10	`//Define the Closed Signal variable`	Declare Tag Signal
11	`var ClosedSignal`	
12		
13	`//Read the HMITag`	Read Tag Signal
14	`ClosedSignal = HMIRuntime.Tags("Valve1_Data_Closed").Read()`	
15		
16	`//Check the Open Signal is TRUE and Closed Signal is False`	Test Acquired Signals
17	`//Store result in "value"`	
18	`value = OpenSignal && !ClosedSignal;`	
19		
20	`//return value to propery`	Return value to property
21	`return value;`	
22	`}`	End Of Function Declaration - Automatically added

Figure 10.19 – Example of a visibility script

Figure 10.19 demonstrates a script that is reading two states of a valve position (`Open` or `Closed`) and, depending on the signals returned, setting the visibility of a screen item. Because this script is looking at two pieces of information in order to set the property, a script is the only option available.

Figure 10.19 also shows different areas that the script contains, with the original `var value;` and `return value;` instances enveloping the rest of the user script.

Reading/writing tags

In order to read an HMI tag in a script, the **HMIRuntime** object must be accessed, and the **Tag** object used to obtain tag information. *Figure 10.19* demonstrates this as `HMIRuntime.Tags("Valve1_Data_Open").Read()`.

The HMI tag to be read must exist in the **HMI Tag** list in order for the script to execute successfully. If a tag does not exist, the default value for the property associated with the script will be returned.

Writing tags can also be achieved by using `.Write(value)`, where `value` is the value to be written to the tag.

> **Note**
> Right-clicking in the script window accesses the **Snippet** menu, which allows quick access to pre-written code. These snippets can be used to quickly set up read and write access to tags; however, they are often bulkier methods across multiple lines and with potentially unnecessary variable declarations.

Compilation errors

A common issue with the Unified HMI platform when using scripts is a The configured tag is invalid: compilation error message, as illustrated in the following screenshot:

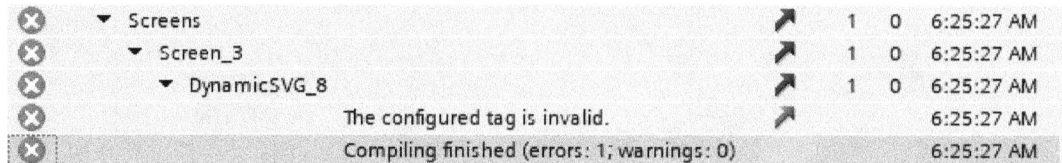

⊗	▼ Screens		↗	1	0	6:25:27 AM
⊗	▼ Screen_3		↗	1	0	6:25:27 AM
⊗	▼ DynamicSVG_8		↗	1	0	6:25:27 AM
⊗		The configured tag is invalid.	↗			6:25:27 AM
⊗		Compiling finished (errors: 1; warnings: 0)				6:25:27 AM

Figure 10.20 – Compilation error

This error suggests that a tag in the script is incorrect; however, that is not strictly true. When TIA Portal first creates a script, the *trigger* for the script is automatically set to **Trigger on tag**. This means that the script is only actioned when a particular tag is accessed by the **HMIRuntime** object. Since TIA Portal does not know which tag a programmer wants to use, it leaves it blank, which causes a compile error.

Unfortunately, the compile error message does not link to the correct **Add trigger** window but, rather, displays the script window only.

By clicking the **Triggers** icon above the script window (the icon is a clock), the following window is displayed:

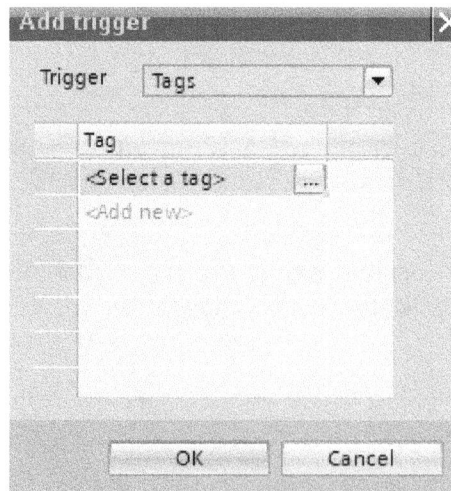

Figure 10.21 – Add trigger window

This window contains the missing tag that is causing the compilation failure. The programmer can then select an appropriate tag or change the trigger method to a cyclic update based on time.

> **Note**
>
> It is a good idea to reduce the number of cyclic triggers used in an HMI. Too many cyclic actions can cause the HMI to have poor performance, especially when large scripts are in use.

Global scripts

Scripts can also be written globally as opposed to written against a screen object property. This then allows local instances of the script to be called and parameterized by screen objects.

The advantage of this is that many items can call the same script, and if the global script is updated, all objects calling that script are automatically updated too. If any interface to the script is changed, TIA Portal will fail the compilation of items that use it, making it easy to find and update local instances.

> **Note**
>
> Global scripts can be found in the **Scripts** object of the **Project tree** pane.

Raising events

As well as reading tags and setting properties for the visual display of data, an HMI is commonly used to set data in a PLC via events, such as the pressing of a button. The Unified HMI platform has not changed this approach from nearly all other Siemens environments, with a simple interface for the declaration of the event. You can see an example of an event in the following screenshot:

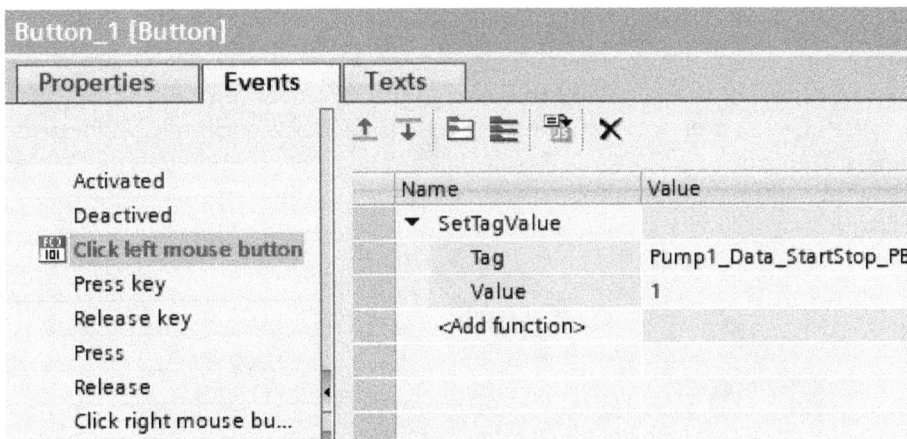

Figure 10.22 – Example of an event

The event configuration can be accessed by clicking on a screen object and selecting **Events** from the **Properties** window at the bottom of the screen. Similar to the **Properties** tab, the **Events** tab lists possible event types on the left and then the configuration of the selected event on the right.

Figure 10.22 is an example of a `SetTagValue` event type, where, when the **Click left mouse button** trigger is raised by the HMI runtime, the `Pump1_Data_StartStop_PB` tag is set to the value of `1` (or `True`).

The `Pump1_Data_StartStop_PB` HMI tag is connected to a PLC variable, and the PLC variable is therefore written to by the HMI, updating the value in the PLC code.

This allows the HMI to directly affect the following logic.

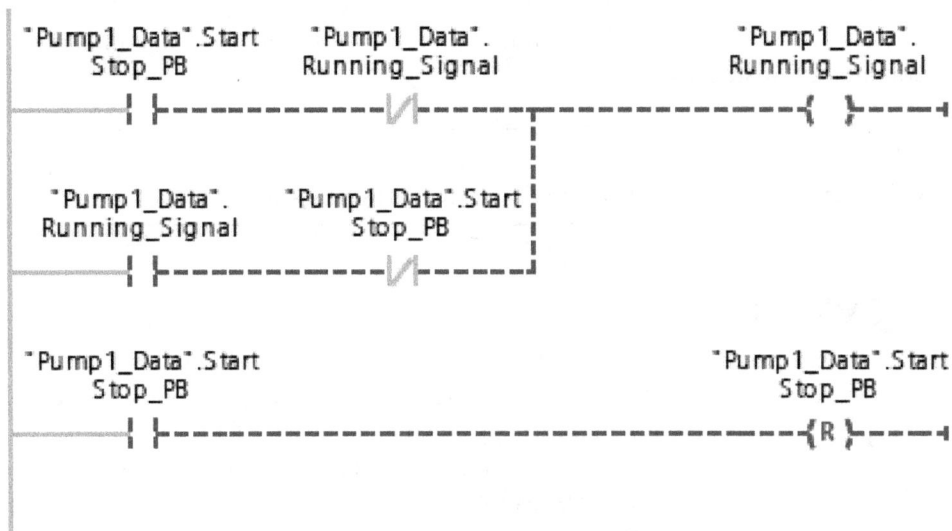

Figure 10.23 – Logic interacted with by HMI event

When the `SetTagValue` event is triggered, the `"Pump1_Data".StartStop_PB` variable in the PLC is written to `1` and the PLC logic then acts accordingly, as illustrated in the following screenshot:

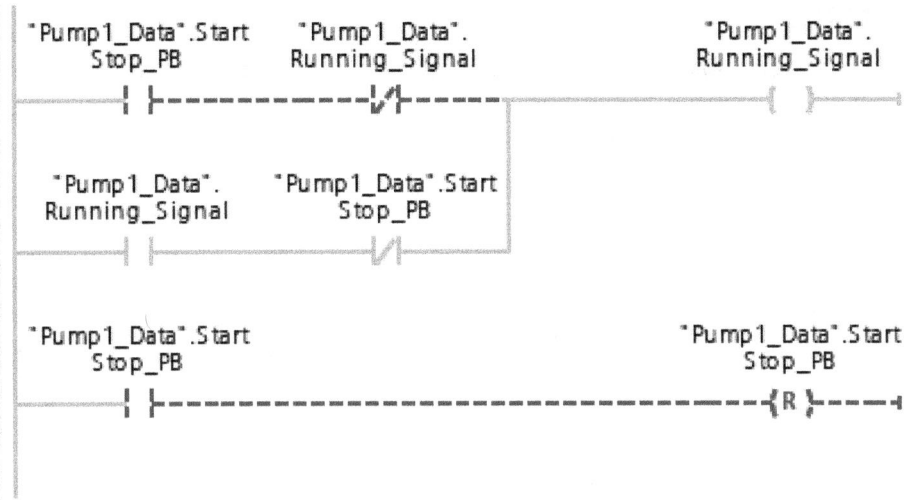

Figure 10.24 – Logic after the HMI event has been actioned

The logic shown in *Figure 10.23* and *Figure 10.24* resets the `"Pump1_Data"`.`StartStop_PB` variable whenever it is set to `True` from the HMI event. It is possible to have the HMI reset a push button variable by making use of more than one event—for example, **Press** and **Release**, where **Press** sets the variable to `True`, and **Release** sets the variable to `False`.

> **Note**
>
> It is important to understand the requirements of the application when setting up events. It is possible, under certain conditions, for events to be missed by the HMI. If it is critical that buttons/events do not *get stuck* with an active value, ensure that the PLC handles this independently of the HMI, as *Figure 10.23* and *Figure 10.24* demonstrate with the *reset coil*.

Event scripts

Events can also be script-based, and TIA Portal offers a quick method to convert an existing event into a script so that customized event data can be added. The following screenshot shows how this is done:

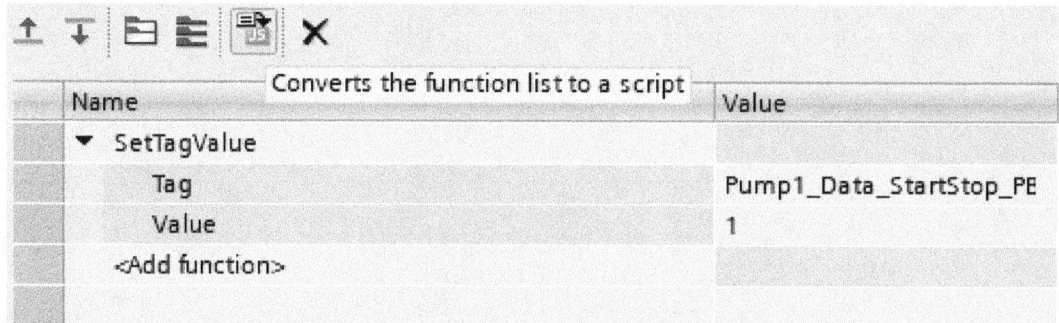

Figure 10.25 – Convert to script button

When the button highlighted in *Figure 10.25* is clicked, the event is converted to a script equivalent, as illustrated here:

```
1   export async function Button_1_OnTapped(item, x, y, modifiers, trigger) {
2   HMIRuntime.Tags.SysFct.SetTagValue("Pump1_Data_StartStop_PB", 1);
3
4   }
```

Figure 10.26 – Equivalent event script

This can then be modified or added to make an event more dynamic. An example use case may be to write to more than one tag dynamically.

Summary

This chapter has covered the setting of properties and the dynamization of properties, as well as scripting and raising events. All of these are required in order to successfully interact with the HMI and any associated PLCs connected to the HMI.

Understanding when to use scripts over the built-in configuration menus is something unique to each programmer and application developed. It is important to remember that scripting should be used only when needed and not *for the sake of it*. Large scripts can cause issues if there are too many running at once.

The next chapter explores connecting structured data from a PLC to predefined **faceplates** in an HMI, allowing large datasets to be retrieved and utilized easily while maintaining a standardized and structured approach.

11
Structures and HMI Faceplates

This chapter expands further on the benefits of using structured data and **faceplates** in the Unified **Human Machine Interface** (**HMI**) environment. By utilizing faceplates, programmers can standardize graphical objects that use standard structures, completing the structured development approach.

In this chapter, we'll learn how to enhance projects even further using structured approaches and how to pass single tags as structures to an HMI faceplate. This approach enables HMI interfaces to stay updated with **programmable logic controller** (**PLC**) interfaces, leading to fast and effective programming.

In this chapter, we'll cover the following topics:

- What are faceplates?
- Creating a faceplate
- Creating interfaces
- Creating and handling events in faceplates

What are faceplates?

Just as with a function block in a PLC, a faceplate is a reusable object that can be instantiated with its own parameters. This allows an HMI to reuse objects with interface properties populated with different values. You can see an example of a faceplate instantiation in the following screenshot:

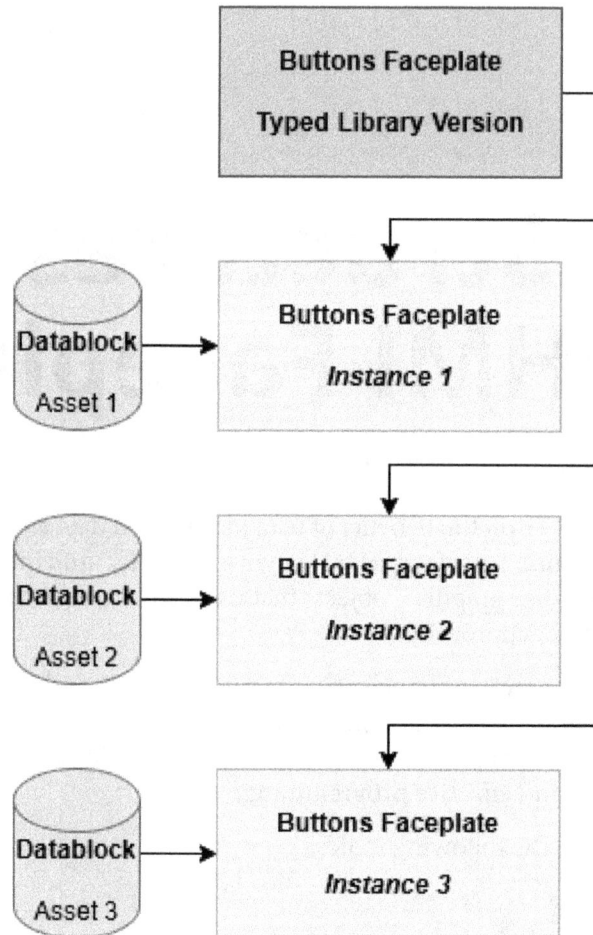

Figure 11.1 – Example of faceplate instantiation

Faceplates are very useful for common object controls due to their reusable nature. They also offer a method by which we can update multiple instances by modifying one library version. In **Totally Integrated Automation Portal** (**TIA Portal**), faceplates must be created at the **Project library** level and then instantiated in the project. This ensures the paradigm shown in *Figure 11.1* is followed at all times.

TIA Portal V17 faceplates

In TIA Portal **Version 17 (V17)**, there are two types of faceplates: one for **Panels / WinCC Runtime Advanced** and another for **Unified Comfort Panel / WinCC Unified PC**, as illustrated in the following screenshot:

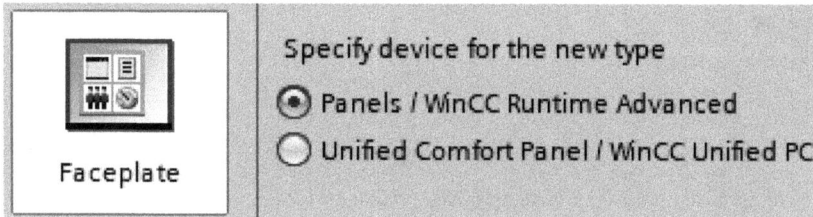

Figure 11.2 – Choice of type of faceplate when creating a new one

The Unified environment is the latest addition from Siemens. For the purposes of this chapter, the **Unified Comfort Panel / WinCC Unified PC** faceplate will be used.

Creating a faceplate

Creating a faceplate starts in the **Libraries** window on the right-hand side of TIA Portal, as illustrated in the following screenshot:

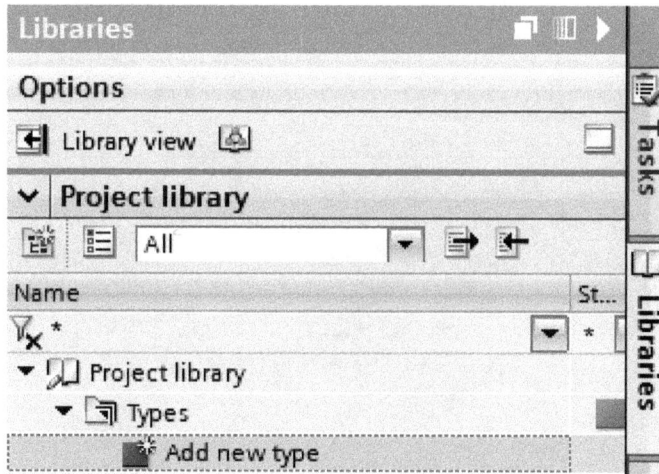

Figure 11.3 – Libraries window

By double-clicking the **Add new type** object, a new window is opened where a faceplate can be named and the type of faceplate can be selected, as illustrated in the following screenshot:

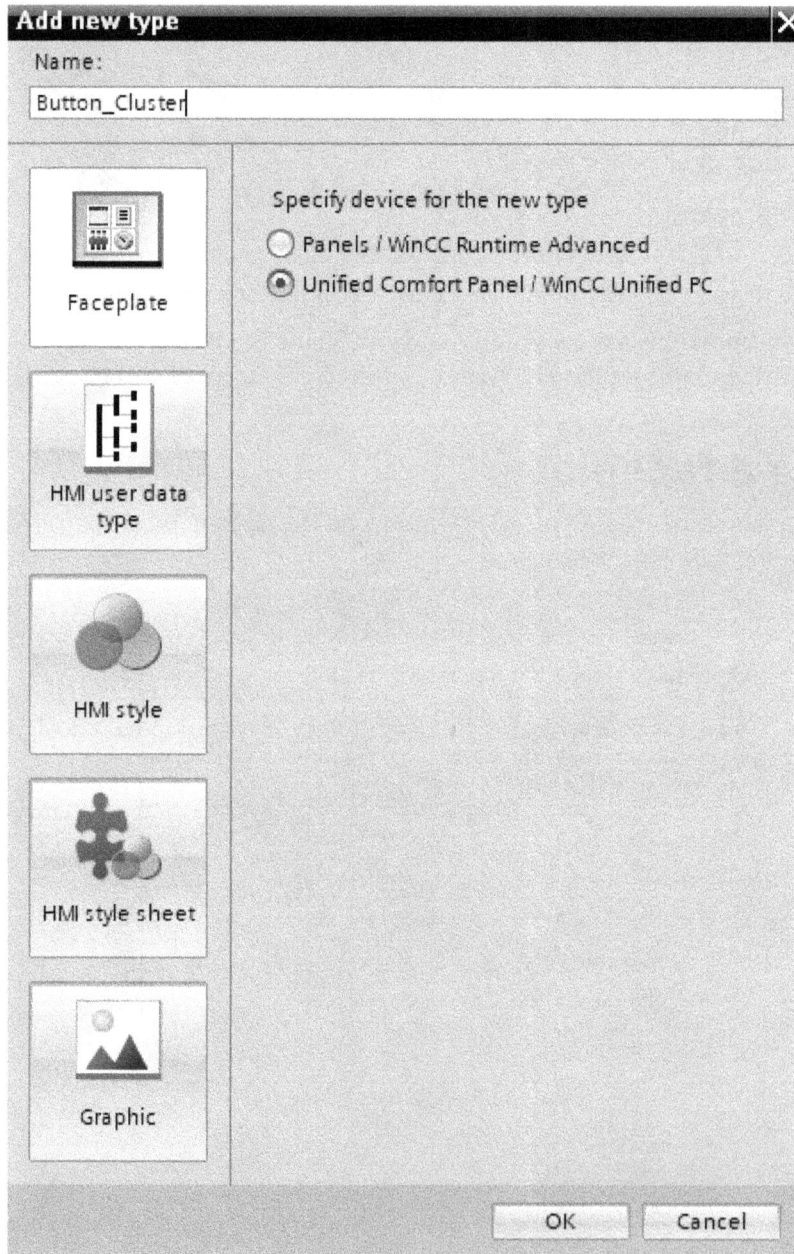

Figure 11.4 – Adding a new faceplate

After clicking **OK, Library view** will open and display version 0.0.1 of the new faceplate. The new faceplate will be in the *in work* status. When a faceplate is *in work*, it is unavailable to be instantiated.

> **Note**
> *In work* is the equivalent of *in test* when editing library objects in a PLC.

Available objects and controls

Not all screen objects and controls are available in faceplates. In the **Toolbox** window on the right-hand side of TIA Portal, the **Controls** and **My Controls** tabs are empty when editing a faceplate, and the **Dynamic widgets** window is not displayed at all.

Faceplates are designed to be basic by design. They are best utilized to serve as repetitive interfaces to standardized data. This is the reason why only **Basic objects** and **Elements** are offered in **Toolbox**, along with items from the **Graphics** section.

Have a look at the following screenshot:

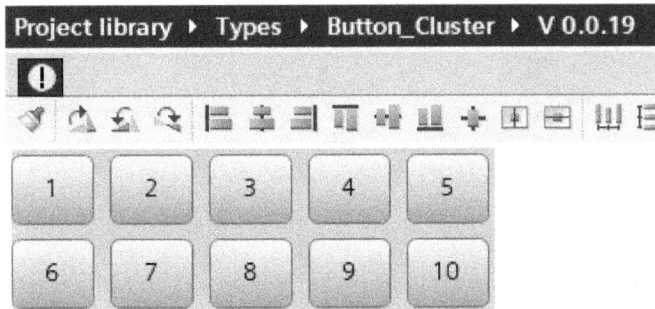

Figure 11.5 – Faceplate example

Figure 11.5 is an example of 10 buttons placed in a faceplate called Button_Cluster. This particular faceplate is on its 19th development iteration, which is evidenced by the incremental version number being 0.0.19.

> **Note**
> Faceplates are created in the same way that screens would be created: dragging objects and elements onto the faceplate and configuring them with properties.

Creating interfaces

Just as with function blocks in the PLC environment, faceplates also have an interface, by which information is passed into the faceplate.

There are two types of interfaces in faceplates, as outlined here:

- **Tag interface**: Connectss HMI tags to the faceplate
- **Property interface**: Sets properties of objects and elements in the faceplate

By using these two types of interfaces, faceplates can be customized both in the data that the faceplate has access to and the behavior of internal screen objects via their properties.

Tag interface

Tag interface is accessed by clicking the tab in the top right of the main window, as illustrated in the following screenshot:

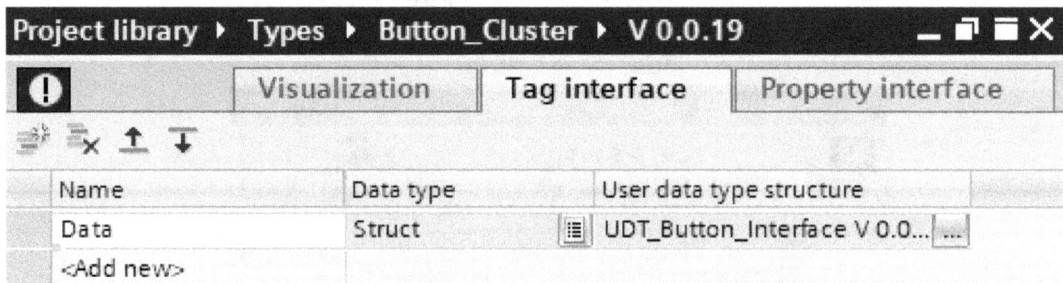

Figure 11.6 – Tag interface tab

This tab contains interface elements that allow the parent screen to pass data types and associated values to the faceplate.

When passing information to the faceplate from the parent screen, the **Faceplate** instance must be selected and the **Interface** property populated with required data, as illustrated in the following screenshot:

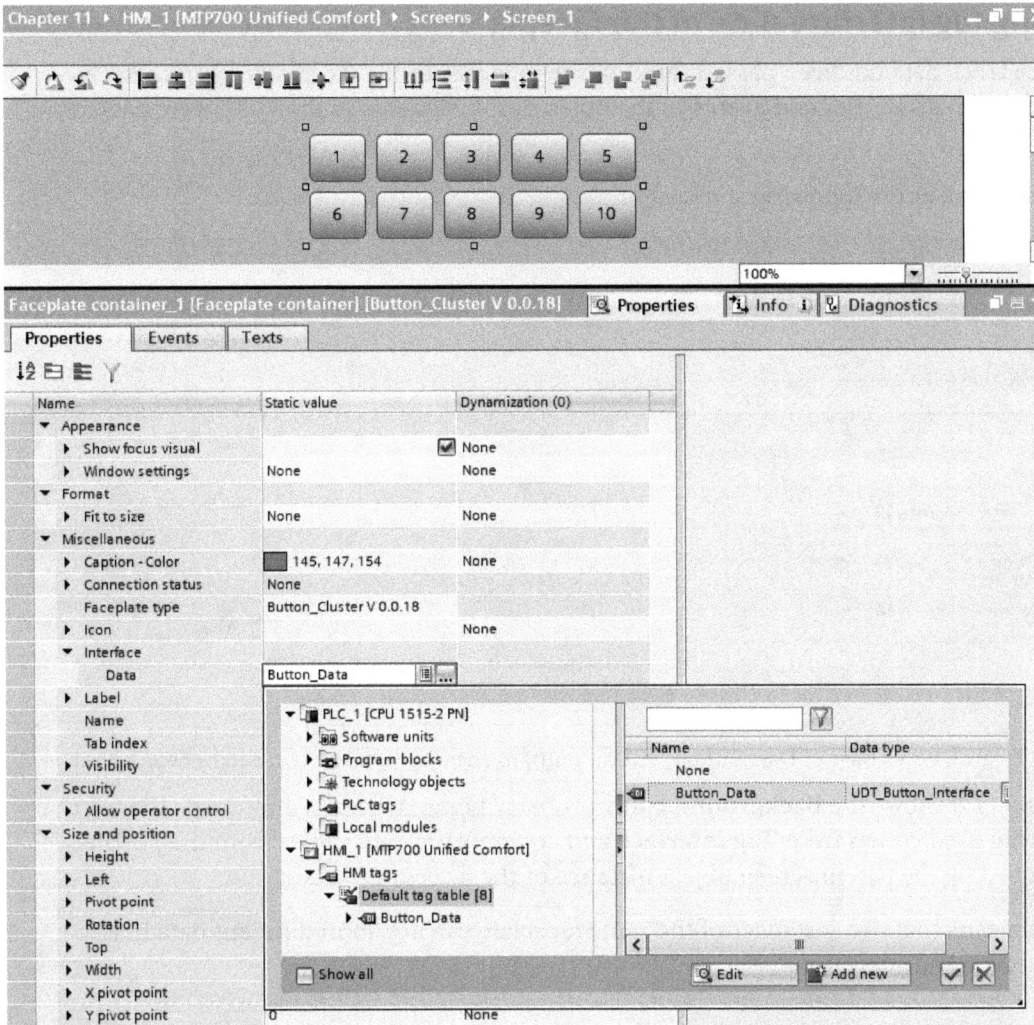

Figure 11.7 – Example of interface data being passed to the faceplate

Figure 11.7 shows an example of passing the `Button_Data` HMI tag to the faceplate using the `Data` tag interface element. By passing the `Button_Data` **user-defined type (UDT)**, the faceplate then has access to all elements within the UDT.

Note

In order for a UDT to be passed as a tag interface element, the UDT must exist as a `Typed` object in the same **Project library** section as the faceplate. A **typed object** is any object that is version-controlled in the `Typed` folder of a **Project library**.

Using tag interface data in the faceplate

Once UDT data has been passed through the interface to the faceplate, it can then be used in the properties and events of the objects as if elements in the UDT were passed individually.

Have a look at the following screenshot:

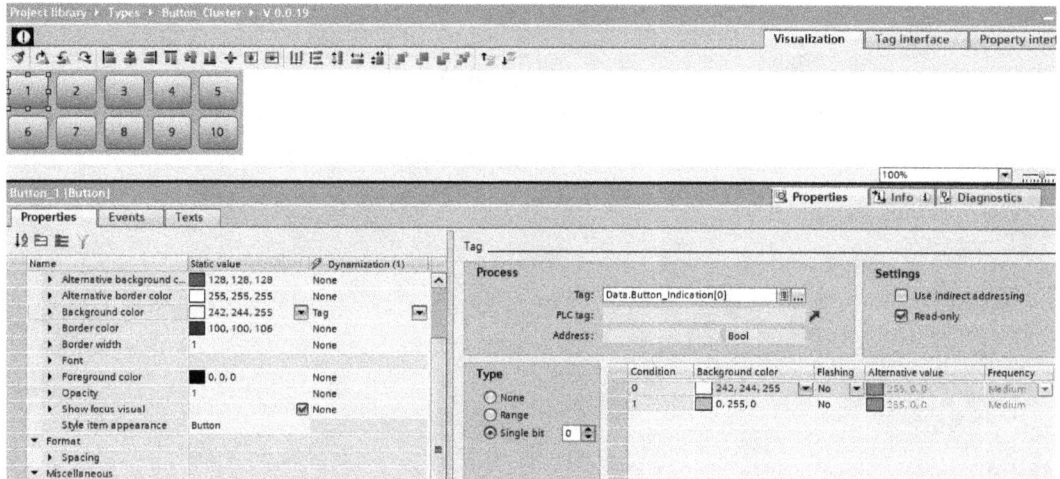

Figure 11.8 – Usage of Data.Button_Indication[0] to change the background color of an object

Figure 11.8 shows the **Background color** property being dynamized by a tag. The tag that is being used comes from **Tag interface** and contains the data values from the HMI tag assigned by the parent screen object instance of the faceplate.

This means that two instances of the same faceplate can act upon different data that is passed via the interface.

Property interface

Property interface is very similar to **Tag interface**, except it allows external access to dynamization for internal faceplate properties.

This means that items such as **Background color** can be exposed at the instance level so that programmers can set the value using the appropriate tools (such as the color picker).

Property interfaces are configured on the **Property interface** tab at the top right of the TIA Portal window when editing a faceplate, as illustrated in the following screenshot:

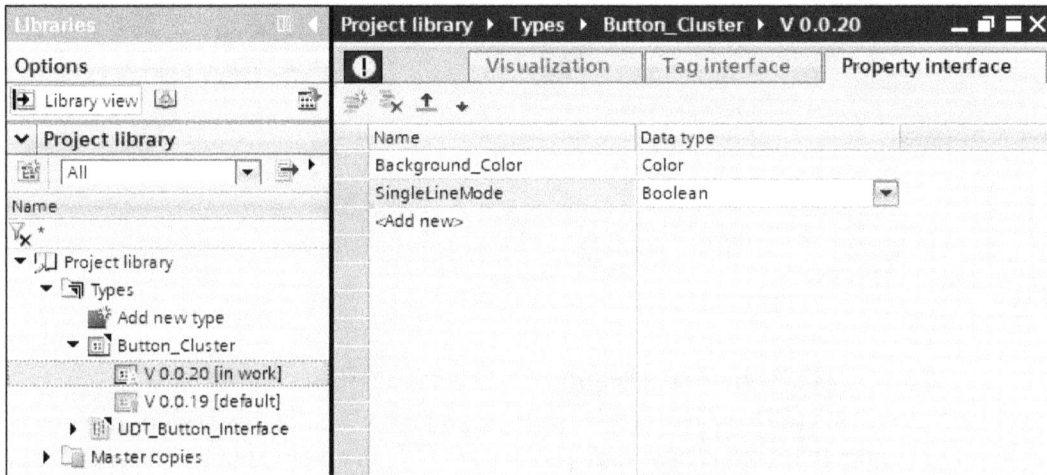

Figure 11.9 – Property interface configuration

Figure 11.9 shows two properties that have been configured for the Button_Cluster faceplate. These properties can be used for the dynamization of objects that exist within the faceplate.

Using a property interface

Once a property interface has been created in the faceplate, the properties can be used by an instance of the faceplate to change the value of a property that the property interface is assigned to.

In order for this to be available, the faceplate must first allocate a *dynamization* method to the property that needs to be accessed at an instance level, as illustrated in the following screenshot:

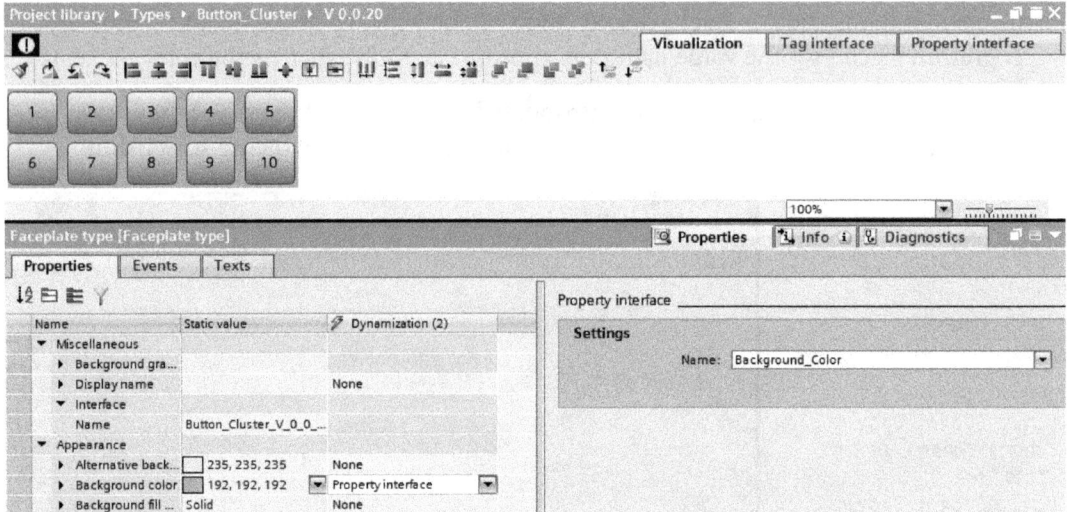

Figure 11.10 – Background color dynamization set as the Property interface

Figure 11.10 shows that the property called **Background color** has had its **Dynamization** method set to **Property interface**. The area to the right, which allows configuration of the **Dynamization** method, simply shows a dropdown of property names that come from the **Property interface** tab.

> **Note**
>
> Only compatible data types are shown. The tag name will not appear in the list if the tag data types do not explicitly match.

Once the dynamization is configured, instances of the faceplate can have the property set at the interface of the faceplate, as illustrated in the following screenshot:

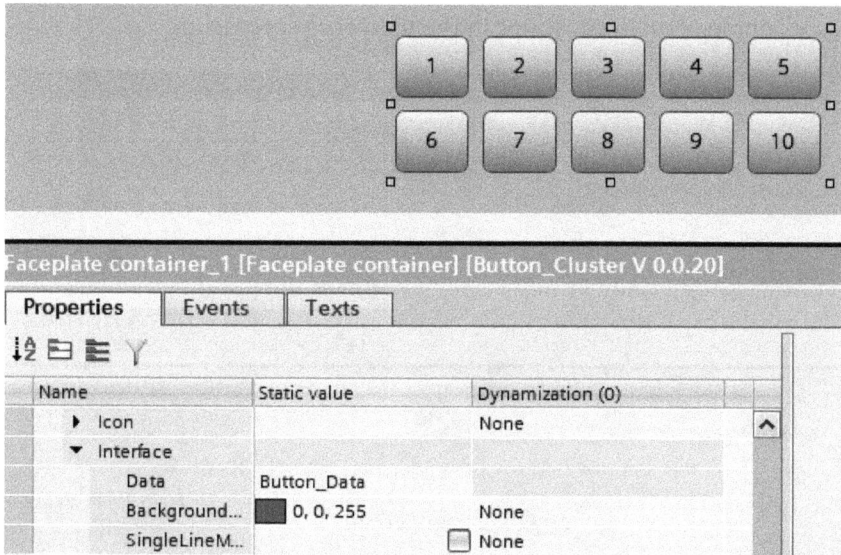

Figure 11.11 – Instance of a faceplate with interface properties

When a faceplate has **Tag interface** or **Property interface** items configured, the configured interface items will appear in the **Interface** property area of the faceplate.

Figure 11.11 shows that the **Background** property and the `SingleLineMode` property both exist in the interface of the faceplate. At this point, these values can be changed in the same way that a normal property value would be changed. The properties can also have dynamization applied to them.

Note

Scripts can also access interface tags. An example would be this:

```
var single;

single = Faceplate.Properties.SingleLineMode
```

The `single` variable then contains the value contained in the **Interface** property.

Creating and handling events in faceplates

Some faceplates may require end users to interact with objects within the faceplate, such as a button or an input field. Unlike normal screen objects, faceplates require events to be handled in **scripts**. This means that programmers will need to write scripts in order to process user events.

You can see an example of an event script in the following screenshot:

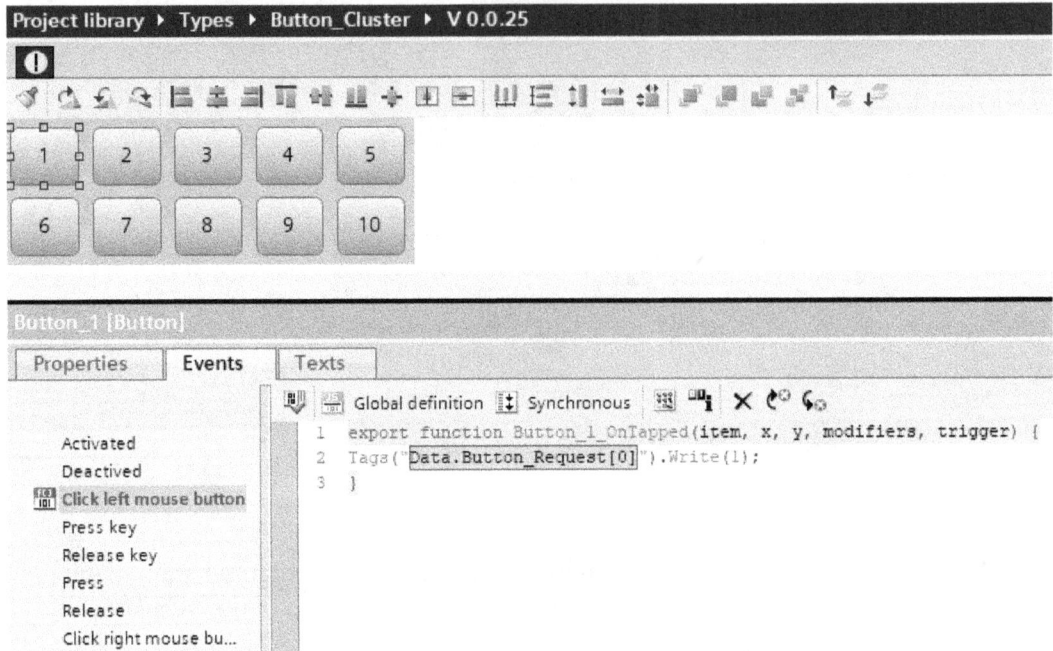

Figure 11.12 – Example of an event script

Events can be configured against a selected object by clicking the **Properties** tab at the bottom of the TIA Portal application and then the **Events** tab.

On the left side of the **Events** tab, *event triggers* are listed. When an event trigger is selected and the **Converts the function list to a script** button is pressed, as illustrated in the following screenshot, an area will appear in which a script can be written:

Figure 11.13 – Converts the function list to a script button

In faceplates, actions cannot be added in any manner other than scripts.

Accessing tags

In order to write a value out of a faceplate, the variable to be written must exist in the **Tag interface**.

Tags can be written using the following script:

```
Tags("TagVariable").Write(Value);
```

Figure 11.12 shows an example of setting up a button click event. When the left mouse button clicks button 1, the `Data.Button_Request[0]` value is written with a value of 1.

Scripting allows programmers to get more dynamic. While hardcoding the name of a tag works perfectly fine, sometimes it's better to derive the name of the tag, which means when the object is copied, the new tag name resolves itself and no editing of the script is required.

Have a look at the following screenshot:

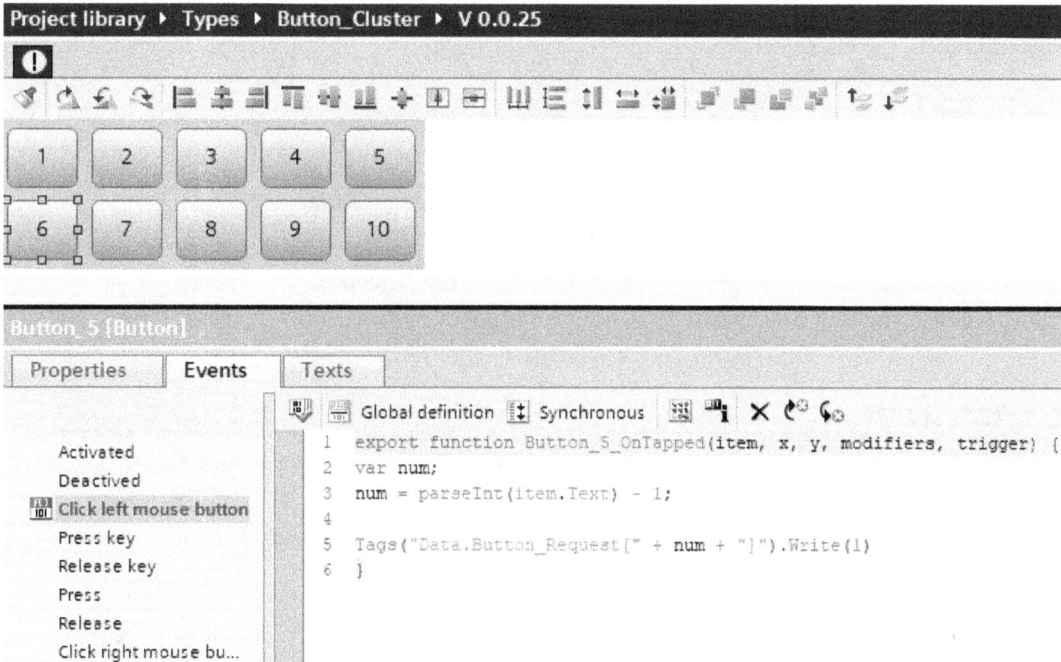

Figure 11.14 – Dynamic script

Figure 11.14 shows a script that derives the numerical index for an array based on the number that is assigned to the button text of the selected object.

Note

Because arrays start at 0, the value in the object's `Text` field is subtracted by 1.

This method allows for any numerical value to be placed in the object's Text field, and the script will dynamically access the correct tag.

This approach also works for reading tags through script dynamization for properties.

Now, have a look at the following screenshot:

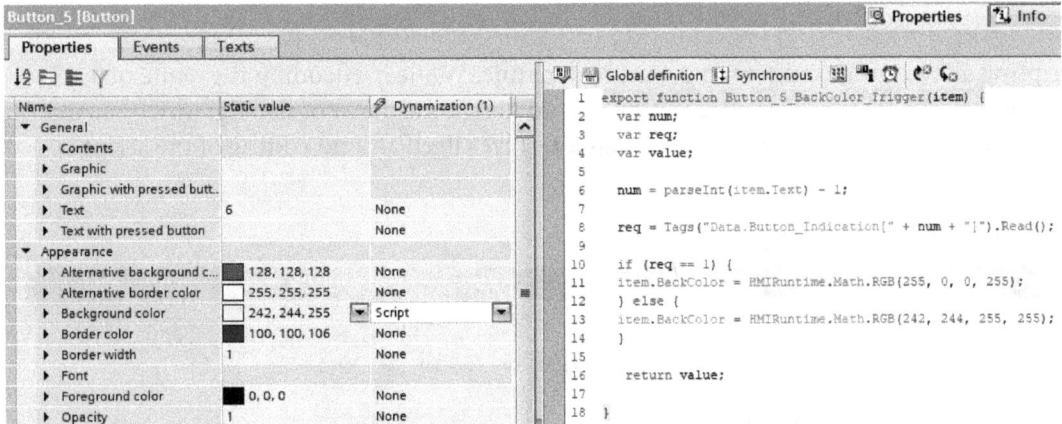

Figure 11.15 – Dynamization with a dynamic script

Figure 11.15 shows the **Background color** property being updated with colors depending on the status of the Data.Button_indication array. The array is indexed by the script, based on the text that is placed inside the object.

Summary

This chapter has explored faceplates and how to create an interface and use a faceplate. By building frameworks and standard objects that use faceplates, UDTs, and standard program blocks, the ease with which projects can be developed increases significantly.

When working with faceplates, it is important to remember that they are encapsulated and have no access to HMI tags or global scripts. In order to use a faceplate, the interface needs to be set up appropriately. It's also worth noting that faceplates must be created in the **Project library**, and only released faceplates can actually be used in a screen. Similarly, in order to modify a faceplate, the typed version must be edited in the **Project library**.

The next chapter covers navigation and alarms in the Unified environment. This includes how to raise, accept, and clear alarms, and the differences between PLC-driven alarming and conventional alarming.

12
Managing Navigation and Alarms

This chapter focuses on how to create navigation between pages in TIA Portal's Unified HMI environment and how to create an alarm management system between the HMI and PLC.

Navigation is important in the HMI, and it is something that an end user interacts with often. There are different styles and approaches to navigation. Alarming is also important as that's how an end user is notified of potential issues. Methods for alarming can also be achieved in many different ways.

In this chapter, the following topics will be explored:

- HMI navigation
- HMI alarm controls
- Alarm tags – accepting and resetting
- PLC-driven alarming

HMI navigation

In order to change pages on the HMI so that more than one page of information can be displayed, a **navigation** system is required.

TIA Portal offers a standard approach to managing page navigation that is both simple and easy to utilize. Depending on the application requirements, pages can be configured in multiple ways, such as *free navigation* or *controlled navigation*:

- Free navigation is a method where every page has a navigation link to every other page:

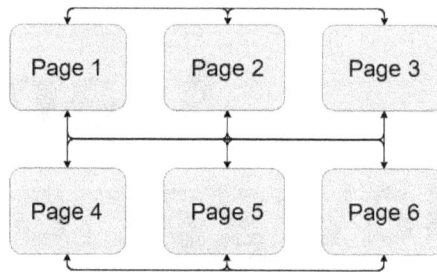

Figure 12.1 – An example of free navigation

- Controlled navigation is a method where page access is controlled via the current page. This allows the segregation of settings pages, for example:

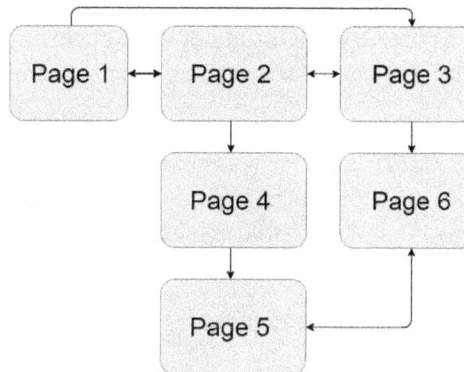

Figure 12.2 – An example of controlled navigation

Free and controlled navigation styles can be mixed also, whereby some pages are accessible at all locations, and some are controlled and can only be accessed in key locations. Mixing navigation styles can also help with the management of page changes.

Managing page changes

TIA Portal Unified HMIs have a built-in event method that triggers a display page change. It is important to understand the dynamics of how the screens can be configured to understand how the event can operate.

TIA Portal offers complete control over navigation, giving options for which screen should be loaded into view and where that view exists. This means that programmers can update the base screen (the first screen that is loaded), or can update a **screen window** object, which is essentially a screen within a screen. *Figure 12.3* is an example of a base screen that is never navigated away from:

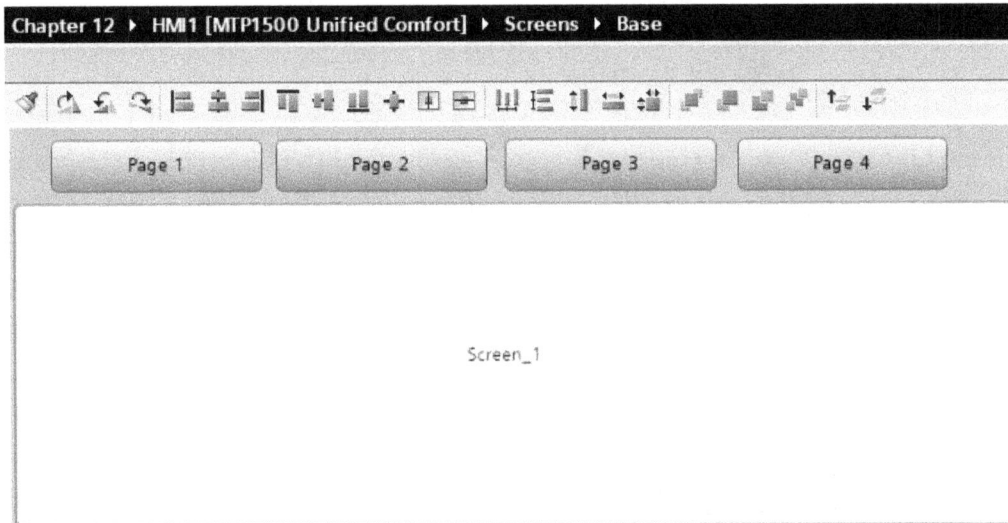

Figure 12.3 – An example of a base screen containing navigation buttons and a screen window object

All of the navigation takes place in Screen_1, which is a screen window object. By following this approach, static items (such as the top navigation menus) can be accessed from any page. Individual screens can then be loaded into Screen_1.

The ChangeScreen event

The **ChangeScreen** event can be used on nearly all objects that allow events to be configured. *Figure 12.4* demonstrates the **ChangeScreen** event being used to update the current screen to the Screen_1 screen:

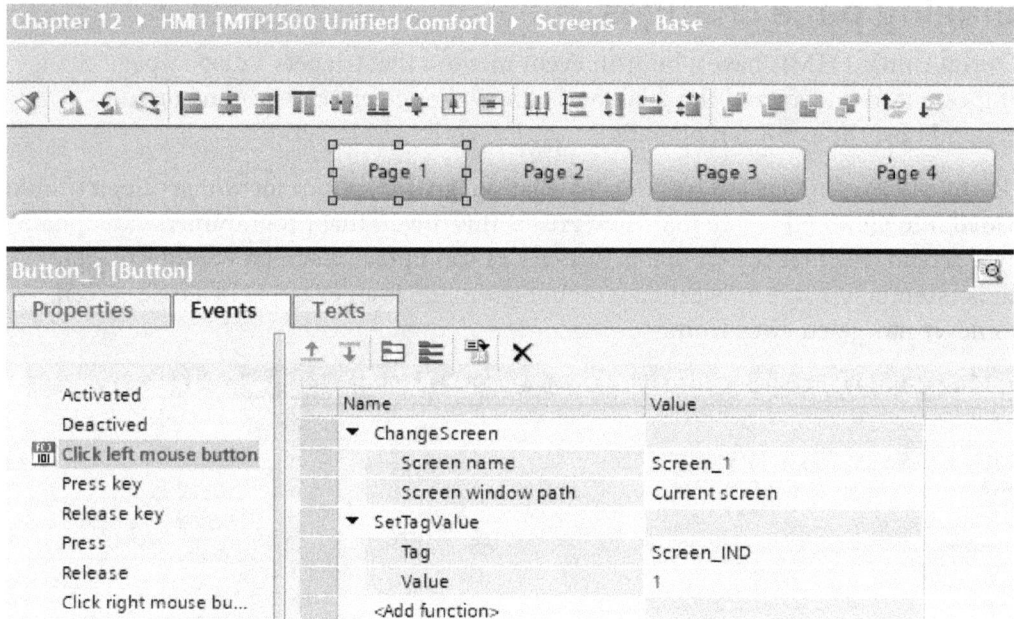

Figure 12.4 – The ChangeScreen event

There is also another event that is setting an HMI tag called Screen_IND to the value 1, which is used to change the background color of the **Page 1** button.

In this particular case, this screen change will result in the navigation buttons for pages 1, 2, 3, and 4 disappearing from view, as Screen_1 would be loaded into the current screen, which is also the **base screen**. By changing the **Screen window path** property, the target of the screen change can be directed to the Screen_1 screen window.

> **Important Note**
> Screen_1 is an ambiguous name in this example and relates to both a screen page and also a screen window. TIA Portal does not raise this as a conflict, as the screen window exists only in the base screen and therefore has a different **call path**.

In order to change the screen window page, the **ChangeScreen** event needs to be modified at the **Screen window path** parameter. This can be done by clicking the small button to the right of the **Value** field and choosing the **Screen window** option:

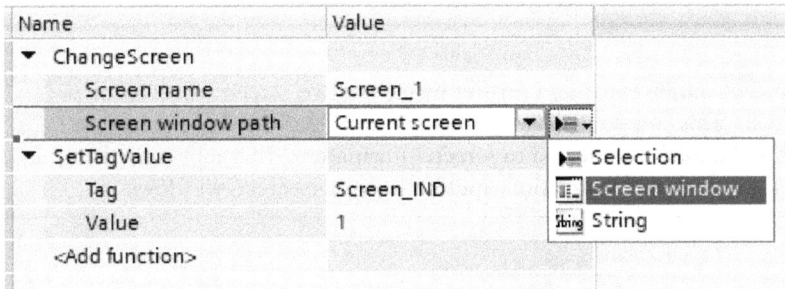

Name	Value	
▼ ChangeScreen		
Screen name	Screen_1	
Screen window path	Current screen	▼ ►☰ ▾
▼ SetTagValue		►☰ Selection
Tag	Screen_IND	🗔 Screen window
Value	1	String String
<Add function>		

Figure 12.5 – The Screen window path options

When this is completed, the **Screen window path** value will be removed, and a new button option will appear. After selecting **Screen window path** as the required screen window, the new path will be displayed in the **Value** column:

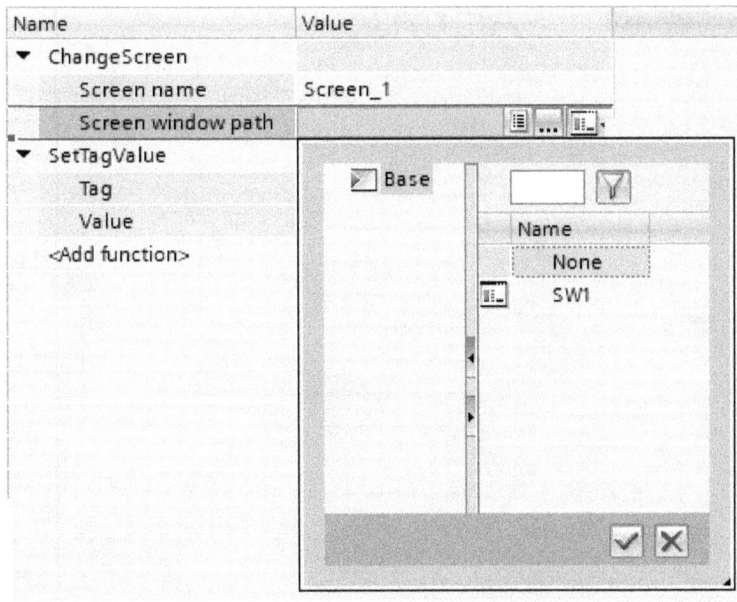

Figure 12.6 – The Screen window path screen window configuration

The button is now configured to change the **SW1** screen window. Because both the button and screen window are on the same base screen, the button is always accessible no matter what page is displayed.

> **Note**
>
> This concept can be expanded further by having two screen windows on a base screen. This can then allow a screen window dedicated to navigation buttons and another dedicated to screen information. This approach allows for complete control over navigation, split between two screen windows.

HMI alarm controls

One of the main reasons for an HMI to be included in a project is so that alarms can be relayed to an operator or end user. TIA Portal has a ready-made control for alarm management called the **alarm control**:

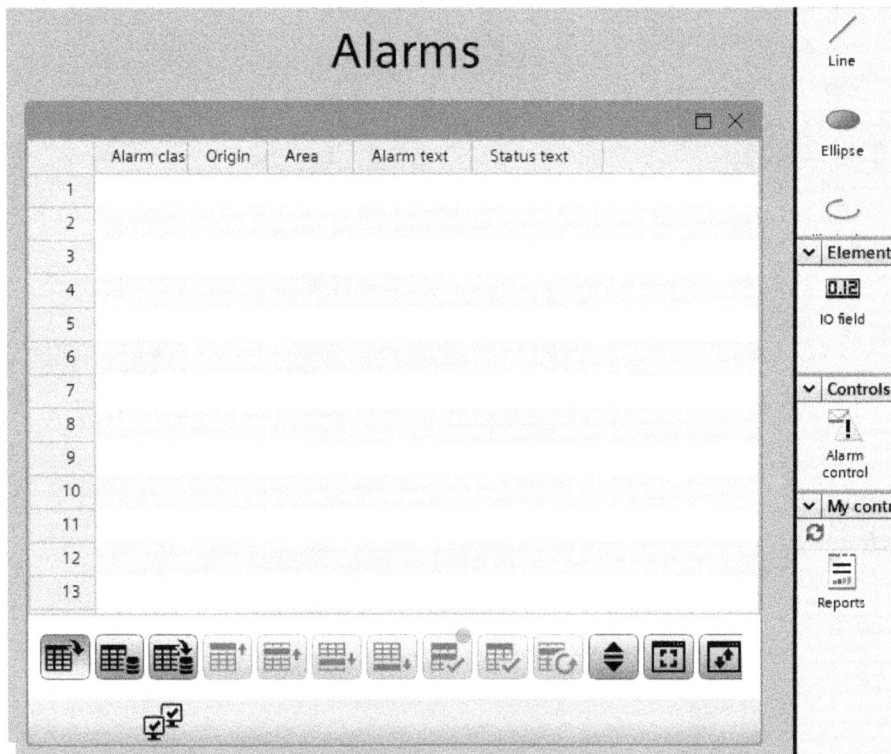

Figure 12.7 – An example of an alarm control on a screen with the alarm control object in the toolbox on the right-hand side

The **Alarm control** option can be found in the **Controls** section of the toolbox. **Alarm control** can simply be dragged and dropped onto a screen as with any other object.

Alarm control connects to many different areas and has many different configuration options for customizability. In most cases, for a quick setup, the preset options are enough to display and interact with alarms.

> **Note**
> The HMI used in this chapter is an *MTP1500*, running *V17 Firmware*.

Configuration of HMI alarms

In the Project tree and under the HMI object, an object called **HMI alarms** exists. It is within **HMI alarms** that the alarm text, triggers, and more are configured. *Figure 12.8* demonstrates the configuration of some discrete alarms:

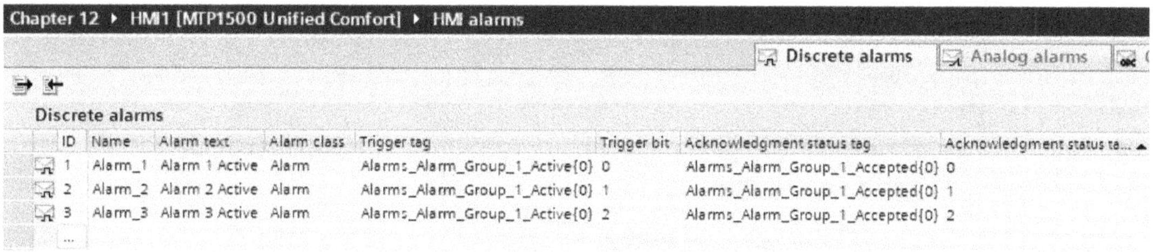

	ID	Name	Alarm text	Alarm class	Trigger tag	Trigger bit	Acknowledgment status tag	Acknowledgment status ta...
	1	Alarm_1	Alarm 1 Active	Alarm	Alarms_Alarm_Group_1_Active{0}	0	Alarms_Alarm_Group_1_Accepted{0}	0
	2	Alarm_2	Alarm 2 Active	Alarm	Alarms_Alarm_Group_1_Active{0}	1	Alarms_Alarm_Group_1_Accepted{0}	1
	3	Alarm_3	Alarm 3 Active	Alarm	Alarms_Alarm_Group_1_Active{0}	2	Alarms_Alarm_Group_1_Accepted{0}	2

Figure 12.8 – Example of discrete alarms configured in the HMI alarms window

These alarms are monitoring an associated trigger tag and acknowledgment status tag. These tags are used to display the alarm's status at various points across an alarm management routine. This routine is managed at a **class** level. The Unified HMI environment allows custom alarm classes to be created, but by default, the **Alarm** alarm class is used.

Trigger tag is the PLC (or HMI) tag that is used to trigger the alarm. In most cases, this is a Word or Dword data type; however, a Boolean is also accepted. **Trigger bit** is used to select the individual bit inside a Word or Dword data type.

The **Acknowledgment status tag** column is used by the HMI in order to hold whether or not the alarm has been acknowledged by an operator. This tag is usually held in the PLC also. Unlike **Trigger tag**, **Acknowledgment status tag** cannot be a Boolean and must be a Word or Dword data type with an **Acknowledgment status tag bit** set.

Analog alarms can also be configured using the **Analog alarms** tab. These alarms use a predefined limit and limit mode. The trigger tag is evaluated against the limit using limit mode; if the condition is `True`, the alarm is raised in the same way that a discrete alarm is raised.

> **Note**
> Discrete alarms are digital alarms, where the trigger tag is driven by a `boolean` data type tag. Analog alarms are driven by a `Numerical` data type tag and have different limit modes that raise alarms.

The `alarm` class is configured by default to follow the following routine, otherwise known as the *alarm with single-mode acknowledgment* state machine:

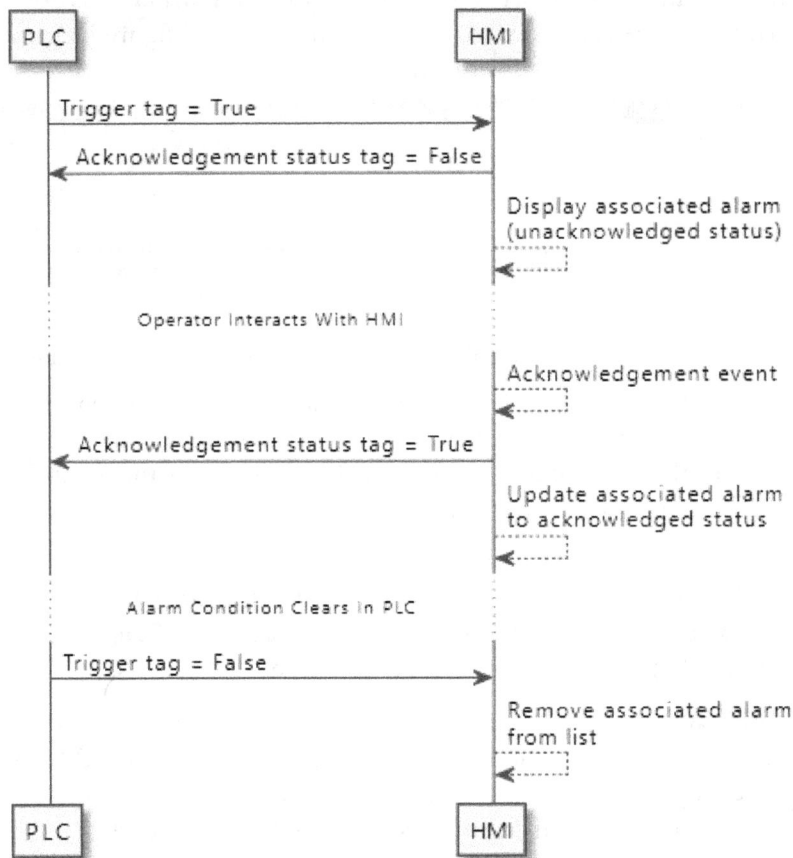

Figure 12.9 – An alarm with the single-mode acknowledgment state machine routine

Figure 12.9 shows how the PLC and HMI interact with each other to manage an alarm. When the alarm's **Trigger tag** Boolean is set to `True`, the HMI performs some management in the form of updating the **Acknowledgment status tag** column and also displaying the alarm on any alarm controls that are configured to display the corresponding `alarm` class.

When an operator interacts with the HMI and acknowledges an alarm, the acknowledgment status tag is set to `True` and the alarm control is updated. When the trigger tag returns to `False`, the alarm is removed from the alarm control list. If the trigger tag returns to `False` before the acknowledgment status tag is set to `True`, then the alarm is immediately removed from the list on acknowledgment.

> **Note**
>
> Alarm controls, by default, display all alarms. However, they can be configured to filter out particular alarms so that only customized lists are displayed. This will be covered later in this chapter in the *Setting filters on alarm controls* section.

The configuration of classes

The **HMI alarms** window contains an **Alarm classes** tab, which allows the configuration of custom alarm classes as well as the configuration of predefined alarm classes:

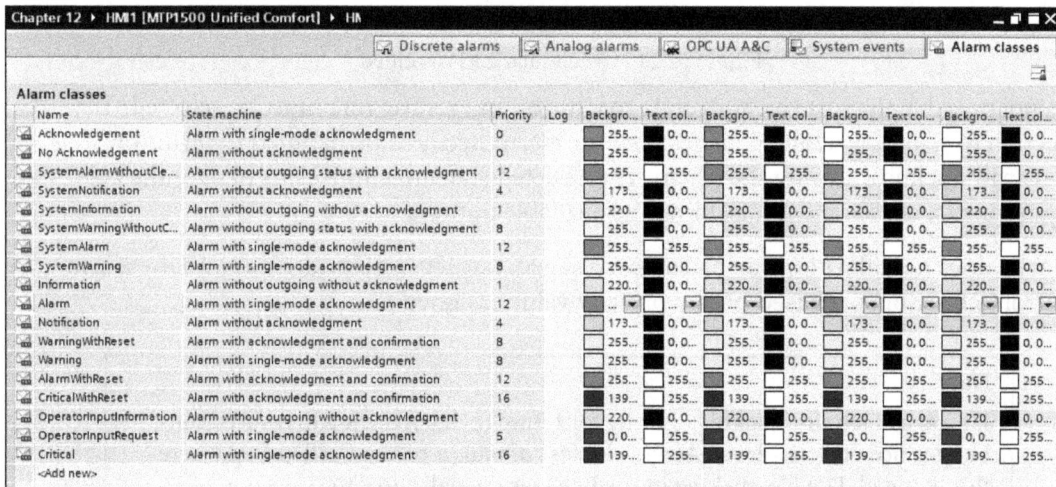

Figure 12.10 – The Alarm classes tab

The predefined alarm classes are set by TIA Portal and the names cannot be edited. In most cases, the colors can be edited, but some are restricted and cannot be edited.

Every alarm configured in the HMI must have an associated class, which is configured on the **Discrete alarms** or **Analog alarms** tab. Depending on the class configuration, the alarm will behave differently.

> **Note**
>
> Alarm controls that have no filters set will display mixed alarm classes, which may result in different alarms behaving differently from each other. It's good practice to either keep the alarm class the same for all alarms in an alarm control or set the **State machine** option for multiple alarm classes that appear in the same alarm control to the same value.

Creating a new alarm class

Whilst TIA Portal offers a wide range of different alarm classes by default, a programmer may want to create a new alarm class for their own system. A custom alarm class can be created by opening the **Alarm classes** tab in the **HMI alarms** window.

Double-clicking the <**Add new**> row will add a new alarm class with the default name Alarm_class_1. A default **Alarm with single-mode acknowledgment** state machine will also be selected:

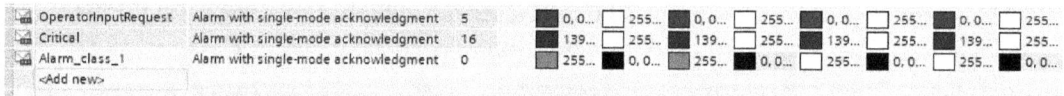

Figure 12.11 – A new alarm class added

It is not possible to create custom state machines; an existing one must be selected. The selection of a state machine changes how the alarm needs to be handled in terms of resetting or acknowledging the alarm. The selected state machine also changes the color requirements, enabling and disabling the available options.

Priority does not affect how an alarm is managed, but setting a *priority* allows the alarm control to filter or sort based on the priority value assigned.

> **Note**
>
> If an alarm requires logging (storing for future review), the log to be used must be specified at a class level. **Alarm classes** contains a column called **Log** where this is added. Double-clicking in the **Log** column will allow the selection of a configured log.

The colors to be displayed in the **Alarm control** are specified at a **Class** level in the **Alarm classes** tab.

Logging alarms

Unified HMIs can log alarms, storing the events related to alarms in a file for later review. In order to log alarms, a log object must be created from the **Logs** window.

The **Logs** window can be found in the Project tree:

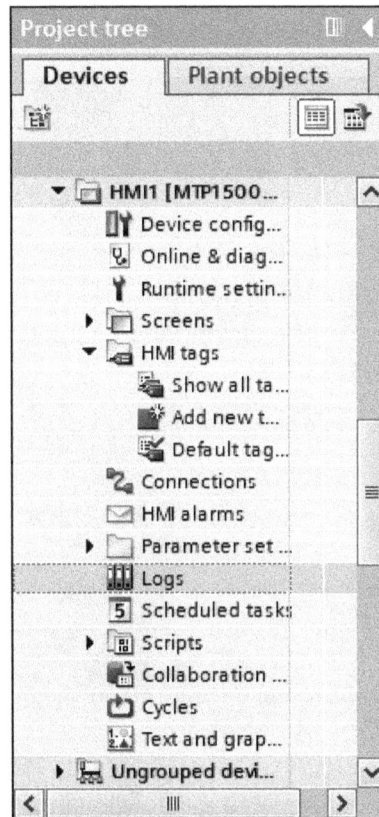

Figure 12.12 – Logs in the Project tree

In the **Logs** window, with the **Alarm logs** tab open, a new alarm log can be added by double-clicking the <**Add new**> row. A new Alarm log_1 object will be created with the default storage medium set to **SD-X51**, which is the memory card medium.

The first time a log is configured, the **Runtime settings** option needs to be modified to allow logging. If this has not been completed, a warning message will be displayed in the **Logs** window:

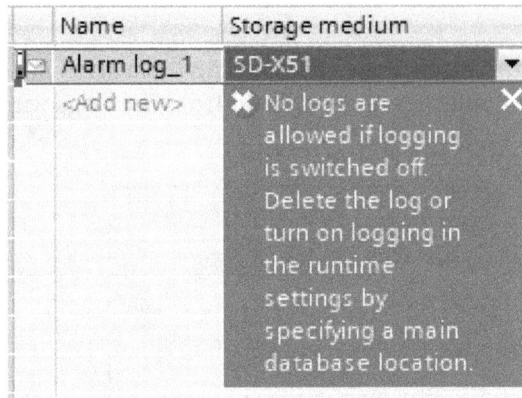

Figure 12.13 – The error if logging is set to disabled

In the Project tree, the runtime settings can be opened and the **Storage system** option selected. The **Main database location for alarm logging** option must be set to an appropriate location, such as **SD-X51**:

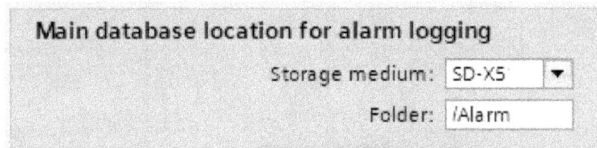

Figure 12.14 – The configuration of the main database location for alarm logging in the runtime settings

> **Note**
> It is a good idea to set up a folder too; this way, alarm logging is segregated from other logging types.

Once the runtime settings have been updated to allow logging, the configuration of the alarm log can continue.

The **Logs** window and the **Alarm logs** tab also require additional options to configure when alarms are logged and how often to save data to the storage medium. A storage directory can also be specified, which further segregates individual logs.

The log time period covers the maximum time period covered by the log. Once the time has elapsed, a new log will be created. The same is true for the maximum log size.

In addition to the log, segments of data within the log also need a maximum time period and size allocated. It is also possible to set a segment start time should it be required to start a segment at a particular time.

> **Note**
> In the Unified HMI environment, it is not possible to set a **backup mode**. The **Backup mode** option is available when using a Windows-based unified environment.

Configuration of alarm controls

In order to actually display an alarm status, an alarm control must be placed on a screen. The alarm control must then be configured to display the required alarms. If a new alarm control is placed on a screen, it will display all alarms from all classes by default.

Alarm controls can be configured to *filter* out alarms that are not required to be displayed by the alarm control. For example, an alarm control with no filter set will show `SystemAlarm` and `SystemNotification` class alarms. *Figure 12.15* shows that an alarm control with no filter shows all alarms from all classes, including those that have not been configured explicitly by the programmer:

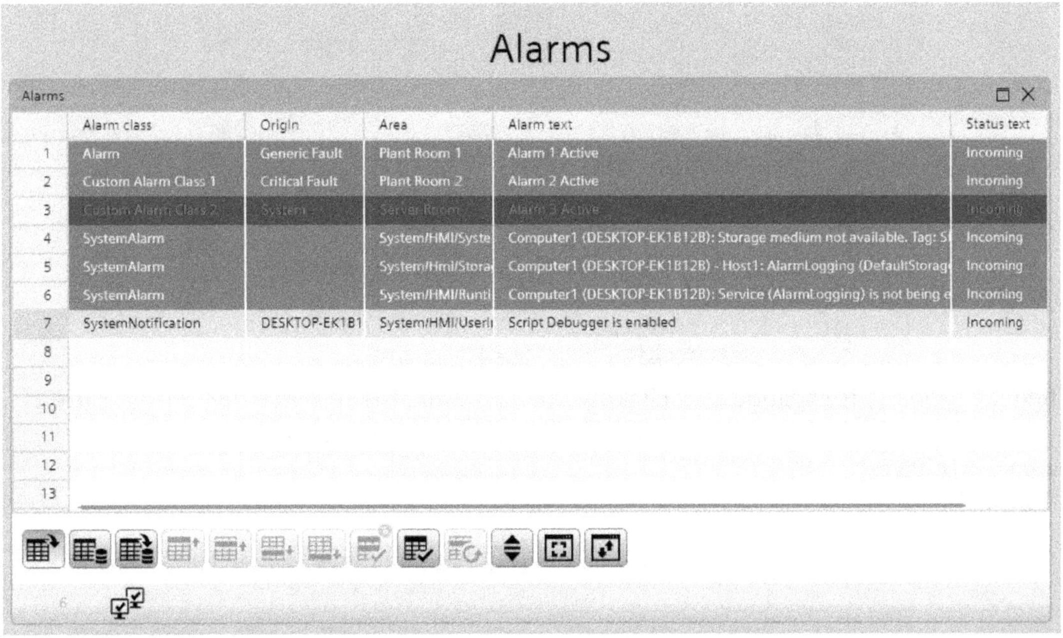

Figure 12.15 – An alarm control showing configured alarms and automatic system alarms

If a filter is not set, system alarms that may be unrelated to the project can be displayed. It is advised that controls are configured with filters so that these alarms are not displayed.

Setting filters on alarm controls

Filters can be configured in the **Properties** tab of the alarm control in TIA Portal or can also (if enabled) be set at runtime on the HMI itself. Allowing HMI operators to set the filter at runtime can be advantageous in some environments, but in most cases, filters are best set in the development environment.

> **Note**
>
> Operators of the HMI will still be able to change/remove filters if the **Selection display** toolbar option is enabled in the **Properties** tab of the alarm control. The **Selection display** button can be removed by selecting the alarm control and changing the property at **Toolbar | Elements | [26] Button (Selection Display) | Visibility** to `False`.
>
> The same can be done with all other buttons, allowing configuration of how the alarm control is to be used.

Filters can be set for an alarm control by expanding the **Filter** property for the alarm control:

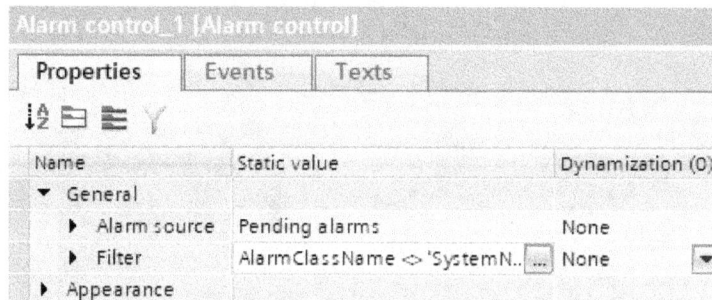

Figure 12.16 – Filtering alarm control

When the property is expanded, by clicking the **...** button, the **Alarm filter configuration** window opens, which allows filters to be created and modified. *Figure 12.17* shows an example of a filter:

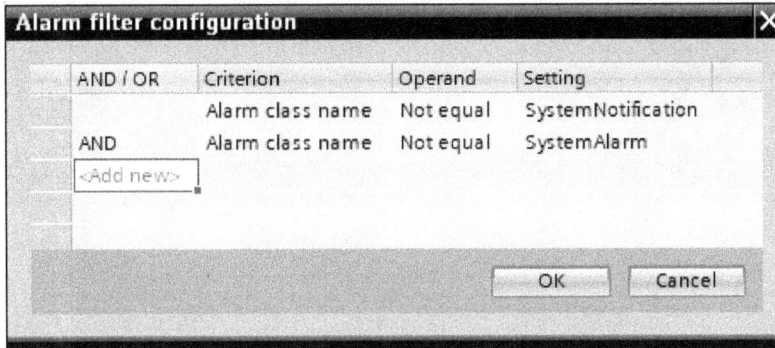

Figure 12.17 – Alarm filter configuration

By declaring **Criterion** as **alarm class name**, the Unified HMI checks which class the alarm belongs to. The **Operand** column shows how the check is performed. The **Setting** column shows the value that will be filtered.

The example in *Figure 12.17* only displays alarms that do not originate from a system class.

> **Note**
> Remember that the filter sets what is being displayed, not what is being removed.

Filters can also be modified or created at runtime if the configuration of the control allows it. To call the filter window, the following button is used on the alarm control. This button is available by default but can be removed if required:

Figure 12.18 – The filter button

When clicked, a window displays the configuration of the filter. This window behaves in a similar way to the development window:

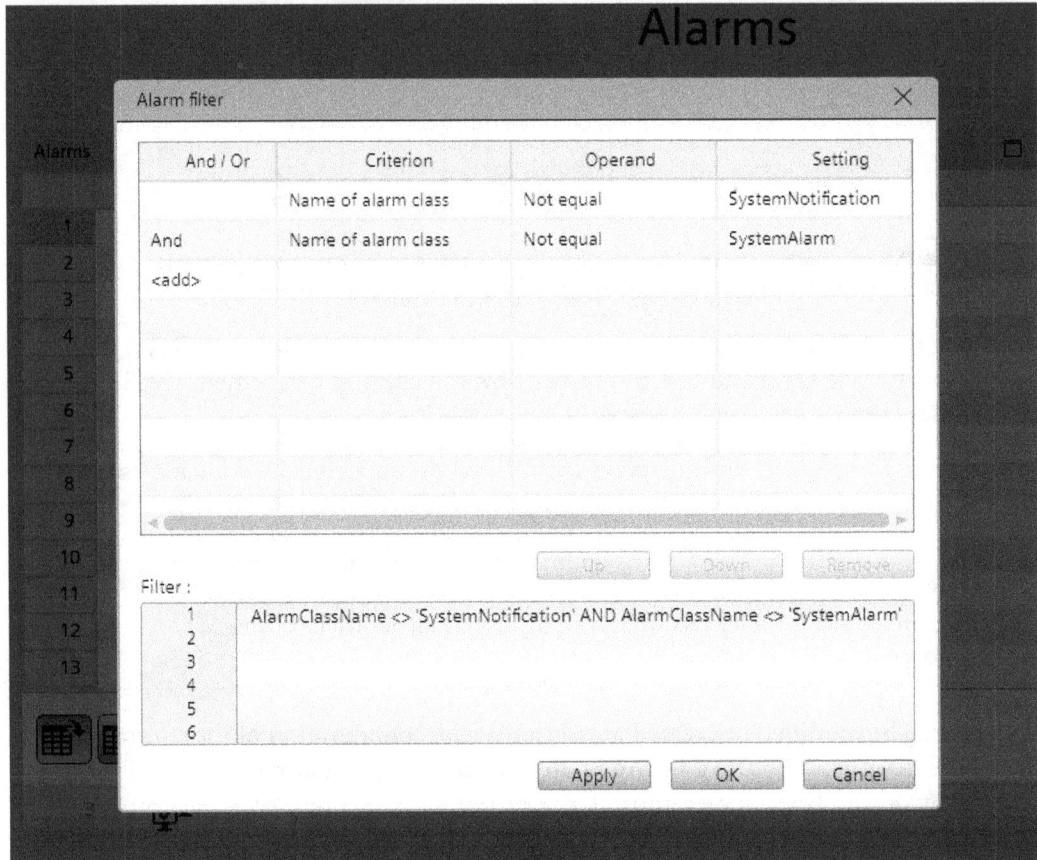

Figure 12.19 – Configuration of filters at runtime

By clicking **Apply** or **OK**, the filter is configured and set. The alarm control will then display the latest changes to the filter.

Note

Once the page containing the alarm control has been reloaded (such as navigating away and returning), the changes to the filter will have returned to those defined in the development environment; the filter is not retained.

Alarm tags

In the PLC, dedicated tags or data block addresses are required to be used in order to tell the HMI that an alarm has been raised. These exist in the HMI as the trigger tag and acknowledgment status tag.

Logic is developed against these tags in order to react to operator acknowledgment of alarms and to raise the instance of the alarm. *Figure 12.19* shows a basic example of ladder logic controlling a latching interlock:

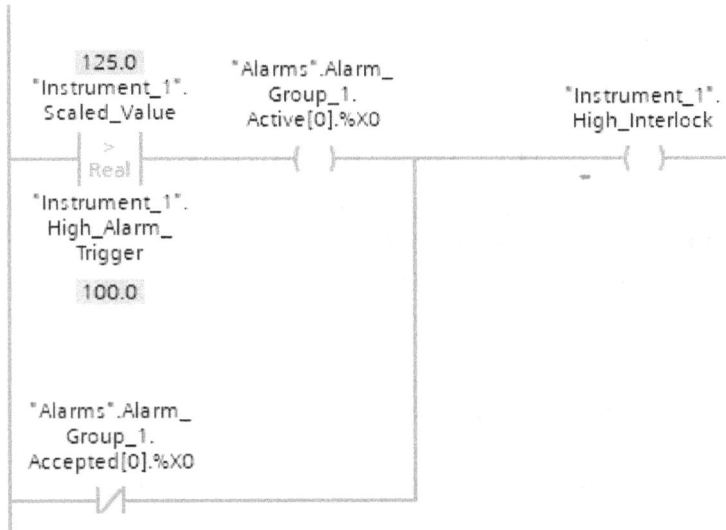

Figure 12.20 – An example of a basic alarm interface

When `Scaled_Value` for `Instrument_1` is above `High_Alarm_Trigger`, both the trigger tag (`"Alarms".Alarm_Group_1.Active[0].%X0`) and `High_Interlock` are set to `True`. When this occurs, the HMI writes to the acknowledgment status tag (`"Alarms".Alarm_Group_1.Accepted[0].%X0`) and sets it to `False`, which then holds on to the trigger tag:

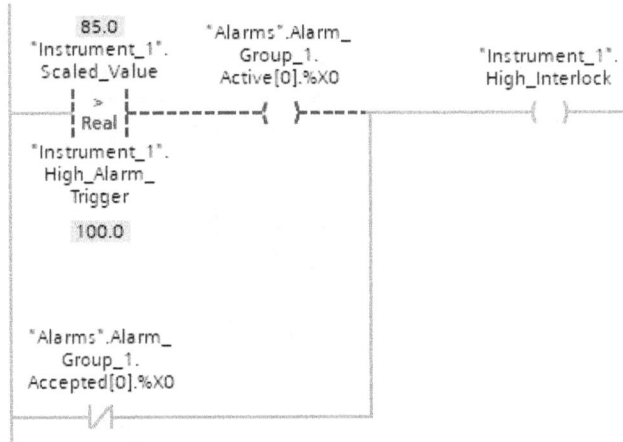

Figure 12.21 – A latched alarm in the HMI

If the trigger tag returns to a `False` value, the acknowledgment status tag is still set to `False`, as the operator has not acknowledged the alarm yet. In this scenario, the logic continues to set `High_Interlock` to `True`:

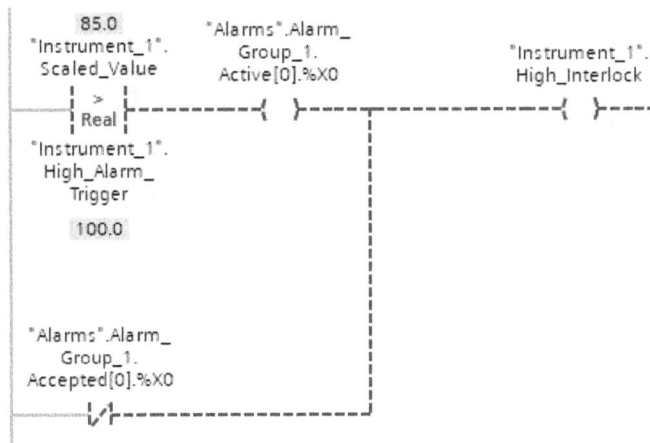

Figure 12.22 – Resetting the alarm

Only when the HMI acknowledges the alarm does `High_Interlock` reset to `False`. It is possible that the operator may acknowledge the alarm before the trigger tag becomes `False` again. If this occurs, then as soon as the alarm's trigger tag is set to `False`, the alarm will clear from the alarm control.

> **Note**
>
> There are many different methods to control alarming; each needs to be considered on the state machine in use by the alarm class.

The PLC logic that controls the alarms is best created as a function block or function that works with the `alarm` class in use. By standardizing the alarm management, alarms become quicker to deploy, and less testing is required to ensure that the alarms work correctly.

> **Note**
>
> *Figure 12.21* shows an example of accessing `bits` in a `Dword` data type. This can be done by dragging the required `Dword` (or any other numerical data type) into the instruction and then adding `.%X<b>` to the end, where `<b>` is replaced with the `bit` number required.
>
> This also works for `Bytes` in `Word` by using `.%B<b>` and other variations.
>
> More information about this is available in the TIA Portal **Help** system (press *F1*) at **Information system | Programming a PLC | Programming basics | Using and addressing operands | Addressing variables in data blocks | Addressing areas of a tag with slice access (S7-1200, S7-1500)**.

PLC-driven alarming

When a PLC and an HMI exist in the same project, **common alarm classes** can be configured for use in both devices.

Using supervisions and the ProDiag function block, the PLC can drive alarms to the HMI without having to configure HMI alarms in the HMI itself. One of the biggest benefits of using PLC-driven alarming is that the alarm resides in the function block associated with a supervision. This means that if the function block is standardized in a library, every project that uses the function block will be able to generate the same alarm automatically.

Supervisions can easily be created for any variable that is stored in a data block or as a PLC tag:

Figure 12.23 – Creating a supervision

In the **Supervision** column, right-click and choose **Add new supervision**. The properties window at the bottom of the screen opens on the **Supervisions** tab.

> **Note**
>
> TIA Portal will automatically create a global ProDiag function block and an associated data block if at least one does not already exist to manage the supervisions.

The **Supervision** tab allows for the configuration of the supervision element that monitors the variable and raises the alarm with the HMI. There are a few different properties to the supervision element that need to be configured to change the behavior of the alarm. In addition to properties displayed for the new supervision, there are also project-wide settings that need to be configured for application usage.

When a new supervision is created, the default supervision type is Operand. The **Type of supervision** option sets a preset configuration for the behavior of the supervision. The **Type of supervision** presets, and other configurations, can be modified by opening **Common data | Supervision settings** from the Project tree:

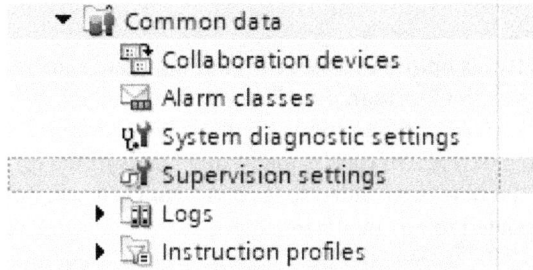

Figure 12.24 – Supervision settings

The **Supervision settings** option configures the global settings for the supervision system (or GRAPH language, in which supervisions are built).

Supervision categories

In the **General** section of **Supervision settings**, **Categories** can be configured. The categories not only help segregate alarms into different types but also allow the configuration of the alarm class:

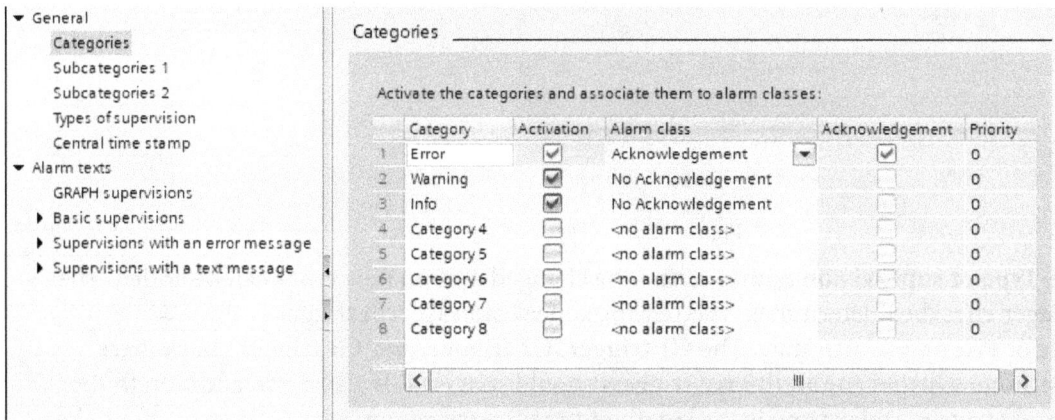

Figure 12.25 – Supervision categories

By default, only three categories are configured, with a maximum of eight available to be configured. Subcategories can also be configured from the menu on the left (as shown in *Figure 12.25*) by selecting **Subcategories 1** or **Subcategories 2**.

> **Note**
>
> The **Supervision settings** option accesses the common data alarm classes, not the HMI alarm classes. In **Common data** in the Project tree, an item called **Alarm classes** exists. Here, configurations for globally accessible alarm classes can be made. When a new alarm class is made, any HMIs in the project will automatically have the new class added to their **Alarm class** tab; however, the name will not be updated to any new name given to the `global` class.

Types of supervision

The **Types of supervision** option relates to how the supervision behaves and the preconfigured options available in the dropdown when configuring a new supervision:

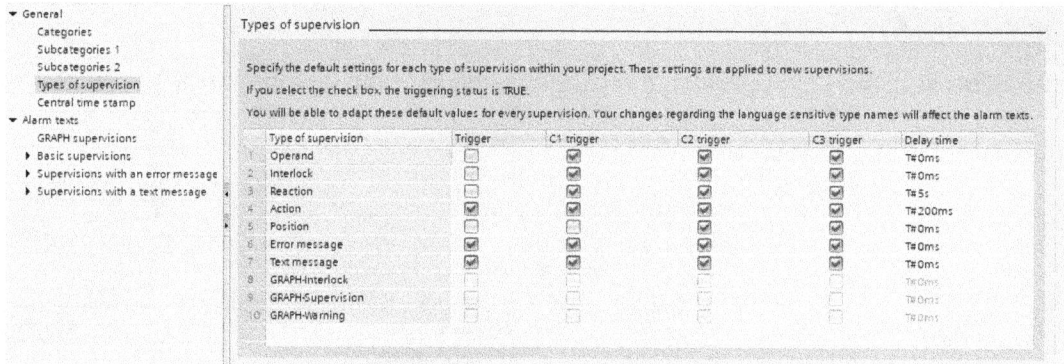

Figure 12.26 – Types of supervision

The **Type of supervision** names cannot be changed and no new types can be added. The **Trigger** checkbox determines whether the supervision's alarm trigger is `True` = alarm state or `False` = alarm state. The **C1 trigger**, **C2 trigger**, and **C3 trigger** checkboxes determine whether the additional triggers should be `True` or `False` in addition to the existing trigger. **Delay time** is the length of time the trigger and any C triggers should be in the required state before an alarm is raised.

> **Note**
>
> Although the **Type of supervision** names cannot be edited, a language value can be applied by inspecting the **Properties** window with a type selected. By changing the default value for a language, **Alarm control** will display the language version instead of simply Operand or the selected type.

Alarm texts

In the supervision settings is a section that allows the setting of alarm texts. By dragging and dropping supported alarm text fields into the **Alarm text** field, a programmer can build an alarm text that is sent to the alarm control. There is more than one type of alarm text, depending on the origin of the supervision. *Figure 12.27* demonstrates how these alarm text values are configured:

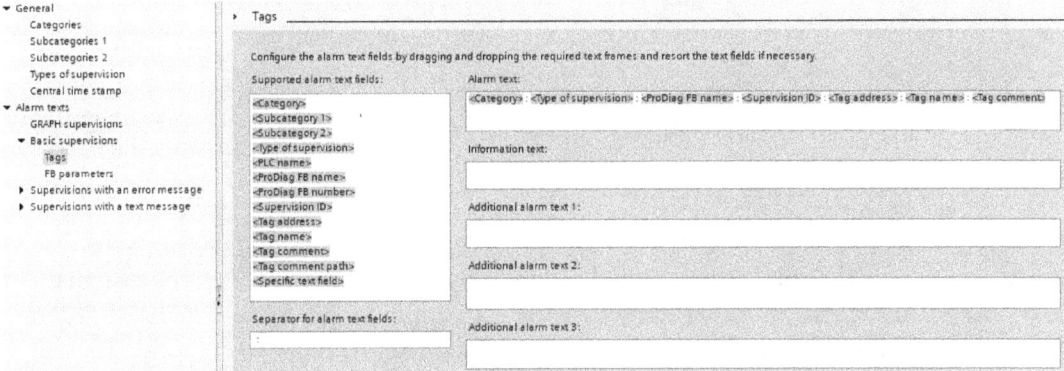

Figure 12.27 – Alarm text configuration

Each of the text fields added to the alarm text is replaced with the relative information from the PLC. Free text is not allowed to be entered in the **Alarm text** area (or any other areas); all text must come from the **Supported alarm text fields** area.

Text fields can be deleted by right-clicking them and choosing **Delete**. Between text fields, the separator for alarm text fields is automatically applied, which can be modified to a different value by editing the **Separator for alarm text fields** value.

The tag comment is most commonly used as the custom alarm text and originates from the comment assigned in the supervision function block in the PLC:

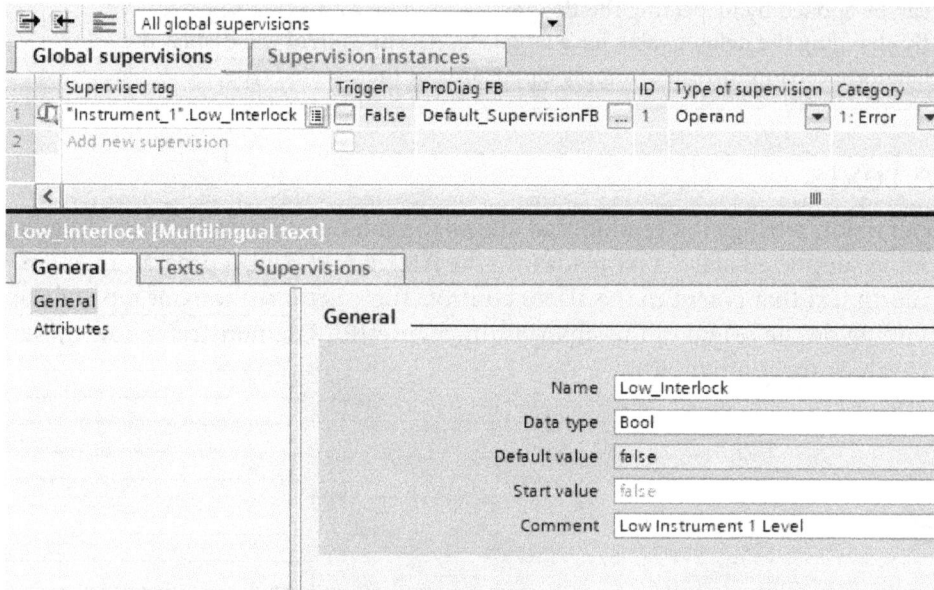

Figure 12.28 – Modification of the tag comment

The tag comment is the comment assigned in the supervision, not the comment assigned to the variable in the PLC. In order to modify the tag comment, open the associated supervision function block and highlight the supervised tag; in **General**, enter the desired tag comment in the **Comment** field.

> **Note**
> When changing the supervised tag's comment, remember to download the changes to the PLC that is driving the ProDiag supervisions for the changes to take effect.

An example of changing the alarm text to a simpler configuration would be the PLC name and tag comment. *Figure 12.28* demonstrates that the PLC name is displayed, followed by the comment assigned in the supervised tag comment:

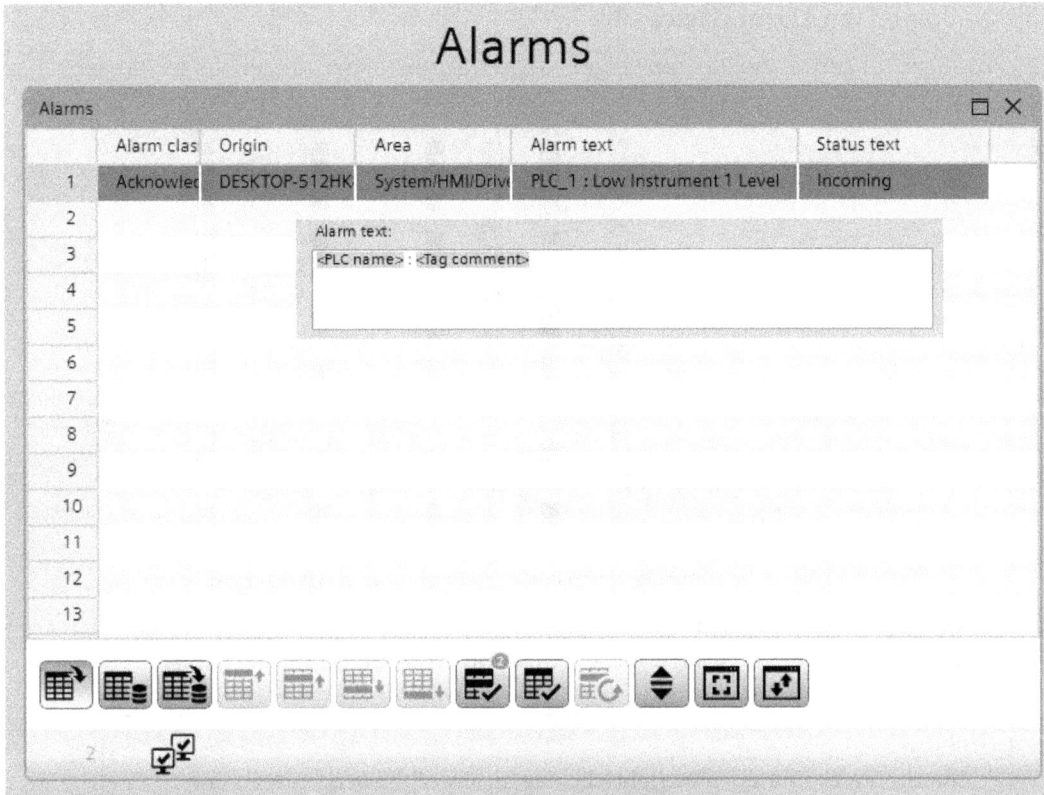

Figure 12.29 – An example of simple alarm text

If the **Alarm text** style is changed (and downloaded to the PLC), the HMI alarm control will update accordingly without any modifications to the HMI.

Setting global alarm class colors

Alarm class colors for global alarm classes cannot be modified at a global level in the common data area. The colors must be assigned on a per-HMI basis in the standard HMI alarms window on the **Alarm classes** tab:

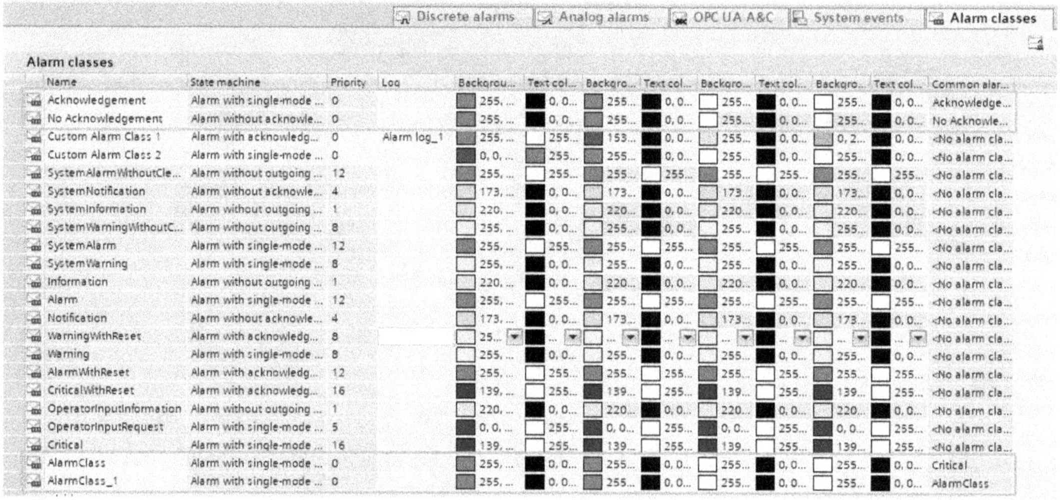

Figure 12.30 – The common alarm color configuration in Alarm classes

By unhiding the **Common alarm class** column, it is possible to distinguish which alarms belong to the global alarm classes from the **Common data** section of the project.

> **Note**
>
> Columns can be unhidden by right-clicking on any of the header items and choosing **Show/Hide** and then placing a checkmark in the box of the column to be unhidden.

Summary

This chapter has explored essential areas of HMI navigation and alarming. These aspects of HMI development are key to providing information to the end user and building a robust and easy-to-use system. Being able to create easy-to-use navigation systems and clear alarming systems can set a project apart from others, ensuring that end users have an easy experience when the overall system is in use.

Using PLC-driven alarming via the supervision tags can help reduce the amount of work required in the HMI by allowing the PLC to interact directly with the HMI's alarm control. The utilization of PLC-driven alarming ensures that function blocks will raise the same alarms no matter what project they are used in (subject to ProDiag supervisions being available), which can have large time-saving benefits as well as standardization.

Supervision tags use the ProDiag system that requires an additional license after the configuration of 25 supervisions. The license can be set in the PLC device properties in the runtime licenses section.

The next chapter focuses on the deployment of the PLC and how to download a project to both a physical PLC and a simulation. The chapter covers areas such as reinitialization and snapshots and what these mean for projects that require downloads.

Section 4 – TIA Portal – Deployment and Best Practices

Learn how to download items to a PLC, including what to watch out for and how to mitigate issues that arise in online PLC downloads. Explore some additional best practices that help to continue the learning experience beyond this book.

This part of the book comprises the following chapters:

- *Chapter 13, Downloading to the PLC*
- *Chapter 14, Downloading to the HMI*
- *Chapter 15, Programming Tips and Additional Support*

13
Downloading to the PLC

Downloading to a Programmable Logic Controller (PLC) is an important and necessary part of any project. It is the point at which an offline project or modification is actually sent to the CPU. A download might consist of just software changes or also hardware changes. TIA Portal does a great job at managing this download process to ensure it occurs fault free. This chapter explores, in detail, how the download process works in TIA Portal and what is new in TIA Portal 17. Before making downloads to a PLC, it's important to understand the actions that TIA Portal needs to take and what this could mean for data in the PLC.

In this chapter, we will cover the following topics:

- Downloading to a PLC
- Retaining data in optimized and non-optimized blocks
- Uploading from a PLC
- Considerations

Downloading to a PLC

At some point, a download to the PLC is required to move the development code into the PLC runtime environment. When a download is initiated from TIA Portal, a sequence is initiated that compiles the project and checks the conditions for download. During this sequence, the programmer is prompted for actions (if any are required) and to confirm the PLC download action before the PLC is affected:

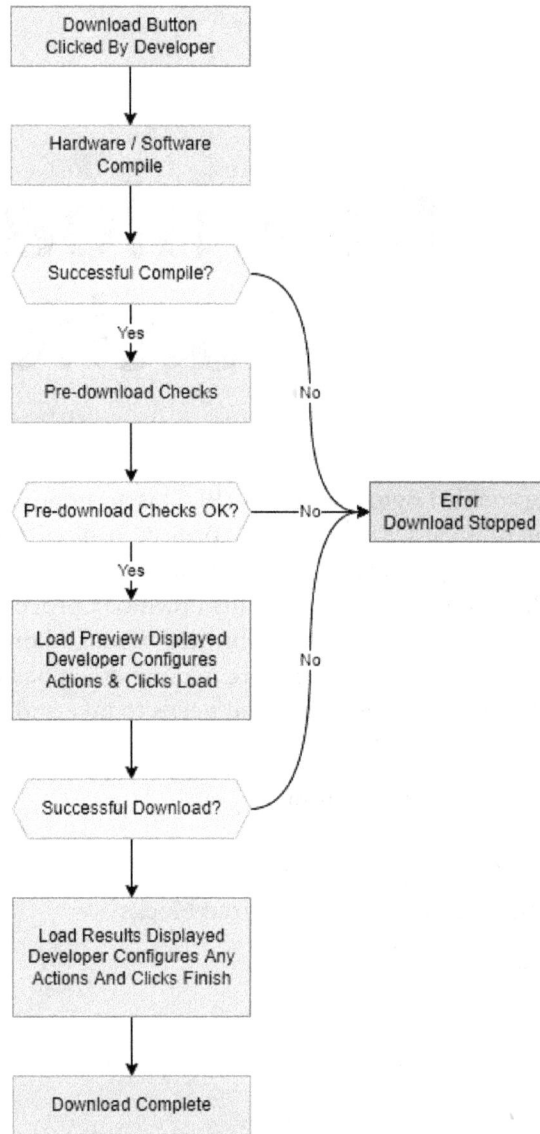

```
          ┌──────────────────────┐
          │   Download Button    │
          │ Clicked By Developer │
          └──────────┬───────────┘
                     │
                     ▼
          ┌──────────────────────┐
          │  Hardware / Software  │
          │       Compile         │
          └──────────┬───────────┘
                     │
                     ▼
          ⬡ Successful Compile? ⬡──────┐
                     │                  │
                    Yes                 │
                     ▼                  │
          ┌──────────────────────┐     │ No
          │  Pre-download Checks  │     │
          └──────────┬───────────┘     │
                     │                  │
                     ▼                  │
      ⬡ Pre-download Checks OK? ⬡──No──┤   ┌──────────────────┐
                     │                  ├──▶│      Error       │
                    Yes                 │   │ Download Stopped │
                     ▼                  │   └──────────────────┘
          ┌──────────────────────┐     │
          │ Load Preview Displayed│     │
          │ Developer Configures  │     │ No
          │ Actions & Clicks Load │     │
          └──────────┬───────────┘     │
                     │                  │
                     ▼                  │
      ⬡  Successful Download? ⬡─────────┘
                     │
                     ▼
          ┌──────────────────────┐
          │ Load Results Displayed│
          │ Developer Configures  │
          │ Any Actions And       │
          │ Clicks Finish         │
          └──────────┬───────────┘
                     │
                     ▼
          ┌──────────────────────┐
          │   Download Complete   │
          └──────────────────────┘
```

Figure 13.1 – The PLC download event sequence

The download sequence ensures that the PLC is up to date before any downloads occur.

> **Note**
>
> When the **Download** button is clicked on, a compilation of the hardware and software occurs. This means that if there are errors anywhere in the logic or the hardware configuration for the PLC being downloaded to, the download will not occur.

Initiating a download

There are multiple methods that you can use to start a download to the PLC; all of them initiate the same download procedure.

The most common method to start a download is to click on the *Download* button from the toolbar at the top of TIA Portal:

Figure 13.2 – The download button (highlighted second from the left)

There are other places from where a download can also be started:

- Clicking on **Online** from the toolbar and selecting **Download to device**.
- Right-clicking on the PLC in the project tree, choosing **Download to device**, and then choosing **Hardware and Software**.
- Additionally, it's possible to download just the software or hardware via the same right-click menu.

In all of these cases, the PLC device (or a child of the PLC device) must be selected in the project tree.

Once a download to the PLC has been requested by one of the preceding methods, TIA Portal will complete a compilation of the software. If any errors are found during this compilation, the download will be aborted, and the **Load preview** window will display the following error:

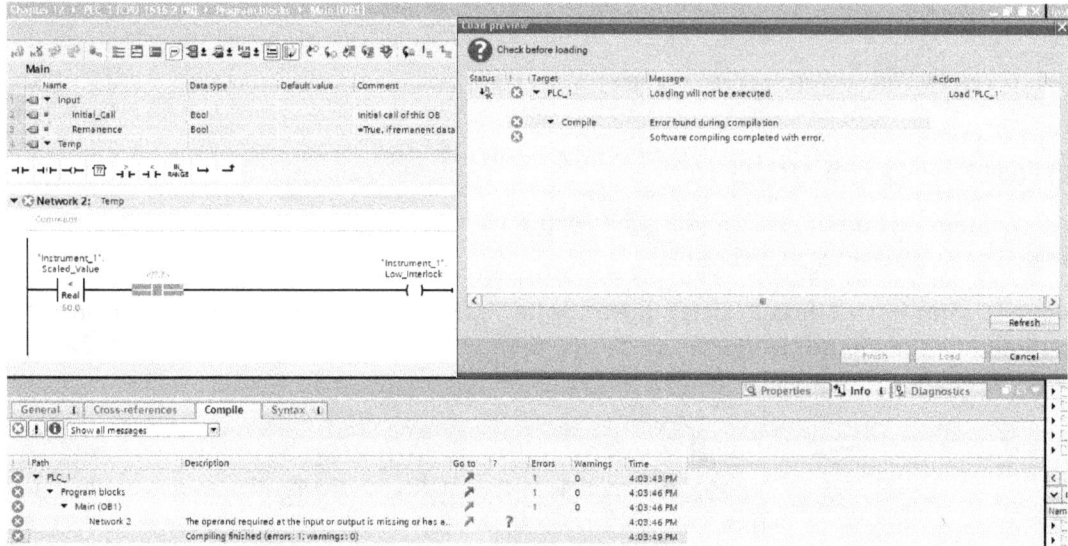

Figure 13.3 – An example of a compilation error during a download

Figure 13.3 displays a compilation error that has occurred during a download. The **Load preview** window displays that an error was found during compilation. The error is displayed in the **Info** panel, which automatically opens at the bottom of the TIA Portal project.

In this example, the issue is caused by an instruction that has not been completed with a variable.

Setting load actions

Once the project has been compiled successfully, the **Load preview** window displays options that are required to be set before the download occurs. These options can vary depending on what has changed in the project since the last compile and download.

Typically, when the **Load preview** window is displayed, one of the following three scenarios occur:

- The download is consistent, and no actions are required.
- The download is inconsistent and actions to confirm reinitialization are required.
- The download requires the PLC to be stopped and restarted.

Depending on the changes made to the program, and how those changes were made, it can be determined which actions are required before the download commences.

Consistent downloads

When a download is consistent, no actions are required before the load takes place. The programmer can simply click on **Load** and the download will commence:

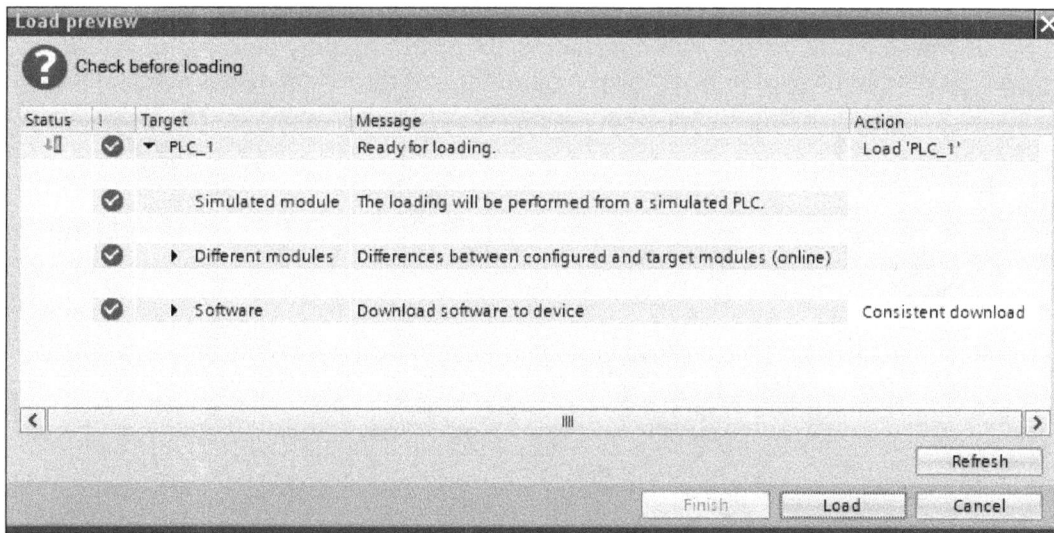

Figure 13.4 – A consistent download

A consistent download means that the PLC does not need to stop, and no data will be lost or reset to their starting values. Although the **Load preview** window displays an action and a drop-down menu for consistent downloads, there is no other option available. Some examples of consistent downloads are listed as follows:

- Logic changes that do not change the declaration section of a **function** or **function block**

- Adding a new **program block** and calling it with a **single instance data block** (global)

- Changing variable comments

- Changing accessibility options for variables (accessible via OPC UA/an HMI)

> **Note**
>
> *Figure 13.4* also demonstrates how the **Load preview** window displays messages that do not have any options for actions, such as *The loading will be performed from a simulated PLC* – this is a message that is displayed when PLCSIM is in use.

Inconsistent downloads

An inconsistent download occurs when the PLC is required to reinitialize or reset data as part of the download. When inconsistent changes to the PLC require downloading, TIA Portal will ask for confirmation from the programmer that the loss of data is acceptable:

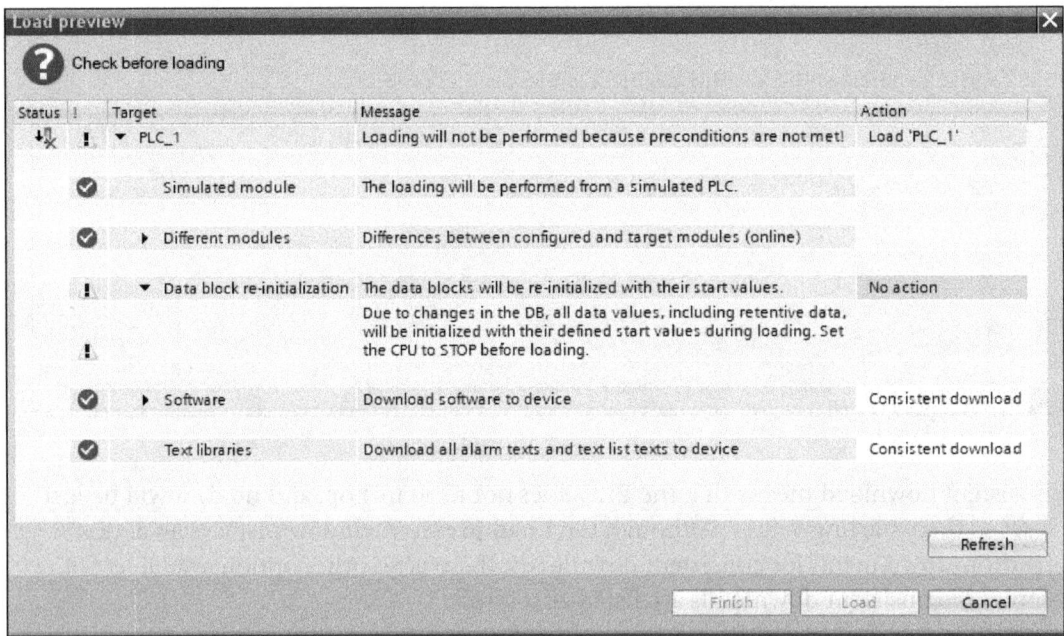

Figure 13.5 – Load preview with inconsistent data block(s)

When this occurs, the option of either *No action* or *Re-initialize* is offered as an **Action** option for the data block reinitialization. The message (when expanded) explains how all data, *including retentive*, will be initialized back to the **start values**. Additionally, it suggests that the CPU be set to STOP before continuing; however, this does not stop the download from executing successfully but can place additional risk on the process the PLC is controlling due to a sudden change of values.

By clicking on the **Software** section of the **Load preview** window, it is possible to see what specific areas of the project are changing. For example, the download in *Figure 13.5* is reinitializing due to the Instrument_1 data block being expanded with new variables:

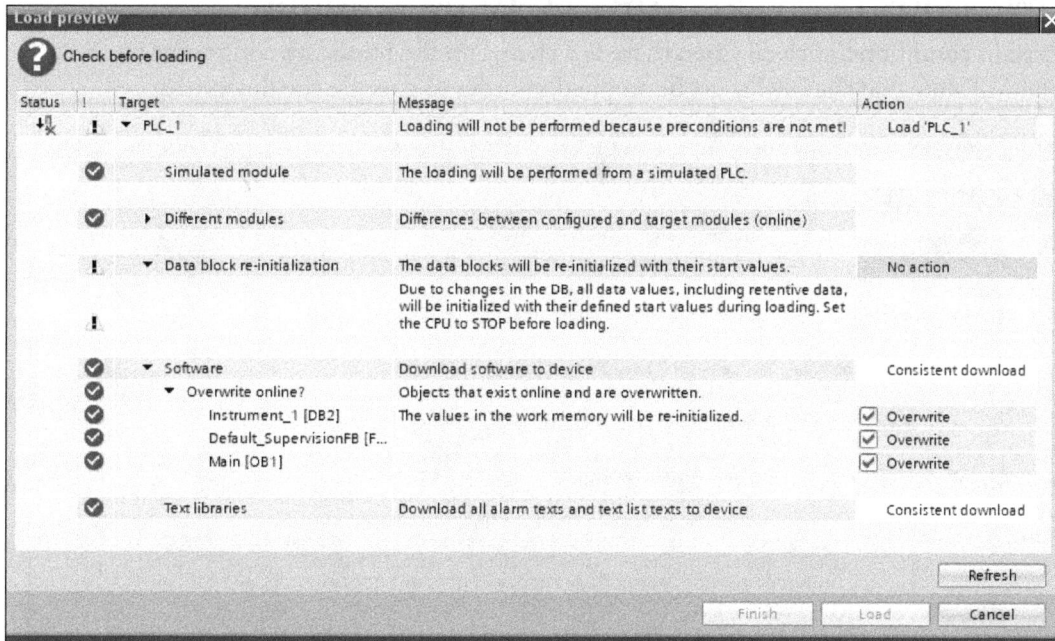

Figure 13.6 – Load preview with detailed changes displayed

The **Software** section, when expanded, details which blocks are changing and how they are changing. *Figure 13.6* shows that Instrument_1 will be reinitialized and overwritten in the PLC. The other two objects will also be overwritten.

Note

To proceed with the download, there must be no actions with the selection of **No action** present. TIA Portal requires the programmer to acknowledge the reinitialization.

Some examples of inconsistent downloads are listed as follows:

- Changing the number of variables in a structure or data block
- The renaming of a variable within a structure or data block
- Changing the **Retain** setting of a variable

Inconsistent downloads do not stop the PLC – they only reset the variable data. This is important to remember as the PLC will continue with the reset data immediately.

Downloads requiring the PLC to be stopped

In certain conditions, such as when there is a change in the hardware configuration, TIA Portal will state that the PLC must be stopped in order to execute the download:

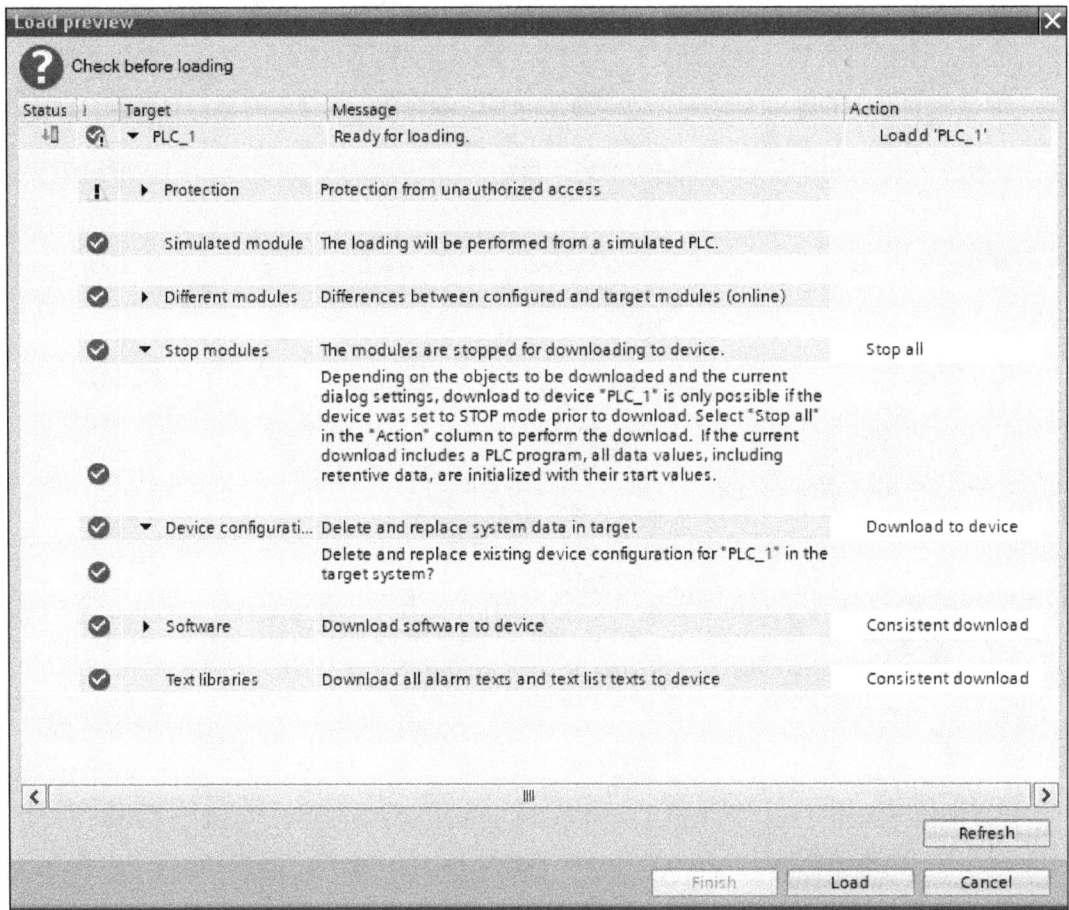

Figure 13.7 – Load preview showing a PLC stop must occur

In *Figure 13.7*, the **Load preview** window displays **Stop modules** with an action already defined as **Stop all**. The only other option for **Action** is **No Action**, which then disallows the download from continuing.

> **Note**
> The PLC program will be stopped. It is important to ensure that any controlled devices are in a safe condition. Data within the PLC will be reinitialized, including retentive data.

Following a stop of the PLC and a successful download, TIA Portal will prompt you via the **Load results** window whether the PLC should be restarted again.

Retaining data in optimized and non-optimized blocks

When a PLC loses power, runtime variable data is lost and reset back to its default values, unless it is checked to be **retained**. Retained data persists after a power fail or a download to the PLC, and it only resets to its default values when a reinitialization occurs or a memory clear of the PLC occurs. If a variable needs to retain information, the **Retain** function needs to be selected in the corresponding data block or interface:

		Name	Data type	Start value	Retain
1	▼	Static			☐
2	■	Scaled_Value	Real	0.0	☐
3	■	High_Alarm_Trigger	Real	45.0	☐
4	■	High_Interlock	Bool	false	☐
5	■	Low_Interlock	Bool	false	☐
6	■	Low_Low_Interlock	Bool	false	☐
7	■	Scaled_Max	Real	0.0	☑
8	■	Scaled_Min	Real	0.0	☑
9	■	Raw_Max	Int	0	☑
10	■	Raw_Min	Int	0	☑

Figure 13.8 – The Retain checkbox

Figure 13.8 shows an example of four values set to **Retain** in a data block. Once the Retain box has been checked for **some given data**, the changes must be downloaded to the PLC. This will cause a reinitialization of the data.

Retaining data in instance data

Function blocks and associated instance data blocks can also have data retention set. However, there are more options involved, and it also depends on whether the function block is **optimized** or **non-optimized**.

Optimized function blocks

Optimized function blocks allow the **Retain** value to be configured within the function block itself:

		Name	Data type	Default value	Retain
1	▼	Input			
2	■	Active	Bool	false	Non-retain
3	▼	Output			
4	■	Status	Byte	16#0	Non-retain
5	■	<Add new>			
6	▼	InOut			
7	■ ▶	Power_Data	Struct		
8	■	<Add new>			
9	▼	Static			
10	■	Line_1_Max	Real	0.0	Set in IDB
11	■	Line_1_Min	Real	0.0	Set in IDB
12	■	Line_2_Max	Real	0.0	Set in IDB
13	■	Line_2_Min	Real	0.0	Set in IDB
14	■	Line_3_Max	Real	0.0	Set in IDB
15	■	Line_3_Min	Real	0.0	Set in IDB
16	■	Rolling_Average	Real	0.0	Retain
17	▼	Temp			
18	■ ▶	Temp_Array	Array[0..20] of Real		

Figure 13.9 – The Retain settings of the optimized function block

Figure 13.9 is an example of a function block that is set to **Optimized** with the **Retain** values configured. **Optimized function blocks** can individually set each variable's **Retain** value. The **Input**, **Output**, and **Static** variables can be retained. However, **InOut** variables cannot be retained. Additionally, **Temp** variables do not have **Retain** configuration data associated with them as they do not persist in any capacity.

Optimized function blocks can choose one of the following three options for **Retain** data:

- **Non-retain**: Data is not retained.

- **Retain**: Data is retained.

- **Set in IDB**: Data is retained, but only if the instance data block is configured to retain data.

> **Note**
>
> The **Retain** option is used to ensure that all instances of a function block will retain the variable, irrespective of the configuration of the instance data block. This has advantages if the information is critical for the continued operation of the function block following a download or a power failure.

Setting the configuration in the **instance data block** (which is abbreviated to **IDB** in the **Retain** configuration) is achieved by opening the associated instance data block and setting the **Retain** checkbox in the same way as a standard data block:

		Name	Data type	Start value	Snapshot	Retain
1	▼	Input				☐
2	■	Active	Bool	false	—	☐
3	▼	Output				☐
4	■	Status	Byte	16#0	—	☐
5	▼	InOut				☐
6	■	Power_Data	Struct		—	☐
7	▼	Static				☐
8	■	Line_1_Max	Real	0.0	—	☐
9	■	Line_1_Min	Real	0.0	—	☐
10	■	Line_2_Max	Real	0.0	—	☐
11	■	Line_2_Min	Real	0.0	—	☐
12	■	Line_3_Max	Real	0.0	—	☐
13	■	Line_3_Min	Real	0.0	—	☐
14	■	Rolling_Average	Real	0.0	—	☑

Figure 13.10 – The Retain settings of the instance data block

When a **Retain** option is set, all the variables inside the data block will be set to **Retain**. It is not possible to retain individual variables in the data block:

		Name	Data type	Start value	Snapshot	Retain
1	▼	Input				☐
2	■	Active	Bool	false	—	☐
3	▼	Output				☐
4	■	Status	Byte	16#0	—	☐
5	▼	InOut				☐
6	■	Power_Data	Struct		—	☐
7	▼	Static				☐
8	■	Line_1_Max	Real	0.0	—	☑
9	■	Line_1_Min	Real	0.0	—	☑
10	■	Line_2_Max	Real	0.0	—	☑
11	■	Line_2_Min	Real	0.0	—	☑
12	■	Line_3_Max	Real	0.0	—	☑
13	■	Line_3_Min	Real	0.0	—	☑
14	■	Rolling_Average	Real	0.0	—	☑

Figure 13.11 – The instance data block with all static options

Figure 13.11 demonstrates how all variables set to **Set in IDB** in the function block are checked when any single variable's **Retain** checkbox is checked.

> **Note**
> The `Active` and `Status` variables are not available to be checked because the function block's configuration for them is set to *Non-Retain*. This means that even the instance data block cannot be retained. Similarly, the `Rolling_Average` variable is checked and disabled, disallowing the unchecking of the **Retain** option. This is because the function block configuration for this variable is set to **Retain**.

Non-optimized data blocks

Function blocks that are non-optimized are unable to set the retain data for individual variables, and the **Retain** column does not exist in the function block:

	Name	Data type	Offset	Default value	Accessible f...	Writa...	Visible in ...	Setpoint	Supervision	Comment
1	▼ Input				☐	☐	☐	☐		
2	▪ Active	Bool	0.0	false	☑	☑	☑	☐		
3	▼ Output				☐	☐	☐	☐		
4	▪ Status	Byte	2.0	16#0	☑	☑	☑	☐		
5	▼ InOut				☐	☐	☐	☐		
6	▪ ▶ Power_Data	Struct	4.0		☐	☐	☐	☐		
7	▼ Static				☐	☐	☐	☐		
8	▪ Line_1_Max	Real	10.0	0.0	☑	☑	☑	☐		
9	▪ Line_1_Min	Real	14.0	0.0	☑	☑	☑	☐		
10	▪ Line_2_Max	Real	18.0	0.0	☑	☑	☑	☐		
11	▪ Line_2_Min	Real	22.0	0.0	☑	☑	☑	☐		
12	▪ Line_3_Max	Real	26.0	0.0	☑	☑	☑	☐		
13	▪ Line_3_Min	Real	30.0	0.0	☑	☑	☑	☐		
14	▪ Rolling_Average	Real	34.0	0.0	☑	☑	☑	☐		
15	▼ Temp				☐	☐	☐	☐		
16	▪ ▶ Temp_Array	Array[0..20] of Real	0.0		☐	☐	☐	☐		
17	▼ Constant				☐	☐	☐	☐		
18	▪ <Add new>				☐	☐	☐	☐		

Figure 13.12 – No Retain column is available in a non-optimized function block

To set retainable instance data in a non-optimized function block, the variables must be set in the instance data block:

Line_monitor_DB

	Name	Data type	Offset	Start value	Snapshot	Retain	Accessible f...	Writa...	Visible in ...	Setpoint
1	▼ Input					☐	☐	☐	☐	☐
2	▪ Active	Bool	0.0	false	—	☑	☑	☑	☑	☐
3	▼ Output					☐	☐	☐	☐	☐
4	▪ Status	Byte	2.0	16#0	—	☑	☑	☑	☑	☐
5	▼ InOut					☐	☐	☐	☐	☐
6	▪ Power_Data	Struct	4.0		—	☑	☐	☐	☐	☐
7	▼ Static					☐	☐	☐	☐	☐
8	▪ Line_1_Max	Real	10.0	0.0	—	☑	☑	☑	☑	☐
9	▪ Line_1_Min	Real	14.0	0.0	—	☑	☑	☑	☑	☐
10	▪ Line_2_Max	Real	18.0	0.0	—	☑	☑	☑	☑	☐
11	▪ Line_2_Min	Real	22.0	0.0	—	☑	☑	☑	☑	☐
12	▪ Line_3_Max	Real	26.0	0.0	—	☑	☑	☑	☑	☐
13	▪ Line_3_Min	Real	30.0	0.0	—	☑	☑	☑	☑	☐
14	▪ Rolling_Average	Real	34.0	0.0	—	☑	☑	☑	☑	☐

Figure 13.13 – A non-optimized instance data block

Similar to the **Set in IDB** option, the instance data block checks all the retain values when a single variable's **Retain** checkbox is checked.

> **Note**
>
> In non-optimized instance data blocks, the interface variables will always be available to set to **Retain**. Only optimized blocks can set the default retain value and prevent the programmer from changing it.

Downloads without reinitialization

One of the drawbacks of the TIA Portal environment is the ease with which a reinitialization is triggered. Changing the name of a variable in a data block from `Variable_1` to `Variable1` will cause a reinitialization of the entire data block. This means that if a project has a data block that contains 100 variables for the parameterization of devices, all 100 devices would be affected due to 1 name change. Therefore, each device would be reinitialized back to its starting value.

The methods offered to help retain data and lessen the impact of modifications to a program vary, depending on how the programmer has configured **program blocks** in the project.

The biggest difference between optimized and non-optimized blocks is that optimized blocks can take advantage of the **Download without reinitialization** option. This feature allows the block size to be defined as larger than the variables that occupy it:

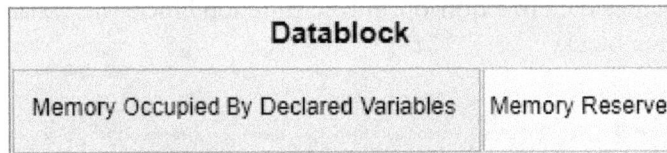

Datablock	
Memory Occupied By Declared Variables	Memory Reserve

Figure 13.14 – An example of a memory reserve

By declaring a fixed number of bytes for the **Memory reserve** option, the block can be expanded into the reserve without a loss of information.

The same can be achieved for any retentive memory by setting the **Retentive memory reserve** option:

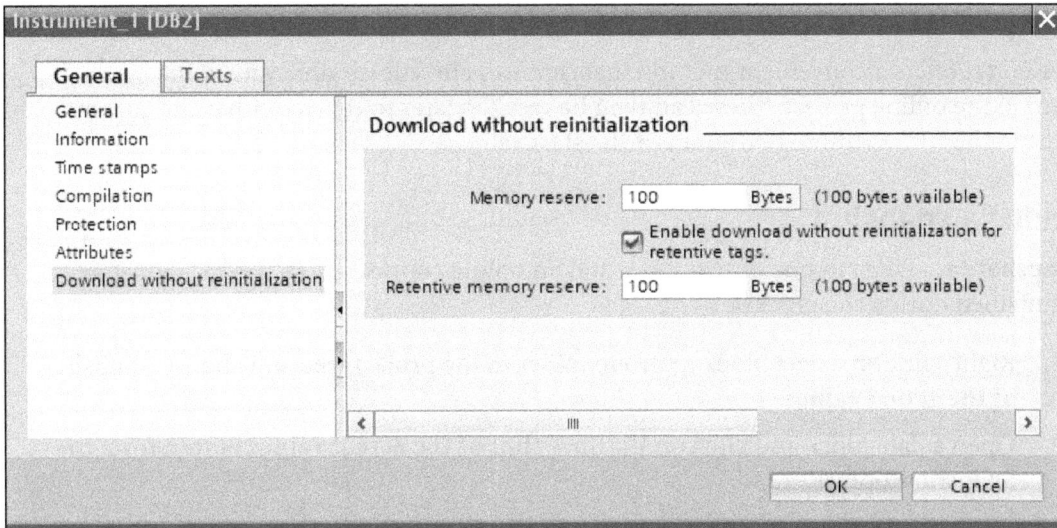

Figure 13.15 – The Download without reinitialization configuration

Figure 13.15 shows the **Download without reinitialization** options. This can be accessed by right-clicking on a program block from the project tree, selecting **properties**, and then navigating to **Download without reinitialization**.

> **Note**
>
> **Download without reinitialization** should be used sparingly. Setting large reserves for multiple blocks is a quick way to fill the PLC memory, leaving little room for further development.

If the **Optimized block access** block attribute is unchecked, the **Download without reinitialization** options will be unavailable for use. Spare variables can be declared when developing the data for later use. However, resizing or renaming these variables will result in reinitialization.

Snapshots

TIA Portal offers a convenient method that captures current variable values and stores them in the offline project. These can then be recalled later to replace either the current value in the PLC's running program or the start value.

Taking a snapshot

Snapshots are taken in one of two ways, and an online connection to the PLC must be maintained during the process:

1. Right-click on a data block or group folder in the project tree and choose **Snapshot of the actual values**.

2. Open a data block and click on the **Snapshot of the actual values** button from the **Snapshot** toolbar:

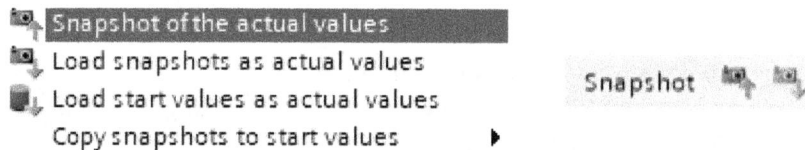

Figure 13.16 – The Snapshot menu items and buttons (left: the right-click menu; right: the data block Snapshot toolbar)

3. Once a snapshot has been successfully taken, a new column will appear when the data block is open, displaying the snapshot values. Additionally, a header text is supplied stating the time and date of the latest snapshot:

Alarms (snapshot created: 1/11/2022 11:45:10 PM)

	Name	Data type	Start value	Snapshot
▼	Static			
■ ▼	Alarm_Group_1	"UDT_Alarm_Group"		
■ ▼	Accepted	Array[0..1] of DWord		
■	Accepted[0]	DWord	16#0	16#0000_0000
■	Accepted[1]	DWord	16#0	16#0000_0000
■ ▶	Active	Array[0..1] of D...		

Figure 13.17 – An example of the Snapshot header text and column

The **Snapshot** column can also be shown or hidden without taking a snapshot.

> **Note**
>
> When snapshots are called from the project tree, all objects beneath the selected object (including the selected object) will have a snapshot taken. If a snapshot is requested from the **Data block snapshot** toolbar, all variables within the data block are captured.

Snapshots cannot be taken if the offline data block does not match the online version in the PLC. This is not highlighted very well by TIA Portal, so it might seem like the snapshot has been successful. It is important to take snapshots before any modification work is required.

Restoring snapshot data

Once data has been captured via a snapshot, it can be reinstated as the current value in the online PLC program. Similarly to taking a snapshot, restoring data can be achieved in more than one way, as shown in *Figure 13.18*:

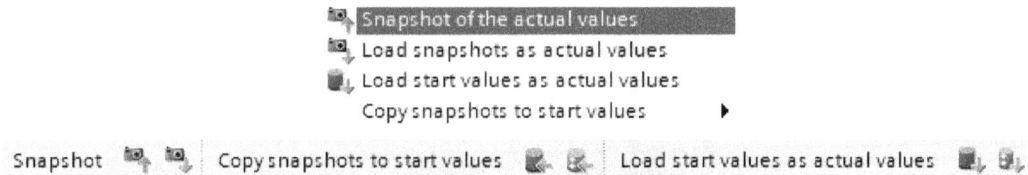

Figure 13.18 – The right-click menu and the data block snapshot toolbar

There are different methods of restoring data into data blocks that have different effects on the data and the PLC program:

- **Load snapshots as actual values**: This option loads the captured values as the current values in the online PLC.

- **Copy snapshots to start values**: This option copies the snapshots to the starting values and filters them according to one of the options selected:

  - All values

  - Only setpoints

  - Only the retain values (only accessible via the right-click menu)

> **Note**
>
> It is important to understand what the PLC program that is running will do when the snapshot data has been loaded, especially if all values have been loaded. For example, sequences might jump from one step to another without determining the intermediate steps due to the sequence step data suddenly being overwritten with data from a previous date.

The **Load start values as actual values** option should not be confused with the snapshots directly but can be used in conjunction with the **Copy snapshots to start values** option to help perform downloads without losing information. When the **Copy snapshots to start values** option is used, the data stored within the snapshots is loaded into the start values. This needs to be downloaded to take effect in the PLC.

When a data block is reinitialized, the new start values from the snapshot are used instead of the default values for the data type.

> **Note**
>
> It is recommended that you use the **Setpoints** checkbox in data blocks to help restrict which variables can be written with new start values and then copied to actual values. If the consequences are not understood, changing variables that control the process equipment or machines should be avoided while the system is running.

Uploading from a PLC

PLC programs can also be uploaded out of the PLC and back into the TIA Portal project, which is useful if the version of the offline project does not match the project in the PLC:

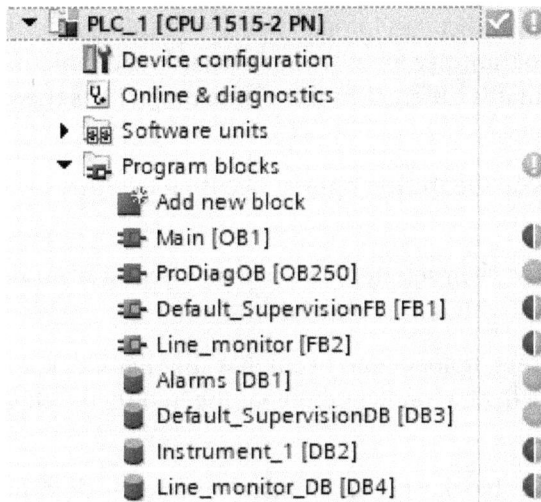

Figure 13.19 – An example of an unmatching online project

The project tree will identify any objects that are not consistent with the online project. By right-clicking on an object and selecting the **Upload from device (software)** option, the **Upload preview** dialog box will open:

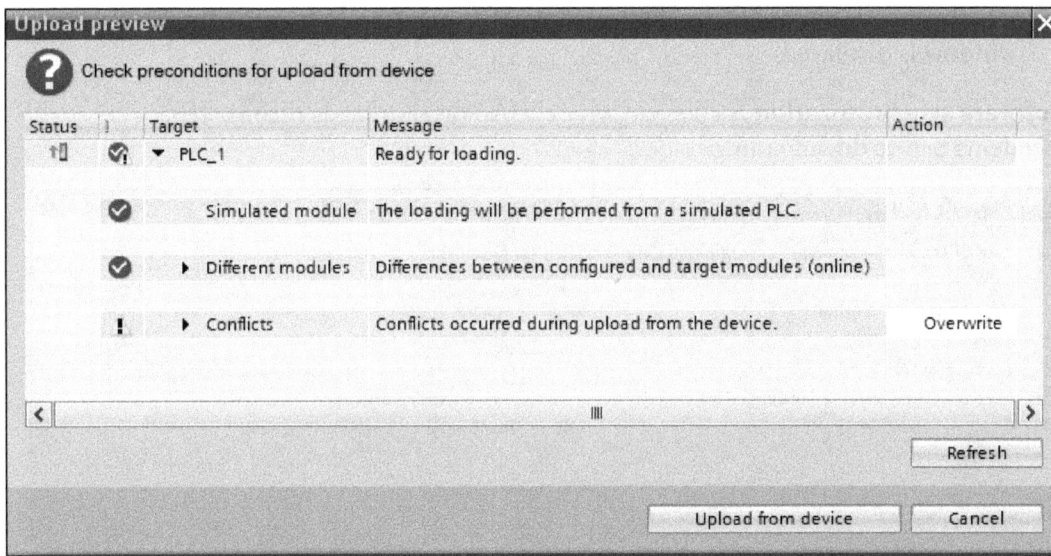

Figure 13.20 – Upload preview

The **Upload preview** window displays information about the upload that is about to take place and also asks the programmer for a decision regarding conflicting objects. A conflict is detected when a block that is different in the offline project has the same name as an object in the online project.

The programmer can choose the **Action** option, deciding to either Overwrite or Insert with different name.

Once the **Action** option has been selected, clicking on **Upload from device** begins updating the offline project with online block data.

Unlike the download process, uploads can occur at a singular block level. By selecting one program block, right-clicking on it, and uploading it, only the single program block is uploaded.

Uploading a single program block, such as an instance data block, can cause an issue where the function block and the associated instance data block do not match. By only uploading the instance data block, TIA Portal remedies the mismatch in one of two ways:

- When the function block is opened, the associated instance data block is regenerated, undoing the upload changes.

- If a download is initiated, the **Software synchronization before loading to a device** window is displayed.

A **software synchronization** is an event where TIA Portal uploads inconsistent dependents before downloading:

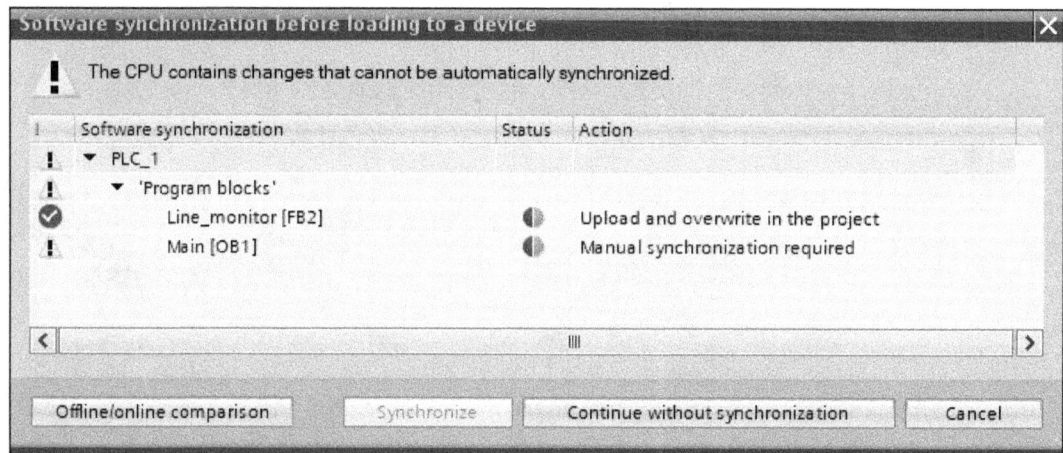

Figure 13.21 – Software synchronization before loading to a device window

Sometimes, TIA Portal is able to automatically synchronize data without having to perform an upload. Where this is not possible, a **Manual synchronization required** message is displayed.

Note

Manual synchronization means TIA Portal is unable to detect the version of the block (offline/online). The message for manual synchronization appears when the PLC's timestamp for the block is newer than the offline project. TIA Portal requests *Manual* synchronization, which means the programmer should click on the **Offline/online comparison** button and review the changes, modifying them appropriately if required.

Click on **Continue without synchronization** to close the **Software synchronization before loading to a device** window and load the normal **Load preview** window. Note that the changes recommended in the **Software synchronization before loading to a device** window will not have been implemented.

At this point, a compilation of the program is executed, and the differing instance data block is regenerated. If this is then downloaded to the PLC, the online changes in the PLC are lost as the older function block has been downloaded.

When TIA Portal displays the **Software synchronization before loading to a device** window, clicking on the **Offline/Online comparison** button loads the **Compare editor online** window:

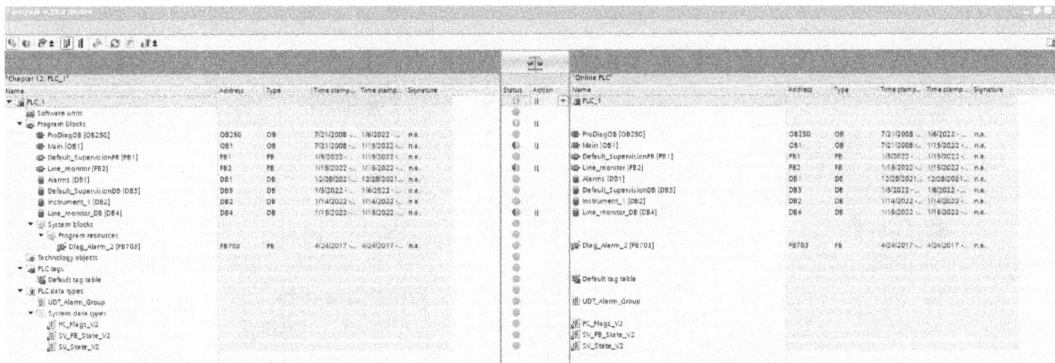

Figure 13.22 – The Compare editor online window

This window allows direct comparison between the offline project (on the left-hand side) and the online PLC project (on the right-hand side). The example in *Figure 13.22* shows that the **Line_monitor_DB** and **Line_monitor** objects are different:

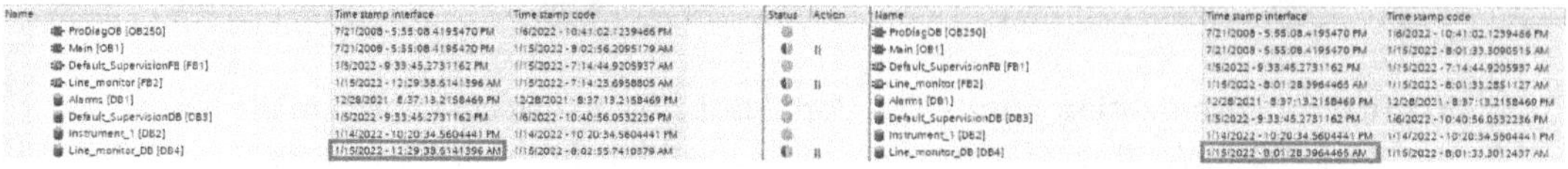

Figure 13.23 – Comparison results

Figure 13.23 demonstrates that the **Line_monitor_DB** object has a newer interface timestamp in the online PLC than the offline project. However, the code timestamp is the opposite, with the offline project being newer than the online project. It is for this reason that a manual synchronization is required.

By clicking on the pause icon between the offline project and the online PLC project for an object, a drop-down menu opens that allows you to select **Upload from device** or **Download to device**, along with **No action**.

> **Note**
>
> It is not possible to set some objects to upload and others to download at the same time.

By setting both the **Line_monitor** and **Line_monitor_DB** objects to upload, it ensures that the function block, which is dependent upon the instance data block, is also kept consistent:

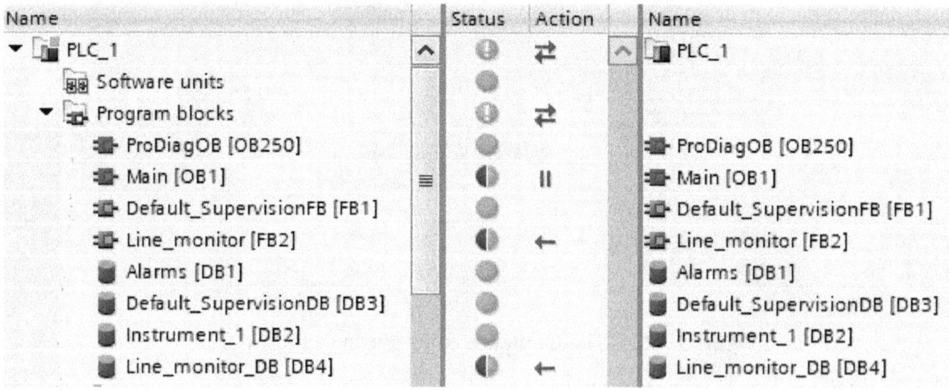

Figure 13.24 – A selection of upload actions

Once the actions have been set, clicking on the **Execute actions** button will perform the synchronization:

Figure 13.25 – The execution button

Depending on the action that has been requested, the appropriate load window will be displayed, and the event (upload/download) is then completed as normal.

> **Note**
>
> When performing uploads, it is important that the dependencies of the objects are examined thoroughly and that manual synchronizations are executed correctly. If manual synchronization is skipped, TIA Portal will download differencing objects, and any online changes will be irreversibly lost.

Considerations

Managing downloads can become more difficult the larger projects become, especially if the process that the PLC is controlling is also critical and PLC downtime is to be kept to a minimum.

Data segregation

A good method of reducing the effect of reinitialization and the chance of it needing to set the data in your project to the starting values is to segregate data into *more than one* data block:

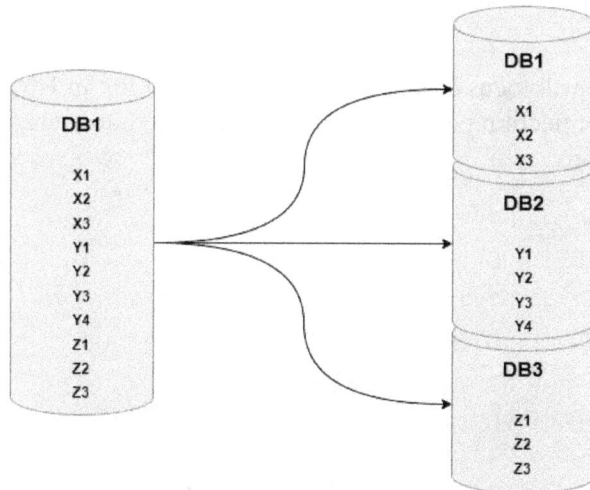

Figure 13.26 – An example of segregating data

By grouping data into singular data blocks, where the data has something in common with the data it is grouped with, reinitialization only affects data that is grouped together.

Figure 13.26 shows an example of mixed data in **DB1** on the left-hand side. If a variable was added, removed, or modified in this data block, the entire data block would be reinitialized. Then, groups of unrelated data would be set to the starting values.

By splitting the relative data into three data blocks, only the data block that contains the added, removed, or modified data will be reinitialized, leaving the segregated data intact.

Using functions

Functions do not utilize their own instance data. This means that any function block that they are placed in does not require its interface to change. When making modifications to a function block, consider whether the modification required can be made in a function, as this will be considered as a consistent download.

Summary

Downloading to a PLC is a fundamental part of PLC programming. However, it is often considered to be a simple click of a button. This chapter has highlighted the important aspects of the download process that are often overlooked, such as the retention of data and methods to keep offline and online projects in synchronization.

TIA Portal always warns programmers of impending changes and will often require the selection of an action in order to proceed. It is important that habits are built by programmers to check the warnings and messages to ensure that the changes about to be actioned are understood.

In the next chapter, we will focus on the same principles, but for an HMI. This includes the configuration of connection parameters, downloading to hardware, and other important considerations.

14

Downloading to the HMI

Human Machine Interfaces (HMIs) found within a project also require a download in a similar fashion to PLCs. During this download, items such as the graphical screens, connection configurations, user administrations, and more are sent to the HMI's runtime system. In order to download to an HMI, additional steps are required in the TIA Portal project than that of a PLC, with more steps required to be confirmed by the programmer performing the download.

In this chapter, we will cover the following topics:

- Connection parameters
- Downloading to an HMI
- Simulating a unified HMI
- Security considerations

Connection parameters

HMIs are designed to offer an interface for a machine or process, hence the name **Human Machine Interface**. Before the HMI can be downloaded to, a *connection* must be made between the **PLC** and the **HMI**. This connection is then used to pass information between the HMI and the PLC.

> **Note**
> Connections must be made before the HMI is able to access variables created in the associated PLC.

TIA Portal offers many different *communication drivers* for HMIs, allowing communication between all Siemens devices and even other vendors such as Allen Bradley and Mitsubishi. This enables a single HMI to talk to multiple PLCs, even if they are not of the same type.

Creating connections

Inside the project tree and within the HMI object, you will find the **Connections** object. By opening the **Connections** object, the **Connections** window opens. Here, connection definitions can be made, including those connections that are not Siemens devices.

A new connection can be added by double-clicking on the <**Add new**> option inside the **Connections** window:

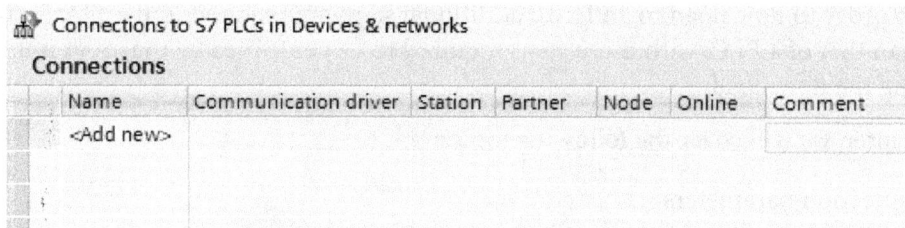

Connections to S7 PLCs in Devices & networks						
Connections						
Name	Communication driver	Station	Partner	Node	Online	Comment
<Add new>						

Figure 14.1 – The Connections window

This will enter a new connection within the **Connections** window. The **Communication driver** field will be set to the last used selection. Alternatively, it will be set to **SIMATIC S7 1200/1500** if no connection has previously been set.

Once a selection has been made for the **Communication driver** field, the **Parameter** tab (which is located below the **Connections** window) will display the configuration properties:

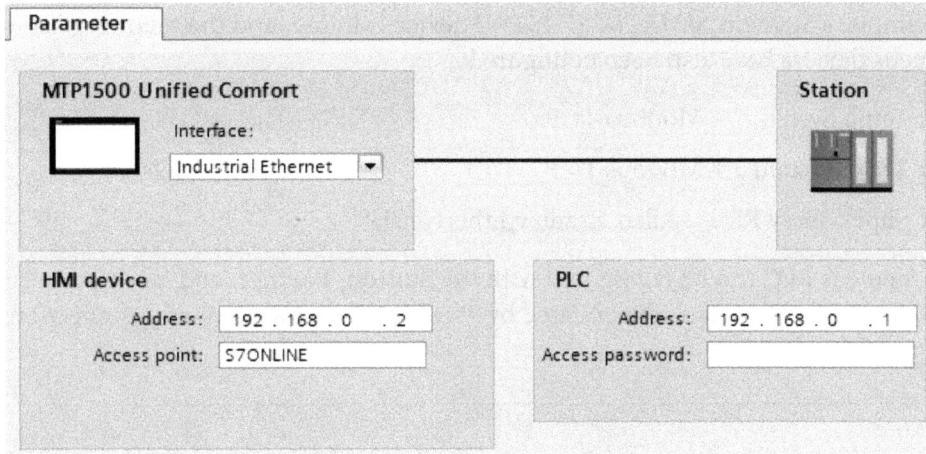

Figure 14.2 – The Connections Parameter tab

Depending on the communication driver that has been selected, different options will appear in this window. For the **SIMATIC S7 1200/1500** driver, the properties displayed in *Figure 14.2* will be used.

> **Note**
>
> In TIA Portal V17, PLCs can have enhanced security profiles configured that deny HMI communication or restrict communications to read-only. Ensure that the PLC's configuration is correct and that the **Access password** value has been entered in the **Parameter** window that is displayed (as shown in *Figure 14.2*).

More than one connection can be configured in the **Connections** window. *Figure 14.3* demonstrates how the HMI can contain more than one connection type across different communication drivers:

Figure 14.3 – The Connections window with four configured connections

In this example, a Siemens *SIMATIC S7 1515-2* device is in use, and the following list of non-Siemens devices have also been configured:

- Air Temp System 1 – Modbus TCP

- Air Temp System 2 – Modbus TCP

- Air Supervisory PLC – Allen-Bradley EtherNet/IP

Only the Siemens PLC can be configured with the **Station**, **Partner**, and **Node** properties. These properties are automatically updated by TIA Portal once a successful connection has been configured.

> **Note**
>
> The **Online** column allows the quick activation and deactivation of connections. This is useful if the HMI is common between different asset types that may or may not be included within the project.

Devices and networks

In the **Connections** window, there is an option that switches the view to the **Devices & networks** window. This button is located above the title of the **Connections** window and is labeled **Connections to S7 PLCs in Devices & networks**:

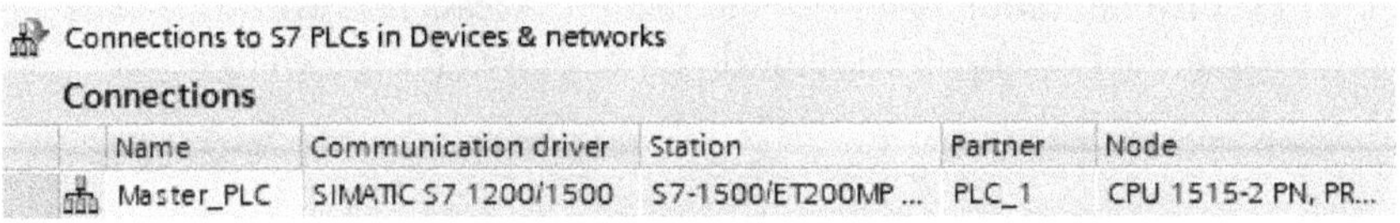

Figure 14.4 – The Connections to S7 PLCs in Devices & networks button

The **Devices & networks** window will be populated with any connections made previously, as shown in the following graphical diagram:

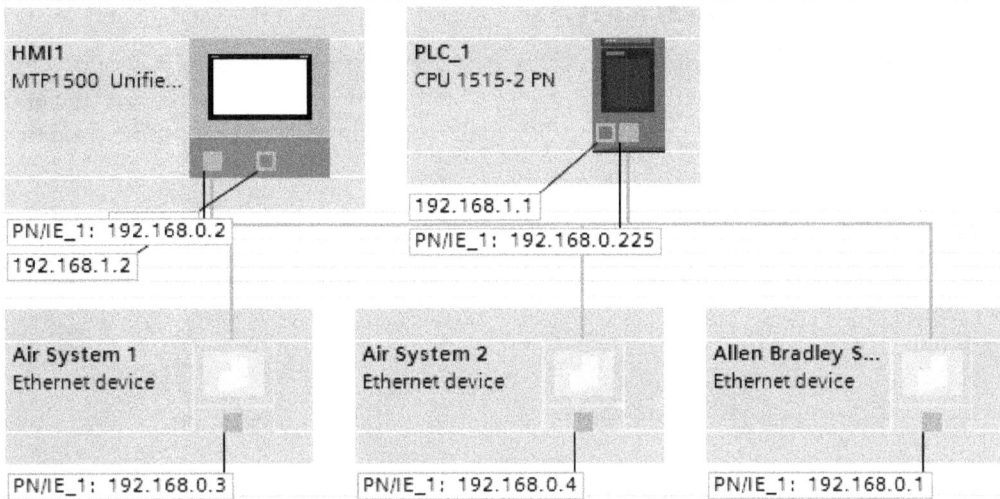

Figure 14.5 – A graphical representation of the configured connections

Additionally, it is possible to modify and change connections from this view by dragging the connection points to the newly added devices or deleting connection points from existing devices.

> **Note**
>
> The programmer might need to reorganize the location of the devices in the **Devices & networks** view. Following the creation of new connections and devices, the objects might all appear on the same line.

Once all the connections have been configured, the HMI can be compiled by selecting it in the project tree, selecting **Compile**, and then selecting **Hardware and software**. If the HMI compiles correctly, it can be downloaded.

Downloading to an HMI

The principles behind downloading to an HMI are very similar to that of a PLC. Starting a download is performed in the same manner – by selecting the HMI in the project tree and then clicking on the **Download** button. Additionally, the right-click menu can be used to initiate a download.

When the **Load preview** window is displayed, the options differ from an HMI to a PLC:

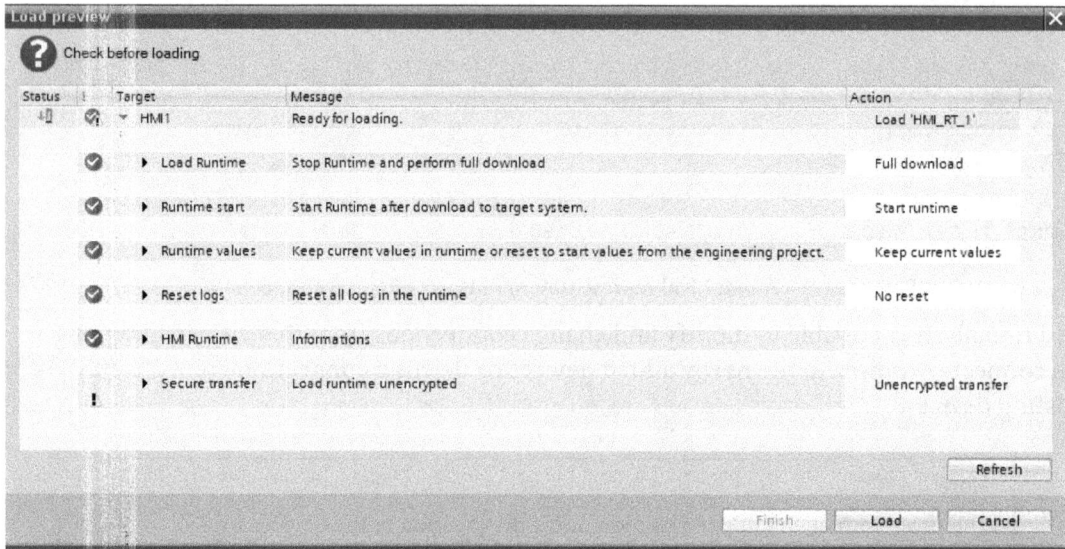

Figure 14.6 – Load preview

The HMI **Load preview** window has more actions available than the PLC **Load preview** window, and these actions directly affect how the download is completed:

- **Load Runtime**: Under certain conditions, a **Delta download** option is available rather than a **Full download** option. **Delta download** allows the HMI to continue running during the download process, and only changes are downloaded as opposed to the entire download. When a **Full download** occurs, the HMI must be stopped temporarily, and this happens automatically.

- **Runtime start**: The action can be set to **Start runtime**, which restarts the runtime on the HMI once the download has finished, or the value can be set to **No action**.

- **Runtime values**: This action has three options:

  - **Keep current values**: All tags, active alarms, and user management data remain the same (if possible).

  - **Reset to start values**: All tags, active alarms, and user management data will be reset to the starting values defined.

  - **Keep selected**: Depending on the selections made when the **Runtime values** section was expanded, the **Keep current values** action will be applied to checked items, and **Reset to start values** will be applied to unchecked items.

- **Reset logs**: Logs, such as tag logs, alarm logs, and other historical functions will be reset in the runtime.

- **HMI Runtime**: This section does not have any actions, but it does highlight important information about the runtime environment that is to be downloaded.

- **Secure transfer**: If **Encrypted transfer** is configured in **Runtime settings**, it will check whether the HMI contains a matching transfer password to the project. If **Encrypted transfer** has been turned on for the first time, at this point, a different message might be displayed explaining that the **Allow initial password transfer via unencrypted download** action needs to be checked to continue.

Once the preceding sections have the appropriate actions applied, clicking on the **Load** button will execute the download. Once the download has been completed, the **Load preview** window will close. Compared to downloading to a PLC, an HMI download does not produce a **Result preview** window after a successful download.

Simulating a unified HMI

TIA Portal V17 can simulate a unified HMI on the *localhost* (that is, the laptop/PC being used to develop the project). The simulation can be started in the same way that a PLC simulation is started – by clicking on the **Start simulation** button from the toolbar.

Unlike with a PLC, no simulation window or program will open. Instead, TIA Portal starts a background service that runs the unified simulation. This can be accessed by opening the **SIMATIC Runtime Manager** window, which can be accessed from the Windows **Start** menu (type in `SIMATIC Runtime Manager` after opening the Windows **Start** menu):

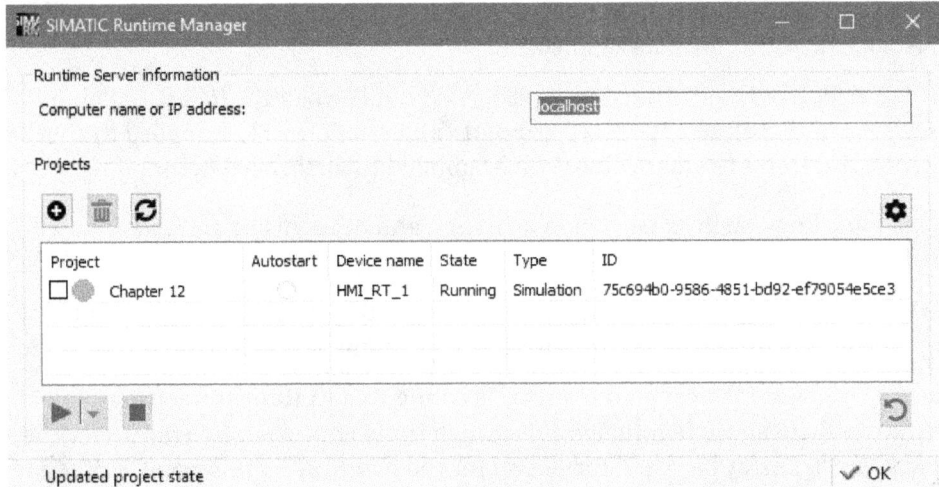

Figure 14.7 – SIMATIC Runtime Manager

The **SIMATIC Runtime Manager** window displays any active runtimes, including simulated runtimes. *Figure 14.7* demonstrates this by displaying a project called **Chapter 12** that is active and running with a **Type** value of **Simulation**. This is the runtime that was used for the *Chapter 12* project.

> **Note**
>
> The **SIMATIC Runtime Manager** window can be used to start and stop a runtime, without the need to open TIA Portal.

Accessing a unified HMI simulation

In order to access the simulation of an HMI, the name of the laptop/PC that is running the simulation needs to be known. This can be found by opening a **Command Prompt** terminal and typing in `whoami`:

Figure 14.8 – The whoami command

This will then display the name of the device followed by \<username>.

In an *HTML5-compatible browser*, open the following URL: https://desktop-512hk8i/.

Replace desktop-512hk8i with the name of the laptop/PC displayed inside the console. If everything is working correctly, the following page will be displayed:

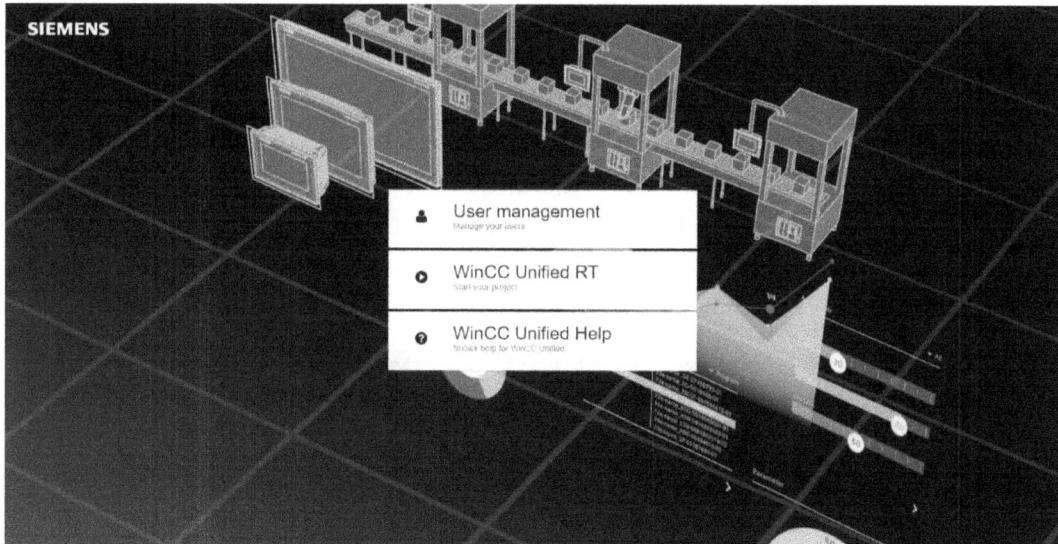

Figure 14.9 – The runtime management menu

Figure 14.9 shows the **runtime management menu**. This is the starting point of a unified runtime system. From here, the user management or the unified runtime can be started, or additional help can be displayed.

By clicking on the **WinCC Unified RT** button, the **User Login** window will be displayed:

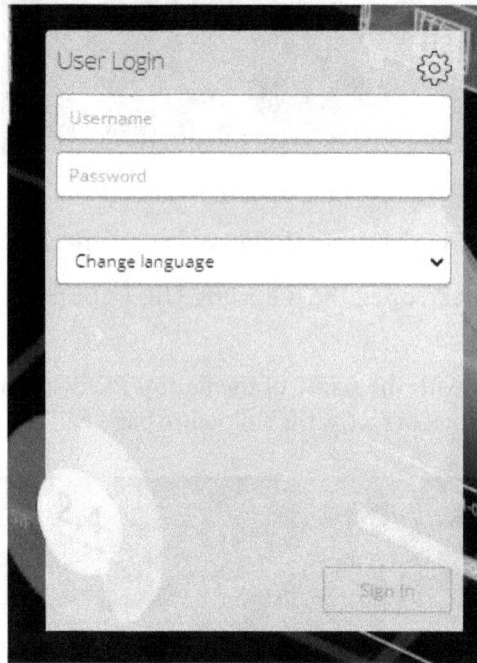

Figure 14.10 – The User Login window

At this point, a user must be logged in to continue. Ensure that the language is set correctly before clicking on the **Sign In** button.

> **Note**
> Users are configured in the **Users and roles** section of the project tree, under **Security settings**. This user configuration applies to the entire project. Ensure that a full download, along with reset runtime values, has been performed to update the runtime with the latest user data.

Once the user has logged in, the runtime will be started and the simulated HMI will begin running in the browser. The behavior of the runtime and how the project is operating can now be tested within a test environment.

> **Note**
>
> Simulated HMI runtimes can interact with physical PLCs via the connections that have been configured. Also, simulated PLCs can be connected to HMIs as if they were physical devices.

Security considerations

In TIA Portal V17, security standards have been implemented and Siemens recommends using security methods whenever possible. These include **encryption transfer** but extend beyond this to items such as OPC UA.

With a vast variety of connection protocols and interfaces available on different hardware platforms that can run the unified environment, Siemens has detailed its security considerations in a freely available document that can be downloaded from the Siemens website or via the following QR code:

Figure 14.11 – The QR code to Siemens Support for HMI security considerations

If the QR code in *Figure 14.11* cannot be read, the same information can be found by searching for *Security guidelines for SIMATIC unified HMI operator devices* on the internet.

On the support site, there is a PDF file spanning over 40 pages that lists, in detail, all the security considerations and steps to enable these security functions. It is unlikely that all projects will need to consider all of the security concerns; however, it is useful for programmers to understand what they are and why some of these options exist.

In nearly all cases, security is disabled by default or, at the very least, has no password set. And it fails the compilation until a password has been set (or the feature has been disabled).

Summary

In this chapter, we covered the steps required to download a unified HMI, including how to simulate and access simulated HMIs. Downloading to HMIs is similar to downloading to a PLC. However, many different options directly affect how the download proceeds and what happens to the runtime of the HMI once the download has finished executing.

The unified environment is capable of running on different hardware platforms, even PCs. The download procedure for all of these should be the same. However, there might be small differences in the **Load preview** window depending on the capabilities enabled (such as OPC UA and other interfaces).

In the next chapter, we will focus on writing tips such as simplifying logic, naming conventions, sequences, and where to find further support.

15

Programming Tips and Additional Support

The final chapter covers programming tips for different areas of TIA Portal. These are items that are useful to know and that may help advance knowledge and writing styles for programmers.

The items described in this chapter serve as a good starting point for any programmer to adapt and modify to suit their individual writing style. This chapter also covers additional support information and where programmers can find more resources to help continue their learning.

This chapter covers the following areas:

- Simplifying logic tips
- Managing sequences
- Naming conventions and commenting
- Additional Siemens support
- Further support

Simplifying logic tips

There are hundreds of different ways of completing the same task when it comes to logic writing. There is no clear approach that is the perfect way to achieve the desired logic output; all that should matter is that it is easy to read, easy to modify, and well documented. In addition to this, programmers should write logic code with their own style that is comfortable to them but be considerate of the fact that it is likely that other programmers will also work on the project.

Good logic will be simple and easy to follow without much deciphering required by those that find themselves working on it.

Delay timers

When using **timers**, such as the **Timer On Delay (TON)** timer, it is important to understand why that timer is being used. Consider a scenario where an output is required to be delayed both before and after a signal is set to `True`. This may look something like *Figure 15.1*:

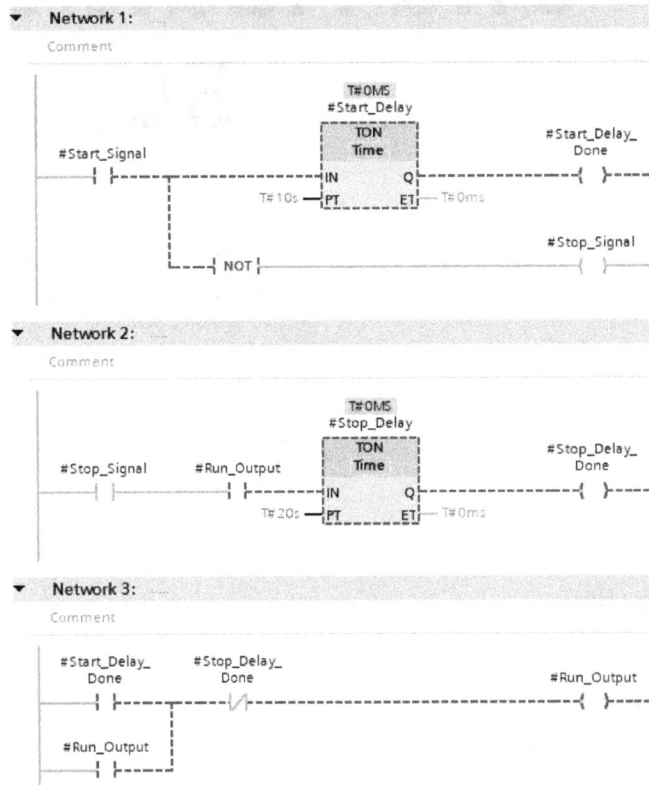

Figure 15.1 – Example of start and stop delay timers

While there is nothing wrong with this approach, it can be simplified, as the Stop_Delay TON timer is being used as a permissive delay instead of an inhibiting delay. When Start_Signal is set to True, the Start_Delay TON timer will start timing for 10s. On completion, the Start_Delay_Done bit is set to True, and it is this variable that sets Run_Output to True on *Network 3*.

When Start_Signal is set to False, then Run_Output is active (as long as the Start_Delay timer has completed). This then starts the Stop_Delay timer on *Network 2*. When this timer completes, the inverted Stop_Delay_Done bit on *Network 3* breaks the **hold on contact** for Run_Output and the system stops.

There is nothing wrong with this approach and all works correctly; however, it can be simplified further, as shown in *Figure 15.2*:

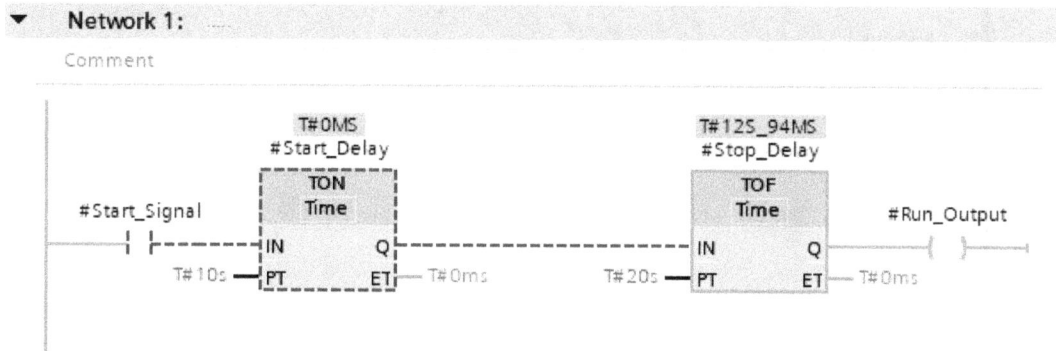

Figure 15.2 – Simplified timer delay solution using a TOF timer

This example shows how using a **Timer Off Delay** (**TOF**) timer in conjunction with a TON timer simplifies the process drastically. No additional variables are required other than Start_Signal and Run_Output, and no new timers are required either.

> **Note**
> The preceding example shown in *Figure 15.2* shows the TOF timer keeping Run_Output in a True state following Start_Signal switching to False.

Remembering what tools are available and how they can be used together is an important practice and one that programmers should try to exercise often.

AT constructor

The **AT constructor** is a special attribute that can be applied when declaring static and temporary variables. It allows memory locations to overlap each other so that the same data can be represented as different data types.

Figure 15.3 demonstrates this functionality:

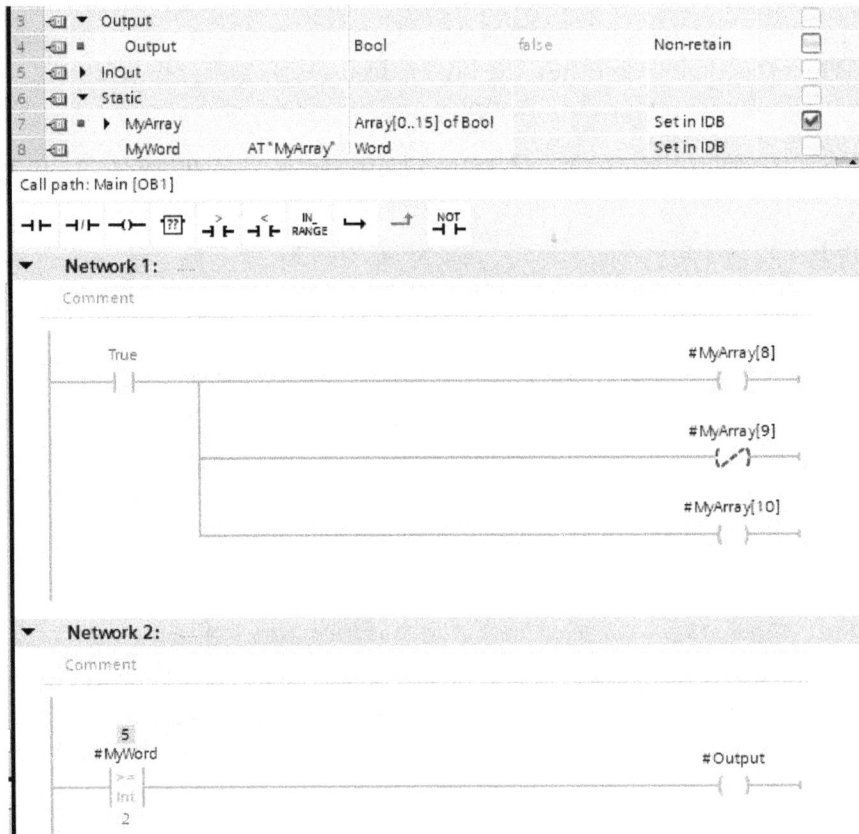

Figure 15.3 – AT constructor example

Network 1 shows a `True` signal setting elements 8 and 10 of a **Boolean array**. Array element 9 is set to `False` as it is an inverted coil.

In the **Interface** section of the block, `MyArray` is defined as a 16-bit `Array` and `MyWord` is defined as `Word`; however, it has the constructor `AT "MyArray"` defined, meaning it is overlayed with the same memory as `MyArray`.

This is demonstrated by *Network 2* having the value of 5, which is the value of the 16 bits represented as a `Word` value.

To use an AT constructor, the preceding variable's **Retain** setting must be set to **Set in IDB** and the word AT must be typed exclusively into the **Data type** field and the *Enter* key pressed. Once this is completed, TIA Portal will prompt for the data type required.

> **Note**
> **Non-optimized** blocks can also use the AT constructor. The **Retain** setting does not affect the availability of using the AT constructor in non-optimized blocks.

IF statements

When writing in the **Structured Text (SCL)** language, it is easy to use IF statements unnecessarily. *Figure 15.4* contains two examples of IF statements that can be simplified:

```
//Example 1
IF #High_level = True AND #System_Running = True THEN
    #High_Alarm := True;
ELSE
    #High_Alarm := False;
END_IF;

//Example 2
IF #Scale_Value < 20 THEN
    #Signal_Healthy := True;
ELSE
    #Signal_Healthy := False;
END_IF;
```

Figure 15.4 – Unnecessary IF statements

Example 1 is simply comparing two boolean variables and setting a third variable to True. *Example 2* is performing a basic comparison between a constant and a variable and setting a variable to True if the variable is less than the constant.

Both examples can be written without the use of an IF statement:

```
//Example 1
#High_Alarm := #High_level AND #System_Running;
//Example 2
#Signal_Healthy := #Scale_Value < 20;
```

Figure 15.5 – Simplified examples

Example 1 and *Example 2* now both store the result of the instructions that were previously within the `IF` statement. This approach writes both `True` or `False` depending on the outcome of the variables being tested.

> **Note**
>
> Set and Reset coils are unavailable in SCL; however, setting a variable to `True` in an `IF` statement that has no `ELSE` statement means that the logic to set the variable back to `False` must be handled elsewhere. This is the same principle as a Set and Reset coil in ladder logic.

Serializing

The **Serialize** instruction is a TIA Portal-provided instruction, found in the **Move operations** folder in the **Basic instructions** window when programming. The instruction converts a `Variant` data type to a numerical sequential representation. The `Variant` data passed can be almost any type, as long as the length of the data is less than 64 KB.

The instruction is most often used to flatten out structured data to either transmit to another device over a communication protocol or to bundle data together for further processing.

An example of this is shown in *Figure 15.6*:

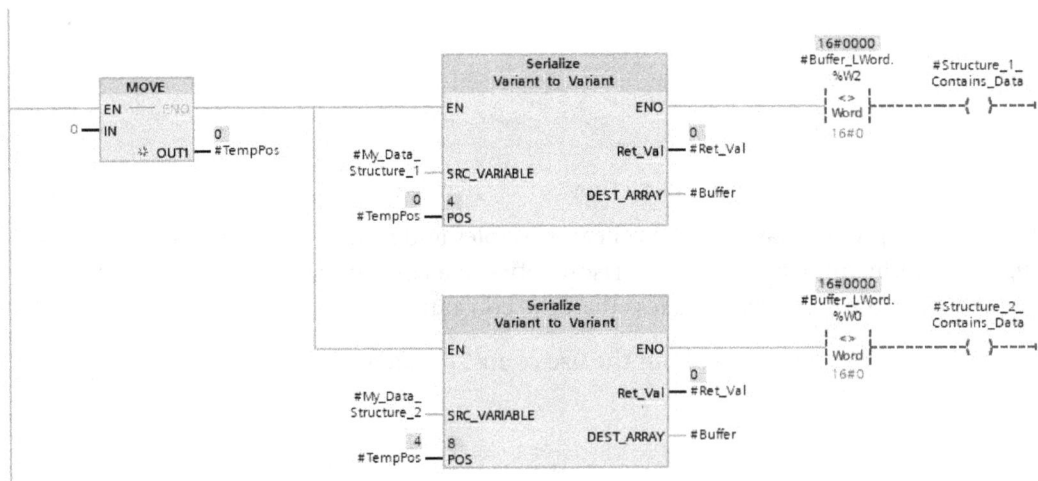

Figure 15.6 – Example of the Serialize instruction

This example is taking two structures of data and serializing them both, and storing them in the `Buffer` output. The `TempPos` variable keeps track of how much data has been processed by sequential `Serialize` instructions so that the data in the `Buffer` variable does not get overwritten by the next dataset.

Once the serialization of the data has been processed, both sets of information are merged and stored in the `Buffer` variable. *Figure 15.7* shows how the data is structured in the interface:

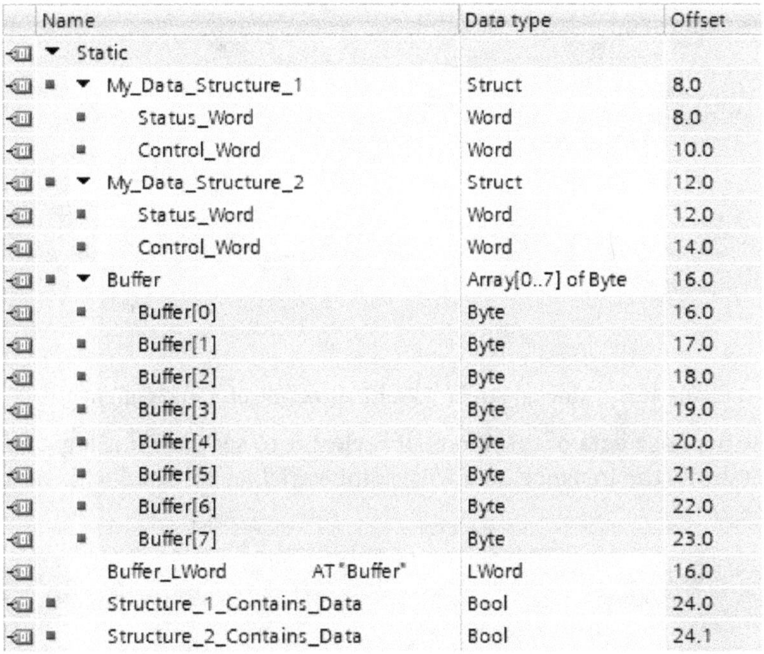

Name					Data type	Offset
◀	▼		Static			
◀	■	▼	My_Data_Structure_1		Struct	8.0
◀		■		Status_Word	Word	8.0
◀		■		Control_Word	Word	10.0
◀	■	▼	My_Data_Structure_2		Struct	12.0
◀		■		Status_Word	Word	12.0
◀		■		Control_Word	Word	14.0
◀	■	▼	Buffer		Array[0..7] of Byte	16.0
◀		■		Buffer[0]	Byte	16.0
◀		■		Buffer[1]	Byte	17.0
◀		■		Buffer[2]	Byte	18.0
◀		■		Buffer[3]	Byte	19.0
◀		■		Buffer[4]	Byte	20.0
◀		■		Buffer[5]	Byte	21.0
◀		■		Buffer[6]	Byte	22.0
◀		■		Buffer[7]	Byte	23.0
◀			Buffer_LWord AT "Buffer"		LWord	16.0
◀	■		Structure_1_Contains_Data		Bool	24.0
◀	■		Structure_2_Contains_Data		Bool	24.1

Figure 15.7 – Interface data

The `Buffer` variable is an `Array` consisting of 8 `Bytes` and an AT constructor is in use, overlaying the `Buffer_LWord` variable on top of the `Buffer` variable. This then means that the `Buffer` variable can be accessed as an `LWord`.

A comparator checks to see if each data structure variable contains any `Control_Word` data being used. This works by accessing each control word directly in the `Buffer` variable. TIA Portal will swap the `Words` in the `LWord` when the AT constructor is used; this is the reason why the first serialize block compares `Word 2` and the second serialize block compares `Word 0`.

> **Note**
>
> This could be simplified by simply comparing `My_Data_Structure_1.`
> `Control_Word`; however, for the purpose of explaining how serialization
> can cause byte/word swapping, the `Buffer` data is used.

Once the serialization of the data has occurred, if both `Control_Word` variables in the structures contain data, the data is output from the block using the logic shown in *Figure 15.8*:

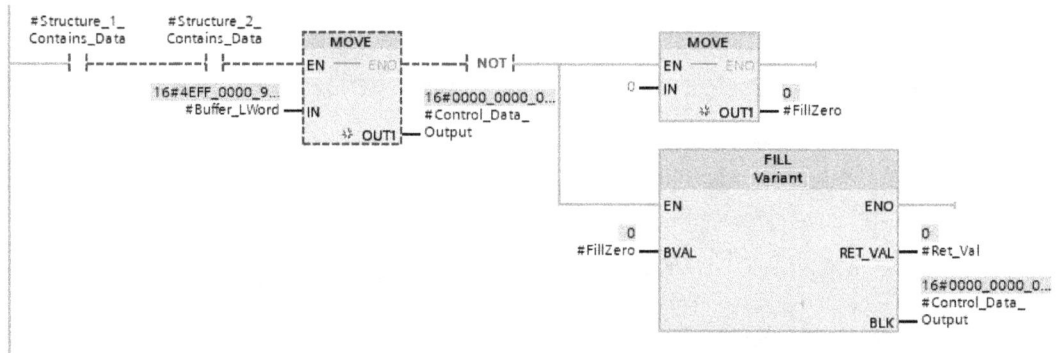

Figure 15.8 – Moving Buffer_LWord into Control_Data_Output

By looking at the instance data of this block, it is clearer to see how this logic affects the data. *Figure 15.9* shows the instance data with some variables updated with values:

Name	Data type	Offset	Monitor value
Input			
Output			
Control_Data_Output	LWord	0.0	16#0000_0000_0000_0000
InOut			
Static			
My_Data_Structure_1	Struct	8.0	
Status_Word	Word	8.0	16#4EFF
Control_Word	Word	10.0	16#0000
My_Data_Structure_2	Struct	12.0	
Status_Word	Word	12.0	16#9024
Control_Word	Word	14.0	16#0000
Buffer	Array[0..7] of Byte	16.0	
Buffer[0]	Byte	16.0	16#4E
Buffer[1]	Byte	17.0	16#FF
Buffer[2]	Byte	18.0	16#00
Buffer[3]	Byte	19.0	16#00
Buffer[4]	Byte	20.0	16#90
Buffer[5]	Byte	21.0	16#24
Buffer[6]	Byte	22.0	16#00
Buffer[7]	Byte	23.0	16#00
Structure_1_Contains_Data	Bool	24.0	FALSE
Structure_2_Contains_Data	Bool	24.1	FALSE

Figure 15.9 – Instance data

At this point, neither of the `Control_Word` variables contain any data. Because of this, no data is output to `Control_Data_Output`, as `Structure_1_Contains_Data` and `Structure_2_Contains_Data` are both `False`.

The `Buffer` variable contains the `Status_Word` variables for both structures decimated into individual `Byte` elements in the `Buffer` array.

> **Note**
> In the `Buffer` array, the data is not **byte swapped**. The data exists in the expected locations for a straight serialization.

Figure 15.10 shows the instance data with values once the `Control_Word` variables contain information:

Name	Data type	Offset	Monitor value
Input			
▼ Output			
▪ Control_Data_Output	LWord	0.0	16#4EFF_1234_9024_5678
InOut			
▼ Static			
▪ ▼ My_Data_Structure_1	Struct	8.0	
▪ Status_Word	Word	8.0	16#4EFF
▪ Control_Word	Word	10.0	16#1234
▪ ▼ My_Data_Structure_2	Struct	12.0	
▪ Status_Word	Word	12.0	16#9024
▪ Control_Word	Word	14.0	16#5678
▪ ▼ Buffer	Array[0..7] of Byte	16.0	
▪ Buffer[0]	Byte	16.0	16#4E
▪ Buffer[1]	Byte	17.0	16#FF
▪ Buffer[2]	Byte	18.0	16#12
▪ Buffer[3]	Byte	19.0	16#34
▪ Buffer[4]	Byte	20.0	16#90
▪ Buffer[5]	Byte	21.0	16#24
▪ Buffer[6]	Byte	22.0	16#56
▪ Buffer[7]	Byte	23.0	16#78
▪ Structure_1_Contains_Data	Bool	24.0	TRUE
▪ Structure_2_Contains_Data	Bool	24.1	TRUE

Figure 15.10 – Instance data with Control_Words containing data

Once the `Control_Words` variables contain data, the `Control_Data_Output` LWord is updated and contains the merged information.

This type of logic can be extremely useful for quickly checking large arrays for a single value. For example, consider an `Alarm Array`. A `Serialize` instruction can be used to pass the `Array` to an LWord and then the LWord compared against a 0 value. If the `Array` contains any information, the LWord will be greater than 0.

Refactoring

Refactoring is a process by which logic and the data used are reviewed and then re-structured or re-written to simplify the process. This is something that all programmers should perform on their projects toward the end of a development cycle. Refactoring provides an opportunity to improve structure and logic without affecting the tested output.

For example, *Figure 15.11* shows logic that repeats the same function a number of times with different datasets:

		Name	Data type	Default value	Retain	Accessible f...	Writa...	Visible in ...
8		▶ Dataset1	Struct		Non-ret... ▼	✓	✓	✓
9		▶ Dataset2	Struct		Non-retain	✓	✓	✓
10		▶ Dataset3	Struct		Non-retain	✓	✓	✓
11		Result	DInt	0	Non-retain	✓	✓	✓

```
1  //Pass1
2  #Result := #Dataset1.A + #Dataset1.B + #Dataset1.C;
3  //Pass2
4  #Result := #Result + #Dataset2.A + #Dataset2.B + #Dataset2.C;
5  //Pass3
6  #Result := #Result + #Dataset3.A + #Dataset3.B + #Dataset3.C;
```

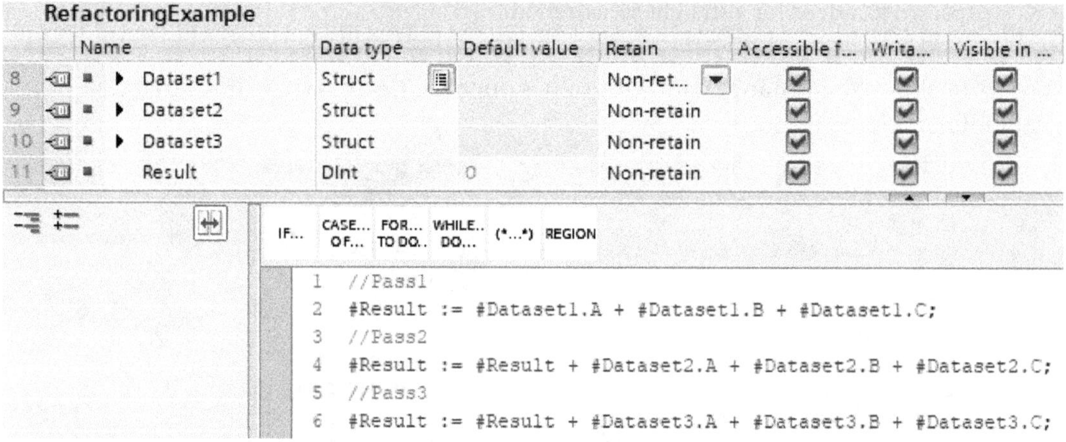

Figure 15.11 – Multiple passes of the same logic with different datasets

While there is nothing wrong with the logic, this approach is not easily expanded, even though it is written in SCL, which makes copy and pasting and replacing easier. If this had to be expanded to include an additional 50 passes, the interface would be busy, and the code would be long. If the logic then needed to add a `Dataset.D` variable into the process, it would become very tedious to update.

By performing a refactoring process on this logic, a few items of potential improvement can be observed:

- The `Dataset` structures, all of which contain the same data, would be better configured as an `Array of struct`.

- The logic could loop the same method instead of having to write the logic for every pass. This would mean the logic could easily be expanded.

Figure 15.12 demonstrates how the logic and data have changed following the refactoring:

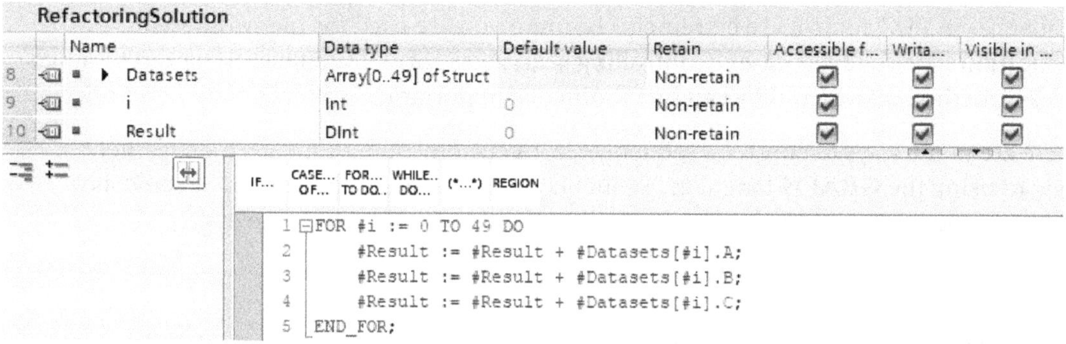

RefactoringSolution							
	Name	Data type	Default value	Retain	Accessible f...	Writa...	Visible in ...
8	▶ Datasets	Array[0..49] of Struct		Non-retain	☑	☑	☑
9	i	Int	0	Non-retain	☑	☑	☑
10	Result	DInt	0	Non-retain	☑	☑	☑

```
1 ⊟FOR #i := 0 TO 49 DO
2       #Result := #Result + #Datasets[#i].A;
3       #Result := #Result + #Datasets[#i].B;
4       #Result := #Result + #Datasets[#i].C;
5  END_FOR;
```

Figure 15.12 – Refactored code and data

This refactored logic performs the same logic 50 times, from `Array element 0` through to `49`. This is achieved as part of the `For` instruction. The `Datasets` variable now holds all of the necessary data and can be easily expanded if required. If the `Datasets` array is expanded, the `For` statement would also need to be increased in the loop count to ensure the new data is processed.

> **Note**
>
> The `Datasets` elements are populated outside of this block. This could also be an important reason for refactoring, as large interfaces are cumbersome to connect to variables. Very large interfaces may even reach a limit, whereby TIA Portal will not compile the block as the interface is too large.

Consolidating blocks

Consolidating code and wrapping it into dedicated **functions** and **function blocks** is a quick way to create re-usable code that can speed up the development process. It is very easy to leave loose code in program blocks that span across multiple lines of code, but these lines of code achieve a common functionality. When this occurs, other programmers may copy the multiple lines of code and use them elsewhere, or in different projects. When this section of code gets updated, it needs to be updated in multiple places and it becomes very difficult to control. Building a habit of consolidating logic into functions and function blocks ensures that code is locked away in a block that can have many instances. When the block is updated, all instances are updated. All instances are also cross-referenceable, making the code easier to find no matter where it is used.

Sequences – best practices

Sequences in PLC control are extremely common and are used for many different application types and use cases. Controlling sequences correctly, efficiently, and retaining an easy method by which they can be modified is important.

There are many ways to control a sequence, from custom-built sequence management logic to using the **GRAPH** language. Sequences are inherently application-specific, but their management does not have to be and can be standardized to some degree.

> **Note**
>
> TIA Portal's GRAPH has a pre-built method to create advanced sequences. However, in many circumstances, GRAPH may not be suitable or desired, especially for sequences that interact with other proprietary code. GRAPH is flexible enough to be programmed to do what needs to be done, but it comes with additional functions that may not be desired.

Using constants instead of numerical values

It's very common to see sequences that use numerical values to manage the sequence step. This may look something like the example shown in *Figure 15.13*:

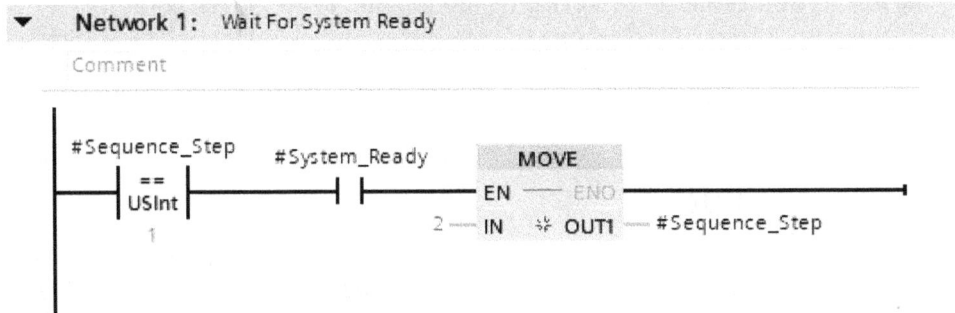

Figure 15.13 – Sequence step management using a USInt value

While there is nothing wrong with the logic behind how this works, the 1 and 2 values do not represent anything significant to the sequence itself. The network comment suggests that it is waiting for the system to be ready and when the System_Ready variable is True the 2 value is moved into Sequence_Step. This does not help another programmer or maintenance engineer understand what sequence step two is.

A better idea is to use **constants** to represent sequence steps. This way, the 1 numerical value is given a symbolic name that can be used to explain the sequence step. *Figure 15.14* demonstrates this approach:

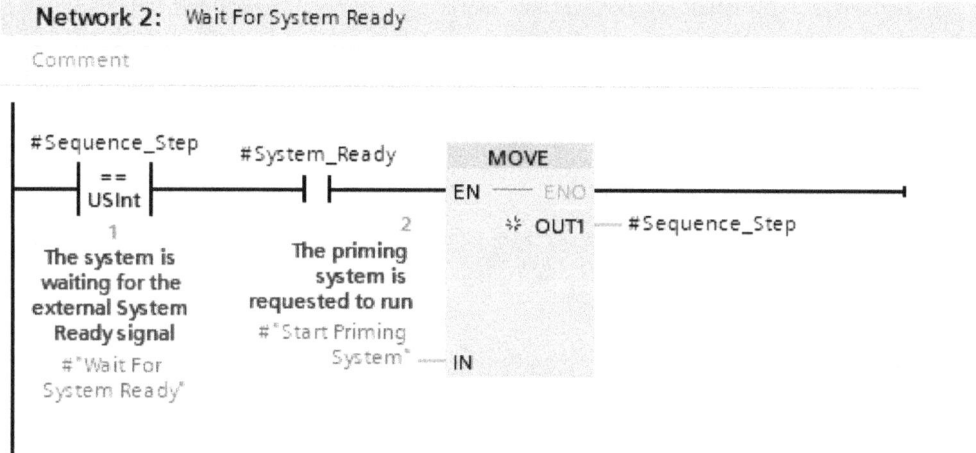

Network 2:   Wait For System Ready

Comment

```
#Sequence_Step      #System_Ready              MOVE
    ==                                   EN ─── ENO
   USInt          ┤ ├                               ┤
    1                           2        ❖ OUT1 ─── #Sequence_Step
The system is         The priming
waiting for the       system is
external System     requested to run
Ready signal        #"Start Priming
 #"Wait For            System" ─── IN
System Ready"
```

Figure 15.14 – Sequence step management using constants

By allocating constants, the current step and the step that is loaded next are much more self-explanatory. Not only can a constant carry a meaningful symbolic name, but it can also carry a comment that adds more context to the meaning of the constant's name. This allows programmers and maintenance personnel to view a section of a sequence and very quickly identify what each step represents.

> **Note**
>
> Another advantage of using constants for sequence step management is that constants can be cross-referenced. This means that programmers and maintenance personnel can quickly see where steps are used and which steps transition to other steps.

Managed transitions

Every sequence needs to transition from one step to the next. These transitions can be managed by a dedicated function or function block. By providing a managed sequence transition, it can be guaranteed that all transitions follow the same approach and additional logic can be applied to all steps.

Figure 15.15 demonstrates a managed transition:

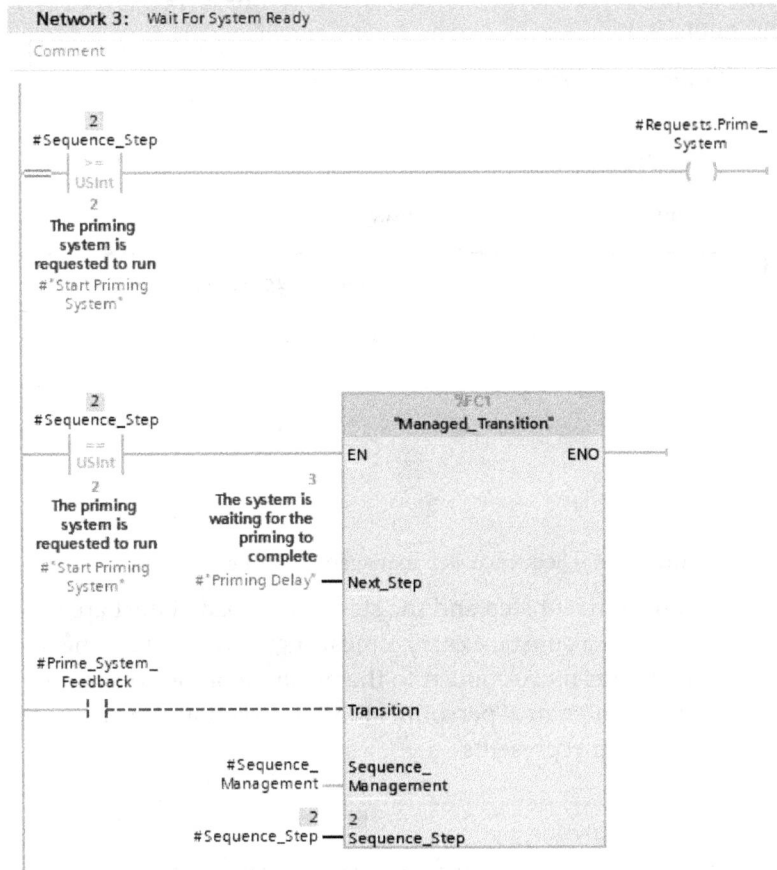

Figure 15.15 – Managed transition step

The key difference with this approach is that the Managed_Transistion dedicated function is being used to change the Sequence_Step value. The Managed_Transistion object accepts the next step, a transition input, a dedicated Sequence_Management structure, and finally the sequence step.

These variables are all used together to control how the sequence is allowed to transition between different sequence steps.

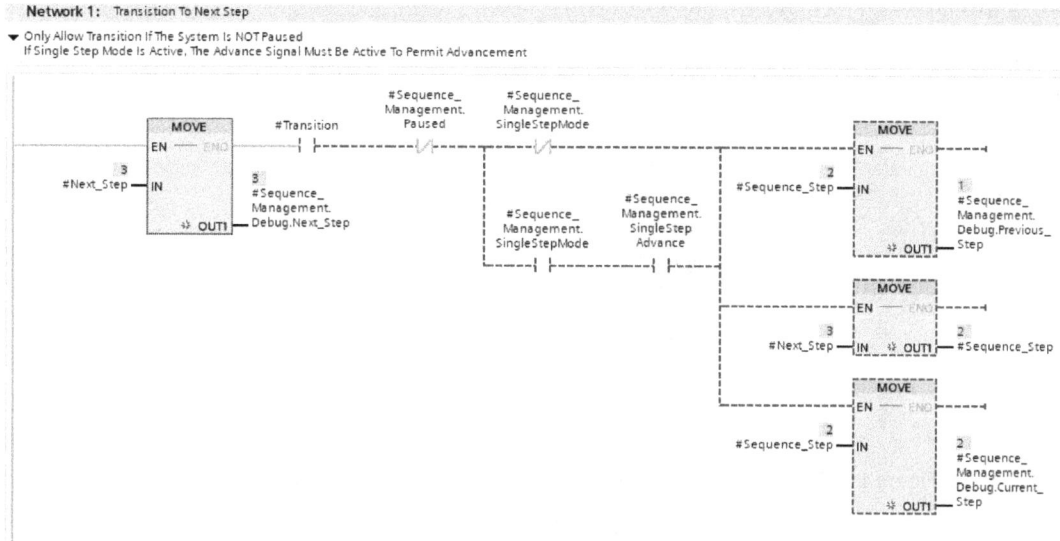

Figure 15.16 – Managed transition function logic

Figure 15.16 shows the logic inside the `Managed_Transition` function. The logic is simple but offers greater control over the transition of all steps that utilize the function. The following controls have been added to every transition that uses this object:

- Ability to block the transition because the system is paused

- Ability to single-step through the sequence if *Single Step Mode* is active and *Single Step Advance* is operated

- Ability to track the next step, previous step, and current step within the `Sequence_Management` structure

In order for this function to work, the parent object must pass the `Sequence_Management` structure to each `Managed_Transition` object. This then holds the relative information as demonstrated in *Figure 15.17*:

4	▼ Static			
5	Sequence_Step	USInt	0 —	2
6	System_Ready	Bool	fal: —	TRUE
7	Prime_System_Feedback	Bool	fal: —	FALSE
8	▼ Sequence_Management	Struct	—	
9	Paused	Bool	fal: —	FALSE
10	SingleStepMode	Bool	fal: —	FALSE
11	SingleStepAdvance	Bool	fal: —	FALSE
12	▼ Debug	Struct	—	
13	Previous_Step	USInt	0 —	1
14	Next_Step	USInt	0 —	3
15	Current_Step	USInt	0 —	2

Figure 15.17 – Sequence_Management structure containing transition data

This methodology can be expanded to include any number of variables or information that a particular sequence may need. For example, a sequence that is tasked with testing widgets may also record if a step passed or failed upon transition to the next step.

> **Note**
>
> The debug section could be expanded further to utilize an array. This would mean that more than one previous step could be held in memory, which means the entire step chain could be held, improving the debugging experience.

Managing output requests

Sequences often control outputs, which usually is not a problem as outputs can be written to within the sequence. Output management can get complicated when a PLC project contains more than one sequence that controls the same outputs, or sequences that run concurrently that use the same outputs.

Figure 15.18 demonstrates an example of this, where three sequences make requests to a common output:

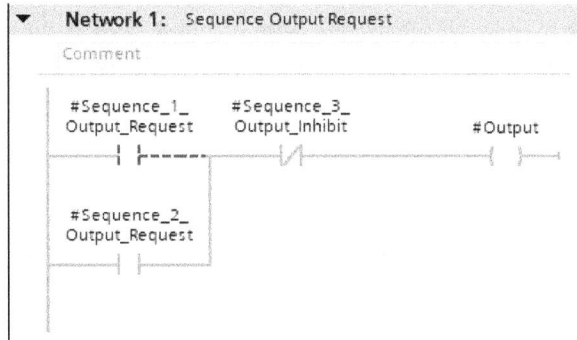

Figure 15.18 – Request example

This logic allows three different inputs to manage a single output. The three request inputs come from three different sequences that all run at the same time but need to interact with a common output.

Sequence 1 and Sequence 2 have the same level of priority. Either can turn on the Output variable as long as Sequence 3 is not inhibiting the output. This means that Sequence 3 has a higher priority and will always stop the output from being True, no matter what Sequence 1 and Sequence 2 are requesting.

This approach is managed inside a dedicated function, as shown in *Figure 15.19*:

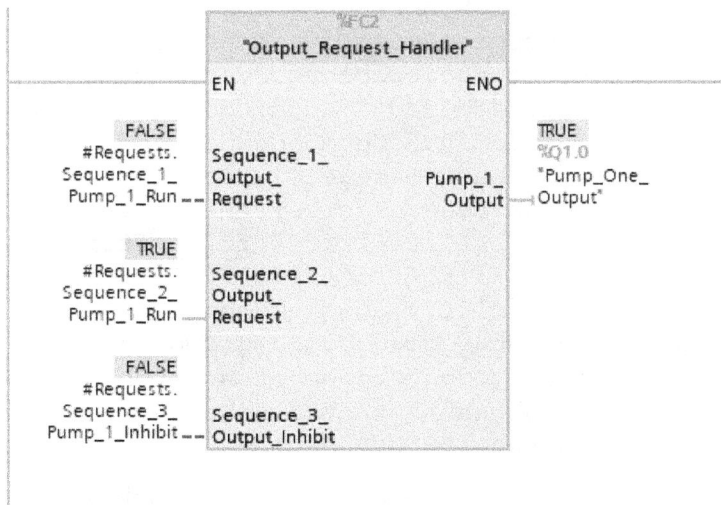

Figure 15.19 – Output request handler

Output_Request_Handler is used to accept requests for the Pump_1_Output from each of the three sequences.

This approach removes the need for using **Set coils** in sequences, which can become very difficult to manage once multiple sequences need to manage the same output. This approach also lends itself to a hierarchical approach; if more sequences are required, they can all be handled in one place.

> **Note**
>
> This approach can handle more than one output. All outputs that require more than one point of control can be handled in a single function. This approach can also apply to general variable updates that originate in a sequence, for example, updating interlocks and alarms.

Naming conventions and commenting

TIA Portal offers a huge 128 characters per symbolic name for a variable across eight available nesting depths. This means that variables can have a combined maximum length of 1,024 characters. To visualize this, the following would be a perfectly acceptable variable name:

```
#aaaaaaaaaaaaaaaaaaaaaaaaaaaaaaaaaaaaaaaaaaaaaaaaaaaaaaaaaaaaa
aaaaaaaaaaaaaaaaaaaaaaaaaaaaaaaaaaaaaaaaaaaaaaaaaaaaaaaaaaaaaa
aaaaa.bbbbbbbbbbbbbbbbbbbbbbbbbbbbbbbbbbbbbbbbbbbbbbbbbbbbbbbb
bbbbbbbbbbbbbbbbbbbbbbbbbbbbbbbbbbbbbbbbbbbbbbbbbbbbbbbbbbbbbb
bbbbbbbbbb.cccccccccccccccccccccccccccccccccccccccccccccccccc
cccccccccccccccccccccccccccccccccccccccccccccccccccccccccccccc
ccccccccccccccc.ddddddddddddddddddddddddddddddddddddddddddddd
dddddddddddddddddddddddddddddddddddddddddddddddddddddddddddddd
ddddddddddddddddddddd.eeeeeeeeeeeeeeeeeeeeeeeeeeeeeeeeeeeeeeee
eeeeeeeeeeeeeeeeeeeeeeeeeeeeeeeeeeeeeeeeeeeeeeeeeeeeeeeeeeeeee
eeeeeeeeeeeeeeeeeeeeeeee.fffffffffffffffffffffffffffffffffffff
ffffffffffffffffffffffffffffffffffffffffffffffffffffffffffffff
ffffffffffffffffffffffffffffffff.gggggggggggggggggggggggggggg
gggggggggggggggggggggggggggggggggggggggggggggggggggggggggggggg
ggggggggggggggggggggggggggggggggggggggggghhhhhhhhhhhhhhhhhhhhh
hhhhhhhhhhhhhhhhhhhhhhhhhhhhhhhhhhhhhhhhhhhhhhhhhhhhhhhhhhhhhh
hhhhhhhhhhhhhhhhhhhhhhhhhhhhhhhhhhhhhhhh
```

This essentially means that there is no real reason why projects developed in TIA Portal should use abbreviated tags and variable names.

For example, the `PMP1_RASR` tag requires the comment *Pump 1 Raw Actual Speed Reference* to make sense to any programmer or maintenance technician who doesn't know what the abbreviation stands for. A better naming convention would involve `Pump 1` being declared as its own data block and the `Raw_Actual_Speed_Reference` variable being created within the `Pump1` data block. The final variable name would appear as `Pump1.Raw_Actual_Speed_Reference`, which makes it immediately obvious what information the variable holds, and what asset it is associated with. This approach also follows the convention for using UDTs and leaves the comment free for something meaningful such as *Divide by 4,000* instead of the expanded abbreviation.

This approach becomes even easier when multi-level nesting occurs. Let's say you declared a global variable as follows:

`MPF_1_CHDA_2_SYST_2_PMPSET_A_PMP_3_RNG`

You also included the following comment:

Main Pump Floor 1, Chemical Dosing Area 2, System 2, Pumpset A, Pump 3 Running

A better alternative would be a global data block with the following structure created:

`Main_Pump_Foor_1.Chemical_Dosing_Area_2.System_2.PumpSet_A.Pump_3.Running`

Comments can then be applied at each point of the structure, as shown in *Figure 15.20*:

Main_Pump_Floor_1		
Name	Data type	Comment
▼ Static		
▶ Chemical_Dosing_Area_1	Struct	Location C1-DFA233
▼ Chemical_Dosing_Area_2	Struct	Location C2-UVTA233
▶ System_1	Struct	System 1 - Acid Dosing
▼ System_2	Struct	System 2 - Caustic Dosing
▼ PumpSet_A	Struct	Caustic Delivery Pumpset
▶ Pump_1	Struct	Asset: 123A
▶ Pump_2	Struct	Asset: 456A
▼ Pump_3	Struct	Asset: 789A
Running	Bool	
▶ PumpSet_B	Struct	Emergency Pumpset

Figure 15.20 – Example of structure with a naming convention

In essence, this naming convention is simply *no abbreviations allowed*, paired with valid structured data. When this is followed, finding information is simplified, and keywords can be searched with a better success rate.

> **Note**
>
> Pressing *F1* to search while in a program block, and then *F1* again will open the project search. Searching for a keyword in the program, such as `Running`, will list all the interfaces and declarations where the word *Running* exists. This is easier to keep consistent with good naming practices instead of abbreviations.

When naming variables, full names are used. Comments are only really required to give a useful hint as to how the data is supposed to be used. *Figure 15.21* gives an example of this:

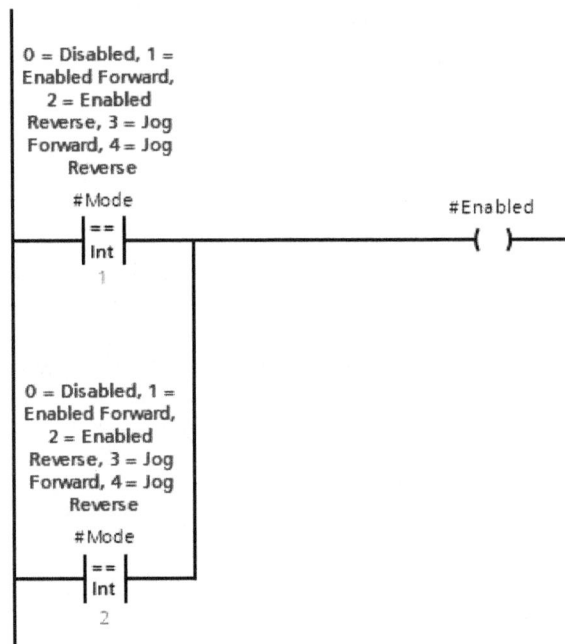

Figure 15.21 – Example of commenting

This example demonstrates how the comment attached to the Mode variable clearly explains what the value within the Mode variable represents. Although comments such as this can make logic look bulky, they are still worthwhile as they help us to understand the exact function the variable is being used for.

Comments in SCL

The SCL language has some additional features for commenting due to the fact that it is a text-based language. In SCL, comments can be added in almost any location of the SCL program. An example is shown in *Figure 15.22*:

```
1   //Call Timers
2
3   //Timer 1
4 ⊟ #Timer1(IN:=#Timer_1_In,
5 │        PT:=T#10s,
6 ⌊        Q=>#Timer_1_DN);
7
8   //Timer 2
9
10 ⊟ #Timer2(IN := #Timer_2_In,
11 │         PT := T#5s,
12 ⌊         Q => #Timer_2_DN);
13
14  //Start Timer 1
15  #Timer_1_In := NOT #Timer_2_DN; //When Timer 2 is not done, run Timer 1
16
17  //Start Timer 2
18  #Timer_2_In := #Timer_1_DN; //When Timer 1 is done, run Timer 2
19
20  //Set Outputs
21  #Dwell_Cycle_Active := NOT #Timer_1_DN;
22  #Run_Cycle_Active := #Timer_1_DN;
23  #Cycle_Complete_Pulse := #Timer_2_DN;
```

Figure 15.22 – Example of comments in SCL

This allows for a commenting convention to be easily established, where all comments in SCL can follow a set pattern. *Figure 15.22* shows comments above and to the side of the SCL code. This is the most common type of commenting found in structured text.

SCL has some additional toolbar buttons for commenting, as shown in *Figure 15.23*:

Figure 15.23 – Comment buttons

These buttons are used to comment out lines of code in one go (or to uncomment). Highlighting the area to be commented and pressing the left-hand button will cause TIA Portal to prefix the line with //, which comments out the line. Using the other button will remove the // prefix.

> **Note**
> Sequential areas of code can also be commented out by using (* *).
> Anything between the * symbols will be commented. This can span many lines and also allows the commented area to be collapsed and hidden.

Figure 15.24 shows an example of using formatted comments. Siemens does not offer a method by which this can be done within TIA Portal, but a simple Excel spreadsheet or pre-formatted note file can allow for quick and easy copying/pasting into TIA:

```
1   //
2   // ‖ Dwell/Run Timer Block - V1.2 - LBEE - 10/02/22                              ‖
3   // ‖ This block runs a dwell cycle (10s), immediately followed by a run cycle (5s)‖
4   // ‖ A cycle complete output is pulsed on a full Dwell and Run cycle             ‖
5   // ‖ Both the Dwell and the Run cycles also have an output to represent when in those states ‖
6   //
7
8   // //====================================================================\\
9   // ||                            Call Timers                            ||
10  // ||]====================================================================[|
11  // || Calls to the TON Function blocks for the timers                    ||
12  // || Note - The IN and Q variables are used elsewhere in the code, cross reference if unsure ||
13  // \\====================================================================//
14
15      // +---------+----------------------------------------------------+
16      // | Timer 1 |                    Dwell Timer                     |
17      // +=========+====================================================+
18      // |         | Called to run when Timer 2 (Run Timer) is not done (Completed) |
19      // +---------+----------------------------------------------------+
20 ⊟   #Timer1(IN:=#Timer_1_In,
21          PT:=T#10s,
22          Q=>#Timer_1_DN);
23
24      #Timer_1_In := NOT #Timer_2_DN;
25
26      // +---------+----------------------------------------------------+
27      // | Timer 2 |                    Run Timer                       |
28      // +=========+====================================================+
29      // |         | Called to run when Timer 1 (Dwell Timer) is done (Completed) |
30      // +---------+----------------------------------------------------+
31 ⊟   #Timer2(IN := #Timer_2_In,
32          PT := T#5s,
33          Q => #Timer_2_DN);
34
35      #Timer_2_In := #Timer_1_DN;
36
37  // //====================================================================\\
38  // ||                            Set Outputs                            ||
39  // ||]====================================================================[|
40  // || Pass the status of the Timers and if the Cycle has completed to the interface outputs ||
41  // \\====================================================================//
42
43  #Dwell_Cycle_Active := NOT #Timer_1_DN;
44  #Run_Cycle_Active := #Timer_1_DN;
45  #Cycle_Complete_Pulse := #Timer_2_DN;
```

Figure 15.24 – Example of formatted commenting

These comments are far easier to read, and the different styles quickly associate them with different types of comments. Indentation is also being used to associate logic under a header comment with the header comment description. This can be seen between lines **15** and **35** in *Figure 15.24*. By indenting the code, it helps distinguish the relationship between logic as the logic is sequentially processed.

Regions in SCL

SCL also contains **regions**, which are similar to single-line comments, but associate code between the region declaration and end of the region to be collapsed. It also allows the navigation of SCL through a dedicated **Regions** toolbar, situated to the left of the main SCL editing window.

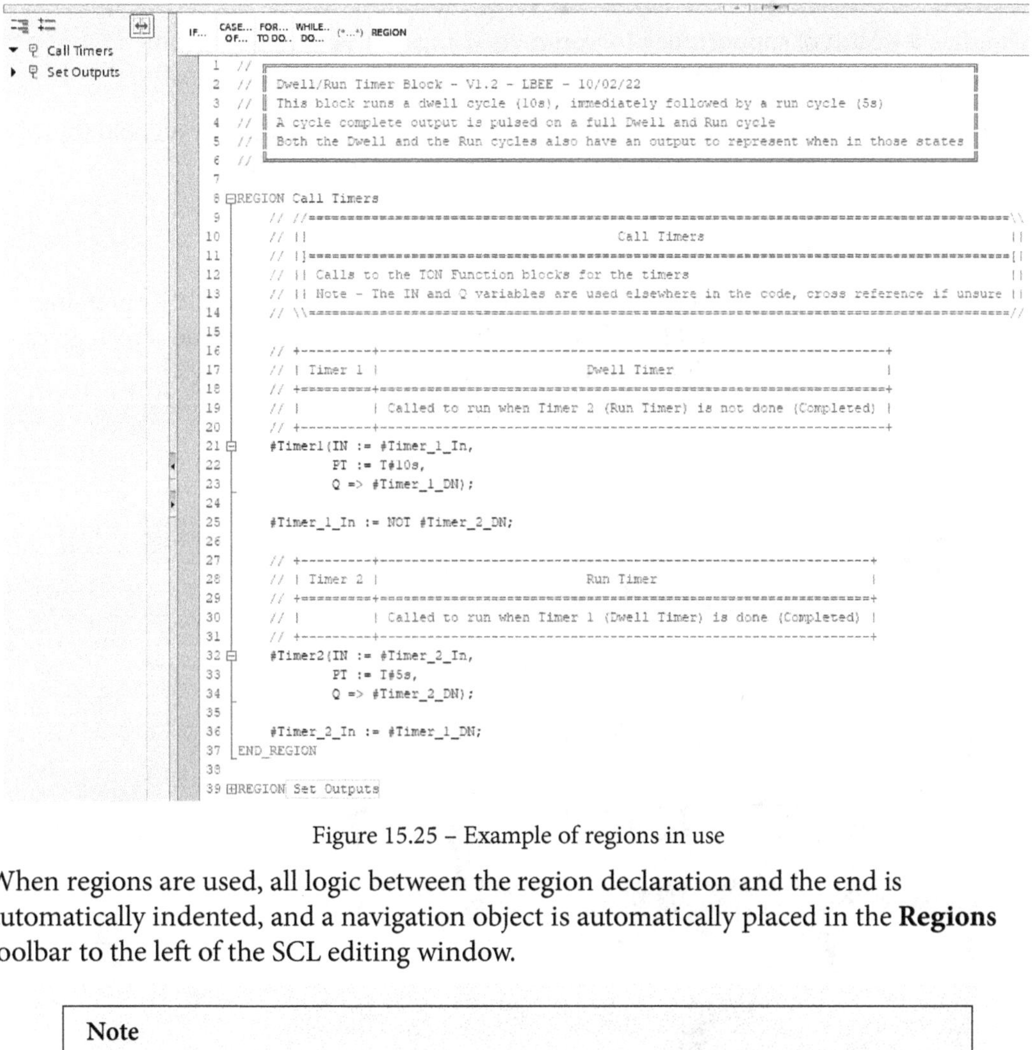

Figure 15.25 – Example of regions in use

When regions are used, all logic between the region declaration and the end is automatically indented, and a navigation object is automatically placed in the **Regions** toolbar to the left of the SCL editing window.

> **Note**
> Regions can also be nested, and the **Regions** toolbar will also show the nesting of regions using collapsible parent objects, similar to the main project tree.

Languages such as **ladder** and **Function Block Diagram** (**FBD**) have these segregation methods built-in by the form of **networks**. With SCL, a similar concept is achievable through well-designed comments and regions.

Additional Siemens support

Siemens has a wealth of support for programmers using TIA Portal. These come in many different forms, from internal TIA Portal help to dedicated forums and websites.

These resources can be a great place to find information quickly, and nearly all help topics have been covered by Siemens or the wider TIA Portal user community.

Using TIA Portal's help system

TIA Portal has an extremely well-documented help system. This is accessible by pressing *F1* or by selecting **Help** and then **Show help**. The **Information System** home screen gives some suggestions for generic help, as shown in *Figure 15.26*:

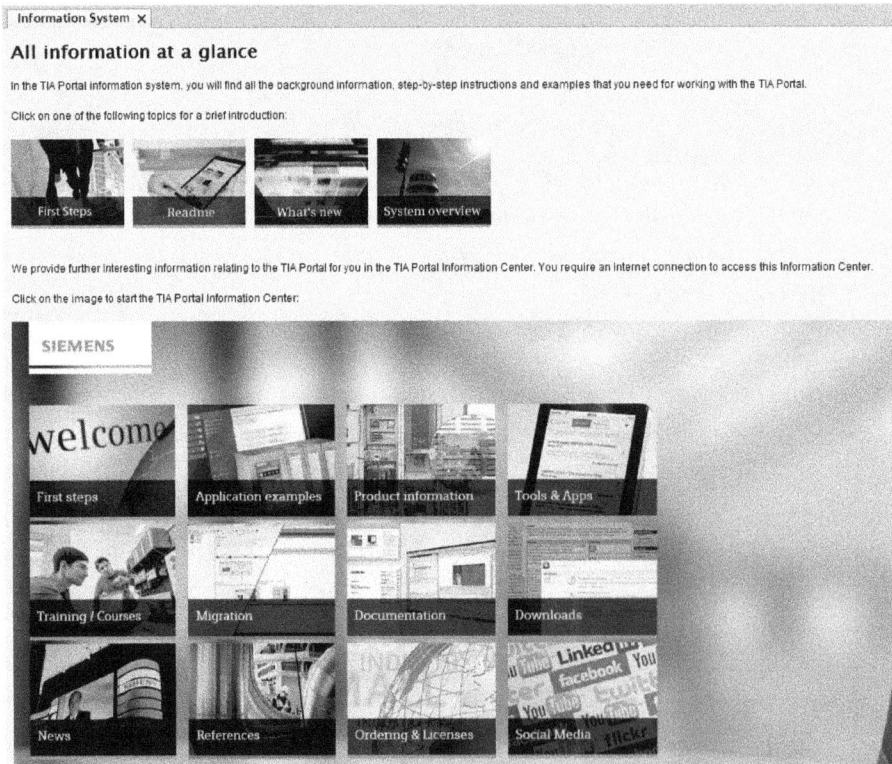

Figure 15.26 – Information System home screen

This is a good place to start for finding items such as generic help, training, and application examples. Most buttons link to the same version of the home page, but on the Siemens website, the buttons link to its relative pages.

> **Note**
>
> Clicking the buttons on the screen will launch a Siemens website page for further drill down for relative help. For example, clicking on **Social Media** will provide links to the **Forum**, **Twitter**, and other social platforms.

Using **Information System** is relatively simple and feels very similar to a modern browser. The system can save favorites (which appear in the navigation panel under the **Favorites** tab), move back and forth between pages, create multiple open tabs for viewing information, and have the ability to print. These tools help make **Information System** an easy-to-use tool when in search of supportive documentation.

> **Note**
>
> General usage and remarks on the **Information System** can be found within the **Information System** itself by navigating to **Information System | Introduction to the TIA Portal | Help in the Information system | General remarks on the Information system**.
>
> Alternatively, the question mark can be clicked to the right-hand side of the screen and **General remarks on the Information system** can be selected.

Navigation

Information System can be navigated using the **Navigation** panel to the left-hand side of the screen. If this panel is collapsed, the word **Navigation** is written where the panel is collapsed. Clicking the expand arrow will open the panel.

Information System is well categorized to make information easy to find. For example, if support information relating to the *Library view* was needed, this can be found at the path **Information System | Using libraries | Using the library view | Overview of the library view**.

The navigation system contains different symbols prefixing the text in the **Navigation** panel. These symbols denote different meanings for the content of the help object, as shown in *Figure 15.27*:

> ▓ Collapsed Category Folder
>
> 📖 Open Category Folder
>
> 📄 Factual Information
>
> 🔧 Operating Instructions
>
> ▶▶ Example
>
> ↗ Reference

Figure 15.27 – Symbol legend for the Information system Navigation panel

These help to identify the type of information before having to open and read the contents.

Searching help

The search bar is located to the left of the **Information System** screen, on any page. By expanding the search bar, the entire TIA Portal help system is available to be searched, as shown in *Figure 15.28*:

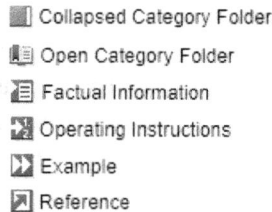

Figure 15.28 – Example of searching the Information system for the term "TON"

All items that contain the **Search for** term will be listed and a selection can be made with a double-click.

> **Note**
> **Information System** may list the same page more than once under different categories and sub-categories.

Advanced searching

TIA Portal's help system also contains an advanced search system that helps narrow the search criteria even further. The difference between the normal search and the advanced search is that the latter allows AND/OR terminology as well as phrase and fuzzy searching.

> **Note**
> A fuzzy search returns all results that match part of the criteria searched for. A good example of this is to search for Boo in the normal **Search for** bar and then Boo~ in the **Advanced search** bar. The normal search returns one result (which is a misspelling of Boolean) and the advanced returns 1,000 results.

To use **Advanced search**, click the button immediately after the normal search bar (the button with three dots). The following window will open:

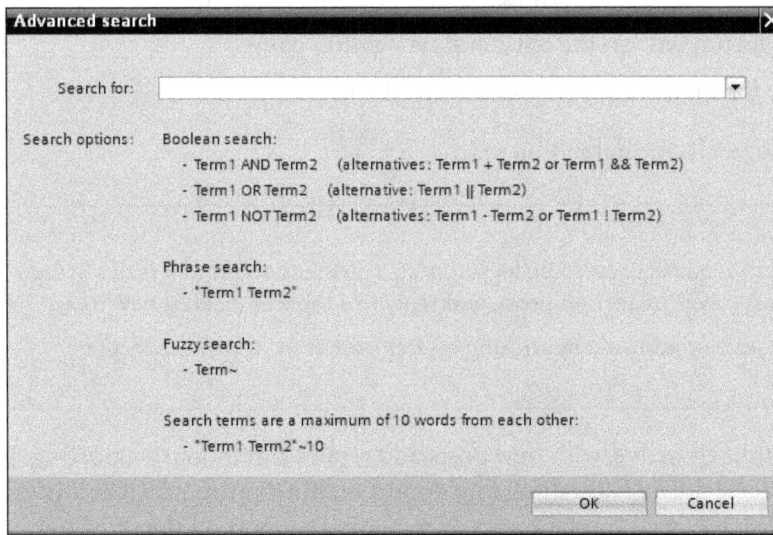

Figure 15.29 – Advanced search

Siemens has helpfully given examples of the types of searches that can be performed, with common alternatives that are supported.

Getting to grips with advanced search methods helps find relative support information far quicker than navigating through the help menus.

> **Note**
>
> It is not actually necessary to open the **Advanced search** window to perform an advanced search. Advanced searches can be performed in the normal search bar too.

Siemens forum

Siemens has an excellent community forum in which users help each other to overcome problems. This is a free and invaluable tool that can be used to obtain information from experts with many years of using Siemens' software and hardware.

The forum can be accessed directly from the **Information System** window by performing the following actions:

1. Press *F1* in TIA Portal.
2. When the **Information System** page loads, click the **Social Media** button from the home page (click it again if the same menu opens in a web page).
3. Click the forum link on the left-hand navigation pane.
4. Select the forum category required from the list.

The forum will open in the default browser.

> **Note**
>
> The forum requires users to be logged in to a Siemens account in order to view some posts, view images on posts, and reply to a topic or create a new topic.
>
> Signup can be completed by clicking on **Register** at the top of the forum window.

The forum is extremely active, with new posts and replies continuously updating. It is always worth visiting the forum and searching for support on the topic required as it is very likely that other people have faced similar issues or requested support on the same topic.

Siemens documentation archive

Siemens has put together a comprehensive list of documentation resources available to end users of TIA Portal. These can be found in a similar method to other resources provided by Siemens:

1. Press *F1* in TIA Portal.

2. When the **Information System** page loads, click the **Documentation** button from the home page (click it again if the same menu opens in a web page).

3. Choose a category from the left-hand menu that matches the documentation category required.

4. A list of available documentation in the category is displayed. Click the **Link** button adjacent to the documentation required.

5. A new browser window will open and will be directed to the Siemens support site that contains a download for the relative documentation.

6. Click on **Download** on the web page presented and the documentation will download.

> **Note**
>
> In most cases, a login is not required to download a manual. However, some pages may ask for a user to log in before the download can commence.

Knowing where to find these resources is half of the battle when searching for documentation. Search engines are a great way of finding documentation, but more often than not, direct the users towards paywalls or non-official sites that are attempting to distribute documentation unofficially.

Accessing documentation by the steps discussed, users can always be sure that the documentation provided is genuine, unmodified, and from Siemens directly.

> **Note**
>
> In some cases, the documentation provided by the documentation links is for older versions of the software. If this is the case, it is still likely that the documentation will offer support, but if the latest version is required, it will need to be searched for on the Siemens Industry Online Support website. This can be done by clicking the link provided, and once you're on the Siemens Industry Online Support website, using the **Search in Online Support** search field.

Further support – Liam Bee

Connect with the author of this book via LinkedIn. Liam is open to questions and discussions on topics of this book and always helps those that want to be helped.

Figure 15.30 – Connect with Liam Bee on LinkedIn

Summary

This chapter has ended the book with some additional information regarding programming tips, some best practices for sequence control, and considerations regarding commenting and naming conventions.

This book has explored many different aspects of TIA Portal and PLC programming in general, as well as HMI development. By covering all topics in this book, you should now have the confidence to apply the learning to projects and help tailor your own personal development. As TIA Portal and the world of automation improves, many of the learned aspects covered in this book will still be valid.

TIA Portal is a very powerful development platform, and when a programmer understands all of the tools that TIA Portal has to offer, almost any automation solution can be developed directly within TIA Portal itself.

Please continue to use this book as a reference document, returning to chapters and content when needed. Use the support content from Siemens, such as the information system and forums, to help learn where required and reach out to connections when needed. The Siemens community is one that is always willing to help.

Congratulations on completing the content of the book. This book was written with a single purpose: to help educate others in the world of Siemens, TIA Portal, and good programming skills. The knowledge contained within this book has hopefully been transferred well and can be used to advance the careers and opportunities of others.

Index

A

advanced simulation
    performing, with standard
        control objects 226, 227
alarm control 310
alarm tags 321-323
analog input 42
anonymous structure 65
Array DB 48
asset manager 39
assignments 95
asynchronous data access
    considerations 249, 250
    correct method 251
AT constructor 374, 375

B

backup mode 317
base screen 262, 308
best practices, sequences
    about 382
    constants, using instead of
        numerical values 382, 383
    managed transitions 383-386
    output requests managing 386-388

B

Boolean array 374
box instructions 21
breakpoints 34
byte
    bit, accessing 115

C

call environment 33
call hierarchy 34, 35
call options 248
call path 308
call structure 42
Cause and Effect - (CEM)
    about 168
    actions 171, 172
    causes 174, 175
    control scenario walk-through 175
    groups 172
    instructions 170
    intersection columns 173
    overview 168, 169
cause and effect matrix (CEM) 98, 99
central processing unit (CPU) 31
code work-memory 189

configuration options, global data
  retain 53
  start value 53
configuration options, instance data
  retain 53
  start value 53
connections
  creating 360-362
  Devices & networks 362, 363
  parameters 360
consistent downloads
  about 339
  examples 339, 340
constants 383
control data
  accessibility, improving with
    UDTs 195, 196
  creating 193, 194, 195
  example 196-200
controlled navigation 306
Controls 268
control scenario
  about 118, 119
  Human Machine Interface
    (HMI), using 123
  inlet valve, opening 120, 121
  overview 120
  stop conditions 122
control scenario walk-through, CEM
  about 175
  filling operation 178, 179
  inlet valve, opening 177
  outlet valve position, calculating 179
  stop conditions, managing 180, 181
  summary 181
  system, starting 176

control scenario walk-through, FBD
  about 138
  aspects 145
  filling operation 142
  inlet valve, opening 141
  outlet valve position,
    calculating 143, 144
  stop conditions, managing 144, 145
  summary 145
  system, starting 138, 140
control scenario walk-through, GRAPH
  about 160, 161
  filling operation 163
  inlet valve, opening 162, 163
  outlet valve position, calculating 164
  parallel stop step 166-168
  stop conditions, managing 164-166
  system, starting 161, 162
control scenario walk-through, LAD
  about 127
  filling operation 130, 131
  inlet valve, opening 129
  outlet valve position,
    calculating 132, 133
  stop conditions, managing 133, 134
  summary 134
  system, starting 127-129
control scenario walk-through, SCL
  about 149
  filling operation 151
  inlet valve, opening 150
  outlet valve position,
    calculating 152, 153
  summary 154
  system, starting 149, 150
CPU operator panel 33

cross-referencing, UDTs/structs
  about 86, 87
  solution 88

D

data
  copying, to instance data 58
  managing, through instance
      parameters 243
  passing, through interfaces 56, 57
  referencing 58-60
  retaining, in non-optimized blocks 343
  retaining, in optimized blocks 343
data blocks 39, 241
data management
  about 38
  example 39
data, managing through
    instance parameters
  principle 244, 245
  TIA Portal example 245-247
data referencing
  about 58-60
  InOut data, drawbacks 61
  InOut interface variables, using 62
  memory advantages 61
data, retaining in instance data
  about 344
  non-optimized data blocks 346, 347
  optimized function blocks 344-346
Details view
  about 15, 18
  uses 18
Download without reinitialization
    option 348

drawbacks, UDTs/structs
  about 81
  cross-referencing 86, 87
  lack of open protocol support 86
  libraries 81-83
  overuse 89
Dword variables
  bit, accessing 115
dynamic properties
  setting 276, 277
  tag dynamization properties 277, 278
  tags, assigning 277
Dynamization feature 276

E

Elements 267
emergency stop (E-Stop) 196
empty box instruction 21
empty folder 27
encryption transfer 369
events
  converting, to script 290
  creating, in faceplates 301, 302
  example 288
  handling, in faceplates 301, 302
  raising 287, 288
example use case, project structure
  dataset 43
  data structure 42
  input mapping 42
  Main (OB1) 40
  output mapping 44
  process area 42, 43
  project tree objects 40

F

faceplates
  about 292
  creating 293, 295
  events, creating 301, 302
  events, handling 301, 302
  interface, creating 296
  objects and controls 295
  tag interface data, using 298
  TIA Portal V17 293
favorite instructions
  adding 22
force table
  versus watch table 223
free navigation 306
function block 24, 78
function block diagram (FBD)
  about 95, 394
  instructions 136
  overview 135
function block diagram
    (FBD) instructions
  about 136, 137
  box instructions 137, 138
  comparators 138
function block interfaces 248
functions 24, 246

G

global data
  about 48
  accessing 51, 52
  configuration options 52
  using 49-51

global data block
  about 49
  structure definition, example 64
global instance data block 49
global library
  about 27
  creating 28
  opening 29
  upgrading 30
  using 30
GRAPH
  about 96, 97, 114, 115, 154
  control scenario walk-through 160
  control summary 168
  instructions 156
  overview 154, 156
graphical languages
  mixing, with textual languages
    in LAD/FBD 110, 111
Graphics and Dynamic widgets 268, 269
GRAPH instructions
  step actions 157
  supervision coils 159
  transitions 159
GRAPH language 382
group of data blocks 27
group of function blocks 27
group of mixed objects 27

H

hierarchy 82
HMI alarm classes
  configuration 313, 314
  creating 314, 315

HMI alarm controls
  about 310
  configuration 317, 318
  filters, setting on 318-320
HMI alarms
  configuration 311-313
  logging 315-317
HMI data
  creating 201, 202
  setpoints/parameters 201, 202
HMI development environment
  overview 259
  Runtime settings 259, 260
  screens 261, 262
HMI navigation
  about 306
  ChangeScreen event 308, 309
  page changes, managing 307
HMI objects
  connections 258
  HMI tags 258
  Runtime settings 258
  screens 258
HMI runtime 272
human-machine interface (HMI)
  about 201, 249, 360
  adding, to TIA Portal project 256, 257
  downloading to 364, 365
  running, in simulate mode 213
  security considerations 369
  using 123

I

inconsistent download
  about 340, 341
  examples 342

InOut interface 79, 227
input mapping 39, 41, 42
input mapping layer
  simulated inputs, modifying
    with 223-226
instance data
  about 48
  accessing 51, 52
  configuration options 52
  using 49-51
instance data block (IDB) 345
instance parameters
  data, managing through 243
instructions
  about 19, 20
  adding, from instructions tab 21
  methods 21
interfaces
  creating 296
interfaces, simplifying with structs/UDTs
  about 74
  functions, passing as single
    input struct 76, 78
  functions, passing as single
    output struct 79
  InOut data, passing as single struct 79
  inputs, passing as single struct 74, 75
  outputs, passing as single struct 78
  static declarations of UDTs/
    structs, creating 80
  structures, in static and
    temporary memory 79
  temporary instances of UDTs/
    structs, creating 80
interface types
  InOut 57
  input 56

output 56
property interface 299
static 57
tag interface 296, 297
Internet Protocol (IP) 14

K

key properties
    appearance 276
    general 276
    miscellaneous 275
    security 275
    size and position 275

L

ladder 394
ladder logic (LAD)
    about 13, 94, 95
    control scenario walk-through 127
    control summary 134
    instructions 124
    overview 123, 124
    Valid networks 113, 114
ladder logic (LAD) instructions
    about 124, 125
    box instruction 126
    comparators 127
    ladder bit logic operations 125, 126
languages
    about 94
    cause and effect matrix (CEM) 98, 99
    function block diagram (FBD) 95
    GRAPH 96, 97
    graphical languages 103
    ladder logic (LAD) 94, 95
    selecting 105

starting with 94
statement list (STL) 98
structured control language
    (SCL), structured 96
textual languages 104
types 102, 104
use case 105
use case, examples 106-109
use case, purpose 110
variations 104, 105
languages, program blocks
    about 99
    function blocks 100, 101
    functions 101, 102
languages, TIA Portal
    Cause and Effect - (CEM) 168
    Function Block Diagram (FBD) 135
    GRAPH 154
    Ladder logic (LAD) 123
    Structured Control Language (SCL) 145
latching circuit 129
libraries
    about 19, 22
    global library 23
    project library 22
libraries, UDTs/structs
    about 81-83
    considerations 86
    dependent types, updating manually 84
    UDT, releasing with dependent
        blocks 84, 85
logic
    general layout 203
    structuring 202
    supportive methods 203, 204
logic tips
    AT constructor 374, 375
    blocks, consolidating 381

delay timers 372, 373
IF statements 375, 376
refactoring 380, 381
Serialize instruction 376-379
simplifying 372

M

manual synchronization 355
master copies
    about 27
    icons 27
    usage 27
memory management 112

N

non-optimized blocks
    about 375
    data, retaining 343
non-optimized data
    about 54
    benefits 56
    versus optimized data 54-56
non-optimized data blocks 346, 347

O

online testing environment
    about 31, 32
    breakpoints 34
    call environment 33
    call hierarchy 34, 35
    CPU operator panel 33
Open Platform Communications
        (OPC) server 260
optimized blocks
    data, retaining 343

optimized data
    about 54
    benefits 56
    integrating, with non-optimized data 56
    versus non-optimized data 54-56
optimized function blocks 344-346
organization block (OB) 49
output mapping 39
output mapping layer
    about 235
    example 236
outputs
    safeguarding, in simulation
        mode 235, 236
Overview view mode 19

P

panels 16
Parameter instance 248
parent object 239, 248
PLC, download considerations
    about 357
    data segregation 357, 358
    functions, using 358
PLC-driven alarming
    about 323-325
    alarm texts 327-329
    global alarm class colors, setting 330
    supervision categories 325, 326
    Types of supervision option 326
PLC, load actions
    consistent downloads 339, 340
    inconsistent downloads 340-342
PLCSIM 214
PLC simulation
    performing 214
    S7-PLCSIM interface 216, 217

support, enabling 214, 215
trustworthy devices 217-219
Portal view
  about 4, 6
  activity area 7
  main menu 7
  submenu 7
Profinet/Industrial Ethernet (PN/
    IE) connection 278
program blocks
  about 81
  languages 99
programmable logic controller (PLC)
  about 4, 277
  download, executing to stop 342, 343
  downloading to 336, 337
  download, initiating to 337, 338
  load actions, setting 338, 339
  performance, measuring 54
  running, in simulate mode 213
  uploading from 352-356
project library
  about 23
  master copies 27
  types 24-26
project library, types
  consistent 25
  default dependent 25
  multiple inconsistencies 25
  non-default version instantiated 25
project structure
  creating 38
Project view
  about 4, 8, 9
  details view 9
  devices, adding 12, 13
  devices, configuring 14, 15
  diagnostics pane 9

info pane 9
main activity area 8
new project, creating 10
project tree 8
project tree, modifying 11
properties pane 9
supportive tools 9
property interface, faceplates
  about 296, 299
  using 299-301
pump
  asset data storage 70
  inputs 70
  outputs 70
  SCADA/HMI exchange 70

R

refactoring process 380
Reference project view
  about 15-17
  uses 17
region 150
Rising Edge Trigger 139
Runtime settings option
  about 259, 260
  Alarms 260
  Language & font 261
  Remote access 261
  Services 260
  Storage system 261
  Tag settings 261
  User administration 261

S

S7-1500 PLC 86
S7-PLCSIM interface 216, 217

screen 261
screen layout 263
screen objects
    about 264
    events 265, 266
    properties 264, 265
screen toolbox 262, 263
scripts
    about 301
    compilation errors 286
    global scripts 287
    interface tags, accessing 301
    script files, constructing 284, 285
    tags, reading 285
    tags, writing 285
    using 283
Secure Digital (SD) card 261
sequences
    best practices 382
Set coils 388
Siemens
    documentation, archiving 399
    forum 398
    languages 112
    support 394
SIMATIC Manager 34
Simple Mail Transfer Protocol
    (SMTP) port 260
simulated inputs
    advanced simulation, with standard
        control objects 226, 227
    managing 220
    modifying, with input mapping
        layer 223, 225, 226
    modifying, with watch tables 220, 222
simulation HMI
    configuring, in TIA portal 229, 231-234

simulation interface
    creating 228
    simulation HMI, configuring in
        TIA portal 229, 231-234
snapshots
    about 350
    data, restoring 351, 352
    taking 350, 351
software synchronization 354
special objects
    about 266, 267
    Controls 268
    Elements 267
standard control interfaces
    planning 189-192
standard control objects
    advanced simulation, performing
        with 226, 227
    creating, considerations
        205, 206, 208-210
standard data
    extending 241-243
standard functions
    extending 238-240
standard interfaces
    InOut variables 186
    inputs 185
    large variables 187-189
    outputs 185, 186
    planning 184
    standard interfaces 186
    variables, defining 184
standard singular objects 27
start screen 262
Statement List (STL) 34, 98
static data 246

static properties
  key properties 275
  setting 272, 273
static value
  about 272
  types 274
step action parameters, GRAPH
  action 157
  event 157
  interlock 157
  qualifier 157
stop conditions 165
structs
  about 64, 231
  best practices 69
  commonalities, finding
    between assets 72
  composure 65
  drawbacks 81
  example, in global data blocks 64
  interfaces, simplifying with 74
  naming conventions 73, 74
  nesting 66
  non-optimized blocks 65
  optimized blocks 65
  requirements 69
  structure variables, defining 70, 71
  variables, accessing 65, 66
Structured Control Language (SCL)
  about 34, 145
  control scenario walk-through 149
  instructions 146, 147
  overview 145, 146
Structured Control Language
    (SCL) instructions
  about 146
  bit logic operations 147, 148

box instructions 148
comparators 148
Structured Text (SCL)
  about 96, 145, 375
  comments 390-392
  regions 393, 394
supervisory control and data
    acquisition (SCADA) 201, 249

T

tag dynamization properties
  conditions, setting 282, 283
  HMI and PLC connection, creating 278
  tags, assigning 280, 282
tag interface data
  using, in faceplate 298
tags
  accessing 302-304
temp data 78
TIA Portal Comfort Panel 256
TIA Portal project
  HMI, adding 256, 257
TIA Portal Version 17 (V17)
  about 4
  bug 264
  Comfort Panel HMIs 256
  faceplates 293
Timer Off Delay (TOF) 373
Timer On Delay (TON) 372
timers 112, 113
Totally Integrated Automation
    Portal (TIA Portal)
  about 238, 273, 292
  Advanced search bar 397, 398
  call structure 45, 46
  dependency structure 46

example 245-247
function block interfaces 248
help system 394, 395
HMI, adding 256, 257
languages 123
naming conventions and
    comments 388-390
navigation 395
parent/child hierarchy 45
parent/child relationships 47
search bar 396
simulation HMI, configuring
    229, 231-234
typed object 297
Types 23
types, modes
  in test 24
  released 24

U

unified HMI simulation
  about 365, 366
  accessing 366-369
Universal Serial Bus (USB) card 261
user-defined types (UDTs)
  about 24, 64, 66, 240, 297
  advantage 67
  best practices 69
  commonalities, finding
    between assets 72
  composure 67
  creating 67
  drawbacks 81
  example, in global data block 66

interfaces, simplifying with 74
naming conventions 68, 73, 74
nesting 69
non-optimized block 68
optimized block 68
requirements 70
structure variables, defining 70, 71
User Management Component
    (UMC) 261

V

variables 64
variable speed drive (VSD) 211
virtual private network (VPN) 33
Visual Basic Script (VBS) 283, 284

W

watch tables
  simulated inputs, modifying
    with 220-222
  versus force table 223
windows and panes
  about 4, 5
  Portal view 6
  Project view 8
Word
  bit, accessing 115

Packt>

Subscribe to our online digital library for full access to over 7,000 books and videos, as well as industry leading tools to help you plan your personal development and advance your career. For more information, please visit our website.

Why subscribe?

- Spend less time learning and more time coding with practical eBooks and Videos from over 4,000 industry professionals

- Improve your learning with Skill Plans built especially for you

- Get a free eBook or video every month

- Fully searchable for easy access to vital information

- Copy and paste, print, and bookmark content

Did you know that Packt offers eBook versions of every book published, with PDF and ePub files available? You can upgrade to the eBook version at packt.com and as a print book customer, you are entitled to a discount on the eBook copy. Get in touch with us at customercare@packtpub.com for more details.

At www.packt.com, you can also read a collection of free technical articles, sign up for a range of free newsletters, and receive exclusive discounts and offers on Packt books and eBooks.

Other Books You May Enjoy

If you enjoyed this book, you may be interested in these other books by Packt:

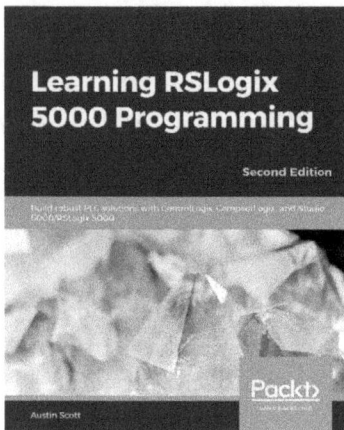

Learning RSLogix 5000 Programming - Second Edition

Austin Scott

ISBN: 9781789532463

- Gain insights into Rockwell Automation and the evolution of the Logix platform
- Find out the key platform changes in Studio 5000 and Logix Designer
- Explore a variety of ControlLogix and CompactLogix controllers
- Understand the Rockwell Automation industrial networking fundamentals
- Implement cybersecurity best practices using Rockwell Automation technologies
- Discover the key considerations for engineering a Rockwell Automation solution

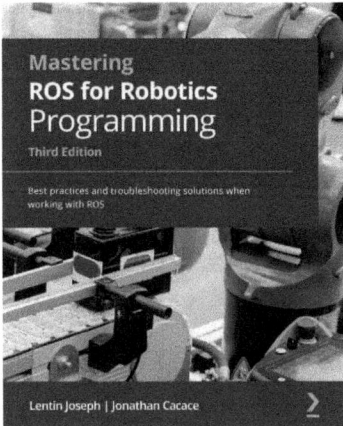

Mastering ROS for Robotics Programming - Third Edition

Lentin Joseph, Jonathan Cacace

ISBN: 9781801071024

- Create a robot model with a 7-DOF robotic arm and a differential wheeled mobile robot
- Work with Gazebo, CoppeliaSim, and Webots robotic simulators
- Implement autonomous navigation in differential drive robots using SLAM and AMCL packages
- Interact with and simulate aerial robots using ROS
- Explore ROS pluginlib, ROS nodelets, and Gazebo plugins
- Interface I/O boards such as Arduino, robot sensors, and high-end actuators
- Simulate and perform motion planning for an ABB robot and a universal arm using ROS-Industrial
- Work with the motion planning features of a 7-DOF arm using MoveIt

Packt is searching for authors like you

If you're interested in becoming an author for Packt, please visit `authors.packtpub.com` and apply today. We have worked with thousands of developers and tech professionals, just like you, to help them share their insight with the global tech community. You can make a general application, apply for a specific hot topic that we are recruiting an author for, or submit your own idea.

Share Your Thoughts

Now you've finished *PLC and HMI development with Siemens TIA Portal*, we'd love to hear your thoughts! Scan the QR code below to go straight to the Amazon review page for this book and share your feedback or leave a review on the site that you purchased it from.

https://packt.link/r/1801817227

Your review is important to us and the tech community and will help us make sure we're delivering excellent quality content.